Astronomy and Astrophysics in Spain (1850-1914)

Astronomy and Astrophysics in Spain (1850-1914)

By

Pedro Ruiz-Castell

Cambridge Scholars Publishing

Astronomy and Astrophysics in Spain (1850-1914), by Pedro Ruiz-Castell

This book first published 2008 by

Cambridge Scholars Publishing

15 Angerton Gardens, Newcastle, NE5 2JA, UK

British Library Cataloguing in Publication Data
A catalogue record for this book is available from the British Library

ISBN (10): 1-84718-439-1, ISBN (13): 9781847184399

TABLE OF CONTENTS

ACKNOWLEDGEMENTS

This book is the result of my doctoral research at the University of Oxford and, therefore, incurs many debts. First of all, I would like to thank both my D.Phil. supervisors, Stephen Johnston, for his reading and insightful comments, and Víctor Navarro-Brotóns, for his crucial guidance. My examiners, Robert Fox and Antoni Roca-Rosell, not only encouraged me to publish the text, but also provided several suggestions in order to improve the quality of my dissertation and which I have tried to include in this book. I am also very grateful for the interest shown in my work to many historians of science at the University of Oxford, such as Jim Bennett, John Heilbron, Faidra Papanelopoulou, and Jessica Ratcliff. This thesis was also possible thanks to the help and comments provided by several Spanish historians of science, like Vicent Salavert and Josep Batlló, who supplied very interesting material and ideas, and also Francisco Pelayo, Alfons Zarzoso, Pepe Pardo, Agustí Nieto-Galán, José Ramón Bertoméu, Antonio García-Belmar, Jesús Catalá-Gorgues, Josep Simón, Tayra Lanuza, Néstor Herran, Xavier Roqué, Aitor Anduaga, and Alfredo Baratas. In addition, I am very grateful to the following professors and lecturers from the University of Valencia: Vicente Muñoz, Ramón Lapiedra, José Ibáñez, and, especially, Vicent J. Martínez and all the staff from the Astronomical Observatory of that university. I am also indebted to the members of the Museo Nacional de Ciencia y Tecnología, particularly to Amparo Sebastián, Pepa Jiménez, Rosa Martín, Ignacio de la Lastra, and Sergio Lozano, as well as to Marcos Villaverde, who made very helpful comments on some technical aspects of astronomy and astrophysics. I am also very grateful to John Woodruff for his proofreading and his very helpful comments of my dissertation, as well as to the archivists and librarians of the institutions where my research took place for their facilities.

Of course, this work would have never been finished without the sympathetic atmosphere provided by my family and friends, indispensable for me to be productive: my mother Julia and my sisters María and Julia, whose support was so important for me, as well as my father Pedro whose advice was extremely useful; Rachael Ker, who shared most of my time in Oxford and whose help and support were the best motivation for this work; Zoë and Andy Rose, and Sandra, Donald, Sarah, Peter and Tim Ker, whose generosity was priceless; my exceptional friends Raeli Bronstein and Ranty Islam, always there for me, like the guys from *Out of the Blue* (Derek, Josh, Seb, Ben S, Tim, James, Ben B, Tom, Gus, André, Vish, Ron, Mike, etc.), Leanne, Marcos and Gabriella, Anupum and Rishta, and Ellen; Raúl Acosta, Germán Freire, Paloma García-Bellido, and Matías Spektor, who always had a word of support in Spanish when I needed it; David, Alistair, Cortney, Assimina, Luca, Isabel, Andrea, Yoon, Jason, Frank, Jan, Jovo, Chad, Marcia and all the people with whom I had the chance to share so many interesting conversations while in England; María Legarda, whose help and love made easier the additional amount of work required to transform the thesis into a book; and my Spanish friends Jordi Gil, Alejandro Marcos, Ximo Català, Javi Agustí, and Carlos Mesejo, whose support was also crucial.

List of Illustrations

LIST OF TABLES

LIST OF ABBREVIATIONS

Abbreviations have been used exclusively in footnotes when dealing with primary sources from the following archives:

OAUV	Archivo Histórico del Observatorio Astronómico de la Universidad de Valencia.
ANF	Archives Nationales de France.
OAM	Archivo del Observatorio Astronómico de Madrid.

CHAPTER ONE

STRUCTURE AND METHODOLOGY

1.1. Structure

The second half of the nineteenth and the beginning of the twentieth century was a time of major change in Spain. In the particular case of the development of science, these changes are very well illustrated in the field of astronomy and astrophysics. The current text aims to explain how these branches of scientific knowledge progressed in Spain during this period. It is divided into two main parts: the first deals with the observatories, institutions, and the scientific community of the period; the second examines the total solar eclipses observed from Spain in the early twentieth century, particularly those of 1900 and 1905. Consequently, the first part takes a more traditional history-of-science approach to present a general overview of the institutional situation in Spain as regards to astronomy and astrophysics during the second half of the nineteenth and the early twentieth century. The second part, however, steps outside the history of the institutions and applies the lessons learned in the first part to the study of specific issues related to the observation of these solar eclipses. It explores the high points of astronomical and astrophysical activity in Spain at the moments of the total solar eclipses of 1900 and 1905.

This separation into two different parts escapes from the typical chronological structure to which many historians are attached. It provides a way for the historian to look at the institutional history through the study of the activity during these solar eclipses. At the same time, it highlights some of the limitations of that more traditional way of doing history and explores new aspects which might help to better understand the processes studied in this work. In sum, it provides a different point of view for more ambitious generalisations and abstractions. What gave rise to the developments in astronomy in Spain during the late nineteenth and early twentieth centuries? Why did astronomy become relevant to Spain? The study of the total solar eclipses of 1900 and 1905 helps to clarify these and other issues, such as the formation of a Spanish astronomical community during that period.

The first part not only deals with the general Spanish context of the second half of the nineteenth century, exploring several social, political, economic, and cultural aspects, but also provides some insights into the situation of astronomy in Spain during that period. An overview of the evolution of astronomy during the second half of the nineteenth century in the country is presented through the developments at the Observatory of Madrid, the main Spanish astronomical institution of the period, and looking at some members of the Spanish astronomical community that was formed during these years. Moreover, the progress of astronomy in Spain during the early twentieth century had the support of several other institutions. Specifically, the institutionalisation of astronomy in Spain during these years had two interesting allies: religious orders and universities. This implied a very heterogeneous scientific community devoted to astronomy, as it will be shown.

The second part of this book explores the state of astronomy in Spain during the late nineteenth and early twentieth century through the study of two particular eclipses whose path of totality crossed Spain in 1900 and 1905. To contextualise these astronomical events, attention is paid to the growth of popular and amateur astronomy during the period studied, as well as to the importance of solar eclipses in the development of solar physics and the rise of astrophysics. The scientific activity developed during the total solar eclipses of 1900 and 1905 is analysed in the subsequent chapters, with some previous eclipses (such as that of 1860) as background, and a comparative analysis of foreign and national programmes and methods of observation is attempted. Consequently, the text not only deals with the organisation and work of the first Spanish official eclipse expedition since the development of astrophysics and with the research pursued by foreign astronomers who visited Spain to observe the solar eclipses of 1900 and 1905, but also explores some social and cultural dimensions surrounding scientific enterprises such as eclipse expeditions. Finally, the study of the solar eclipses of 1912 and 1914 allows an analysis of the situation in which astronomy was left in Spain after the total eclipses of 1900 and 1905 through. When compared with the state of astronomy during the second half of the nineteenth century, one appreciates the qualitative step taken in astronomical research and initiatives in the country.

1.2. Methodology

This book focuses on the astronomers and institutions of Spain in the late nineteenth and early twentieth centuries. This includes not only Spanish individuals and institutions, but also foreign eclipse expeditions sent to Spain to observe the solar eclipses of 1900 and 1905. Of these, more attention is paid to the British and French expeditions because of the numbers of astronomers from these countries who went to Spain to observe the two eclipses. In fact, France and Great Britain were for Spain the two most influential European countries in terms of scientific research, at both the institutional and the amateur level, by the early twentieth century. That is why many of the bibliographical sources, particularly primary sources, cited in this book are French and British publications. Several archives and libraries in these two countries were visited in order to obtain information, including the Archives Nationales de France, the archives of the Observatory of Paris, the Bodleian Library, the archives of the Museum of the History of Science at Oxford, and the British Library. This made possible the systematic study of the main astronomical and astrophysical journals of the period. The following technical periodicals in the field of astronomy and astrophysics were consulted for this book:

- *Astrophysical Journal*, volumes 1 (1895) to 50 (1919).[1]
- *Monthly Notices of the Royal Astronomical Society*, volumes 53 (1892) to 73 (1913).[2]
- *Astronomical Journal*, volumes 1 (1849) to 50 (1944).[3]
- *Astronomische Nachrichten*, volumes 121 (1889) to 210 (1920).
- *Publications of the Astronomical Society of the Pacific*, volumes 3 (1891) to 19 (1907).
- *Journal of the British Astronomical Association*, volumes 1 (1890–1891) to 30 (1919–1920).[4]

---

[1] Storrs B. Barrett (comp.), *The Astrophysical Journal: A General Index by authors and by subjects to volumes I to XXV (January 1895 to June 1907)* (Chicago: University of Chicago Press, 1908); Storrs B. Barrett (comp.), *The Astrophysical Journal: A General Index by authors and by subjects to volumes XXVI-L (July, 1907, to December, 1919)* (Chicago: University of Chicago, 1920).

[2] See: Royal Astronomical Society, *General Index to the Monthly Notices of the Royal Astronomical Society, volumes LIII to LXX, 1892–1910, together with General Index to Illustrations in the Memoirs, volumes I to LIX, and the Monthly Notices, volumes I to LXX, 1822–1910* (London: Burlington House, 1911)].

[3] G. M. Clemence and Louise F. Jenkins, *General Index to the first fifty volumes 1849–1944* (New Haven, 1948).

- *The Observatory*, volumes 13 (1890) to 42 (1919).
- *L'Astronomie*, volumes 9 (1890) to 13 (1894); and the *Bulletin de la Société Astronomique de France*, which absorbed *L'Astronomie* in 1895, volumes 1 (1887) to 24 (1910).

Other astronomical publications, such as the Russian *Astronomicheskii Zhurnal*, did not appear until later, in this case in the early 1920s. Other important journals dealing specifically with astrophysics, like the German *Zeitschrift für Astrophysik* and the French *Annales d'Astrophysique*, were published for the first time in the 1930s (in these two cases, 1930 and 1938, respectively). Some British and French general scientific periodicals, such as the *Proceedings of the Royal Society of London*, *Science*, *Nature*, *Knowledge*, *Annales du Bureau des Longitudes*, and the *Comptes Rendus de l'Académie des Sciences*, have also been methodically consulted. Finally, most of the Spanish primary sources cited in this book, including manuscripts, periodicals, and newspapers (such as *El Imparcial, El Heraldo de Madrid, El Liberal*, *El Mundo Científico*, *Memorias del Observatorio del Ebro*, *Boletín de la Sociedad Astronómica de Barcelona*, *Razón y Fe*, and *Ibérica*), have been obtained not only from the Spanish Biblioteca Nacional but also from the library of the National Museum of Science and Technology of Madrid and the archives of the Astronomical Observatory of the University of Valencia, the Observatory of Madrid, and the Ebro Observatory.

[4] Volumes 1 to 18 can be searched with F. W. Levander, *General Index to the Journal of the British Astronomical Association, volumes I.-XVIII* (Edinburgh: Neill and Co., 1909).

Part I

The development of astronomy in Spain: institutions and individuals

CHAPTER TWO

SPAIN IN THE NINETEENTH CENTURY

2.1. Astronomy and the State

Greenwich, Paris, Pulkovo … National astronomical observatories such as those of England, France, and Russia were established during the seventeenth and eighteenth centuries to investigate navigational and geodetic problems. Navigation was strongly linked to power. It implied not only military authority, but also territorial and economic resources. Several authors, such as William McNeill and Fernand Braudel, have pointed out the relationships between "technological" knowledge, the improvement of oceanic navigation and military techniques, and the growth of national economies.[5] These relationships have been identified as the basis for the development of Western civilisation.[6] Aware of its importance, the most powerful countries supported and financed astronomy during these years. Practical astronomy was to play a significant role in the development of both European imperialistic policies and oceanic commerce. This agenda was familiar to Spain, a country particularly interested in looking after the important market and font of resources that its American colonies provided. Evidence of this may be

---

[5] See for instance: William McNeill, *The Pursuit of Power: Technology, Armed Force, and Society since A.D. 1000* (Oxford: Basil Blackwell, 1983); and Fernand Braudel, Capitalism and Material Life (London: Weidenfeld & Nicolson, 1973). McNeill points out the significant relationships between technology, armed force, and society, while Braudel underlines the importance of the conquest of high seas for modern Europe's world supremacy.

[6] See: J. R. McNeill and William H. McNeill, *Las redes humanas. Una historia global del mundo* (Barcelona: Crítica, 2004); and Samuel P. Huntington, *El choque de las civilizaciones y la reconfiguración del orden mundial* (Barcelona: Paidós, 1997), pp. 54–63.

seen in the publication of astronomical ephemerides in the *Almanaque Náutico* of the Naval Observatory of San Fernando from 1791.[7]

The Royal Observatory of Cadiz, originally conceived by the Spaniard Jorge Juan y Santacilia (1713–1773), was founded in 1753.[8] In 1735 Jorge Juan, together with the Spaniard Antonio de Ulloa y García de La Torre (1716–1795), had participated in the expedition to Peru led by the Frenchman Charles Marie de la Condamine (1701–1774) to measure an arc of the meridian at the equator. On his return in 1744 he devoted himself to the reorganisation of the Spanish Navy. Once appointed Captain of the Compañía de Guardiamarinas (Royal Marine Guards) in 1751, Jorge Juan promoted the establishment of an astronomical observatory. The Observatory was created as an annexe to the Royal Marine Guards Academy with the purpose of teaching astronomy to future naval officers. In 1798 it moved next to the island of León. The Observatory became independent from the Academy six years later and in 1814 changed its name to Real Observatorio de San Fernando.[9]

For political and economic reasons, and in the absence of mature institutional frameworks, navies encouraged and supported the development of astronomy all over eighteenth-century Europe. The equation "economic power = military power" would remain valid during the nineteenth century.[10] In Spain, astronomy was developed during the eighteenth and early nineteenth centuries partly as a result of the needs, stimulus, and protection of the Navy. Nevertheless, by the end of the

---

[7] See the *Almanaque náutico y efemérides astronómicas para el año ... calculadas de orden de S.M. para el Observatorio Real de Cádiz* (Madrid: En la Imprenta Real, 1791–1959); 168 volumes.

[8] For a complete history of the Observatorio de la Marina de San Fernando, currently known as the Real Instituto y Observatorio de la Armada, see: A. Lafuente and M. Sellés, *El Observatorio de Cádiz (1753–1831)* (Madrid: Ministerio de Defensa, 1988); and Francisco José González González, *El Observatorio de San Fernando (1831–1924)* (Madrid: Ministerio de Defensa, 1992). On the scientific instruments of the observatory, see: Francisco José González González, *Instrumentos científicos del Observatorio de San Fernando (siglos XVIII, XIX y XX)* (Madrid: Instituto de Historia y Cultura Naval, Ministerio de Defensa, 1995).

[9] The island of León was renamed the city of San Fernando in 1813. The change of name, made by the liberal courts who met on this island in 1813, was probably in honour of the Spanish king Fernando VII.

[10] See: Paul Kennedy, *Auge y caída de las grandes potencias* (Barcelona: Plaza & Janés, 1994).

eighteenth century new attitudes in favour of a more theoretical and speculative astronomy were developing within the country. Good examples are the attempts of educational reforms to promote astronomy in Seville and Valencia and the foundation of a Royal Observatory in Madrid, in charge of cartographic activities and astronomical research.[11]

The economic problems and political instabilities of nineteenth-century Spain, however, proved to be enormous obstacles to the consolidation of such projects. There is no doubt about the connections between the State and the progress of astronomy during the seventeenth, eighteenth, and early nineteenth centuries. In Spain, these links seem to have been particularly significant. In fact, any attempt to study the evolution of astronomy during the nineteenth century has to take these into account. Some of the most relevant aspects of the social, economic, and cultural development of Spain during the nineteenth century are now considered.

2.2. Nineteenth-century Spain

By the end of the eighteenth century, Spain had a reasonably significant role in the world as a country with moderate power. However, Spain had to begin the nineteenth century relegated to a second order. The Battle of Trafalgar (21 October 1805) confirmed the superiority of the British Navy and the decline of the Spanish *Armada*. The Napoleonic invasion of 1808 heralded a century of political instability. The Spanish Empire also began to disintegrate with the new century. The events of 1808 in Spain created on the other side of the Atlantic a crisis of political legitimacy and power. The process of independence in Chile started in 1810—although it did not finish until 1823. Argentina claimed independence from Spain in 1816, five years before the independence of Mexico, Central America (Guatemala, Honduras, Nicaragua, El Salvador, and Costa Rica), Peru, Venezuela, and Ecuador. By 1824 the conclusive loss of the South American colonies had been formalised.[12]

---

[11] See: Manuel A. Sellés, 'Astronomía y navegación', in Manuel Sellés, José Luis Peset, and Antonio Lafuente (comps.), *Carlos III y la ciencia de la Ilustración* (Madrid: Alianza, 1988); pp. 81–98.

[12] For a more complete understanding of the processes of independence in Latin American countries, see: Leslie Bethel (ed.), *The Cambridge History of Latin America. Volume III, From independence to c.1870* (Cambridge: Cambridge University Press, 1985); and John Lynch, *Las revoluciones hispanoamericanas 1808–1826* (Barcelona: Ariel, 1980).

During these years a sense of national decline and of inferiority alongside European philosophical and scientific progress was felt amongst the Spanish intellectual elites. For many of them the roots of the problem were found in the enormous pressure that the Catholic Church had exerted in the country. Nonetheless, the central years of the century brought new changes. From the 1850s the gap between Spain and Europe was seen by contemporary Spaniards not only as an intellectual problem or a lag in culture, but also as an economic fact[13]—even when during these years foreign trade experienced a vigorous growth in general terms.[14] From 1858 to 1863 the government led by General Leopoldo O'Donnell y Jornis (1809–1867) pursued a *policy of prestige* based on the imperialistic expansion of Spain. Again, the aim to control new lands was strongly linked to economic interests. Although less important than the campaigns of the most powerful European countries, military expeditions were organised to places such as Morocco and Mexico. They did not create the basis of any new Empire, but encouraged and promoted Spanish national identity.[15]

It was in this new context that the elite began to revisit the problems of the country from new points of view. The freedom of thought brought about by the *Glorious Revolution* of 1868 encouraged it even more. One of the most characteristic examples is the public debate that developed during the 1860s and 1870s into the well-known *polémica* about Spanish science—a debate centred on whether there existed a scientific tradition in modern Spain.[16] On the one hand, public figures such as José Echegaray y Eizaguirre (1832–1916), Gumersindo de Azcárate (1840–1917), Manuel de la Revilla y Morera (1846–1881), and José del Perojo y Figueras (1850–1908) denied the existence of such a scientific tradition in the country.[17] Their arguments, again, identified the roots of the problem with

---

[13] See: Raymond Carr, *Spain 1808–1939* (Oxford: Clarendon Press, 1966).

[14] Albert Carreras and Xavier Tafunell, *Historia económica de la España contemporánea* (Barcelona: Crítica, 2003).

[15] José Álvarez Junco, 'La conformación de una identidad', in José María Jover Zamora (dir.), *Historia de España Menéndez Pidal. Tomo XXXVI, La época de la Restauración (1875–1902). Volumen II, Civilización y cultura* (Madrid: Espasa Calpe, 2002); pp. 5–45.

[16] See for instance: Enrique García Camarero and Ernesto García Camarero, *La polémica de la ciencia española* (Madrid: Alianza, 1970).

[17] Their approach was, however, slightly disparate, as they were strongly influenced by different philosophical schools. Gumersindo de Azcárate, for instance, was a recognised *krausist*, while Manuel de la Revilla and José del Perojo were well known for their neo-kantian ideas.

the political and religious intolerance of, and the oppression by, the Church and the State. Others, such as Felipe Picatoste y Rodríguez (1834–1892) and Marcelino Menéndez y Pelayo (1856–1912), claimed that a nationalistic approach to the problem was needed and underlined the importance of Spanish scientific activity during the sixteenth and seventeenth centuries.[18] The discussion was, however, part of a broader ideological and political debate. Even when it dealt explicitly with scientific details, the public debate was infused with more general arguments about freedom of thought and the roles of the Catholic Church and the State in the country.

The *polémica* took different shapes and forms. A good example is the debate precipitated by the Spanish translation of the *History of the Conflicts between Religion and Science*[19] of 1876. The prologue, written by Nicolás Salmerón Alonso (1838–1908), exposed the incompatibility of Catholic faith with cultural and scientific progress.[20] The powerful neo-Catholic movement, however, reacted immediately, and replies to the text were published during the late 1870s, the 1880s, and into the 1890s.[21] A similar debate emerged in Spain during this period around Darwinist ideas. But as Glick has pointed out, those on both sides of the debate defended postures whose origins and objectives were far removed from the scientific problem itself.[22] A glance at some of the names associated with this public discussion confirms this observation. Echegaray, who taught mathematics and physics at the Engineering School of Madrid, was a well-known politician and a famous playwright. Azcárate, Professor of Law at the Universidad Central of Madrid, was also an active politician. Revilla was a recognised writer, like Perojo and Picatoste, who were also distinguished politicians. Menéndez y Pelayo was a prestigious philologist and historian. Salmerón was another prominent politician, President of the Spanish First

---

[18] See the main work of Marcelino Menéndez y Pelayo, *Historia de los heterodoxos españoles* (México: Porrúa, 1995) [First edition: Madrid, 1882].

[19] John William Draper, *History of the Conflict Between Religion and Science* (London: H. S. King, 1875). The book was translated into Spanish by the astronomer and meteorologist Augusto T. Arcimis Werle as *Historia de los conflictos entre la religion y la ciencia* (Madrid: Imprenta de Aribau y Cª, 1876).

[20] For a brief introduction to the conflicts between science and religion in nineteenth-century Spain, see: Francisco Pelayo López, 'Ciencia y religión en España durante el siglo XIX', *Asclepio*, 40 (1988), 187–207.

[21] Enrique Menéndez Ureña and José Manuel Vázquez-Romero, 'El pensamiento y las ideas', in Jover Zamora (dir.), *Historia de España Menéndez Pidal. Tomo XXXVI. Volumen II* (Madrid, 2002); pp. 125–176.

[22] Thomas F. Glick, *Darwin en España* (Barcelona: Península, 1982).

Republic in 1873. This indicates how easily the debates set up around scientific topics moved into a more general arena.

At the same time, as Sala Catalá has pointed out for the particular case of biology, many of those who in the late 1860s began to develop new ideas in Spain on scientific methodology relegated the practice of science to a secondary role because it was seen as an activity subject to rationally constructed philosophical deductions.[23] As a consequence of the terms in which the public discussion was set up, he argues, no experimental biology was practised in the country. Philosophical, political, and educational issues were considered, in many cases, more important. Indeed, public discussions about science, common in nineteenth-century Europe, were greatly politicised in Spain, obviating any examination of how scientific activity had to be pursued. Consequently, many of these debates have to be understood as part of a more general battle between "two Spains": the modern versus the traditional, the progressive against the conservative.

Two of the aspects traditionally linked to the progress of science in nineteenth-century European countries are economic and industrial developments. In Spain, the early years of this century were characterised by the loss of the American colonies and, as a consequence, the increasing debt of the State.[24] However, no government reformed the inequitable and unproductive Spanish taxation system, under which the burden of tax rested much more heavily on the poor. During this period the State was dependent on loans and did not have sufficient means to promote a process of modernisation in a backward economy.[25] The first decades of nineteenth-century Spain saw a strong economic contraction. Nevertheless, a slow and gradual recovery, in terms of industrial and urban development, began in 1840. As Nadal has argued, between 1830 and 1840 a process of industrialisation was initiated in the country, more or less at the same time as in other countries such as Belgium, France,

---

[23] José Sala Catalá, 'Ciencia biológica y polémica de la ciencia en la España de la Restauración', in José Manuel Sánchez Ron (ed.), *Ciencia y sociedad en España: de la Ilustración a la Guerra Civil* (Madrid: El Arquero-CSIC, 1988), pp. 157–177.

[24] Josep Fontana, *La quiebra de la monarquía absoluta* (Barcelona: Ariel, 1971).

[25] Raymond Carr, *Modern Spain 1875–1980* (Oxford: Oxford University Press, 1980).

Germany, Austria, Russia, and Sweden.[26] Steam-powered machines were introduced in the cotton industry during the 1840s, a railway system began to be constructed during the 1850s, gasworks for urban and domestic lighting were set up during the 1860s, and both non-industrial manufacture and industrial production began to co-exist in the food, textile, and metallurgical sectors.[27]

The permanent political crisis that shook Spain during the first two-thirds of the nineteenth century, dominated by successive military *coups d'état*, severely hampered attempts to lead the country along the pathway of economic modernisation. These projects, introduced mainly by the Progressive Party, aimed to create and consolidate the political and institutional bases for an industrial and technologically advanced society. As Tortella has summarised:

> The most notable of these were the legislative program of the Progressive Biennium (1854–1856) and the executive and legislative of the Glorious Revolution (1868–1874). These efforts were not wholly fruitless, and not all nineteenth-century Spain stagnated: Catalonia, and especially Barcelona, underwent considerable industrialization; land reform was carried out; a railroad network was built; improvements were made in the educational system; monetary, banking, and fiscal reforms were implemented; yearly budgets were duly prepared and enacted into law; some manufacturing industries developed and some older industries were partially modernized.[28]

Despite the slow modernisation of the country, the process of growth set in motion during the mid-nineteenth century gained momentum as the century progressed and finally rapidly accelerated in the twentieth century.[29] This process was catalysed by the political stability created by the restoration of the monarchy in the last quarter of the century.

---

[26] Jordi Nadal, *El fracaso de la Revolución industrial en España, 1814–1913* (Barcelona: Ariel, 1975).

[27] Miguel Artola, 'La burguesía revolucionaria (1808–1874)' in Miguel Artola (dir.), *Historia de España* (Madrid: Alianza Editorial, 1990).

[28] Gabriel Tortella, *The Development of Modern Spain. An Economic History of the Nineteenth and Twentieth Centuries* (Cambridge, Mass., and London: Harvard University Press, 2000).

[29] *Ibid.*

2.3. Restoration Spain and the end of century crisis

The restoration of the monarchy was proclaimed in December of 1874, after the *pronunciamiento* (military insurrection) of Arsenio Martínez Campos (1831–1900) in Sagunto (Valencia). The coronation of the King Alfonso XII (1857–1885) took place in January 1875. The political system of Restoration Spain, conceived by Antonio Cánovas del Castillo (1828–1897), was based on two main parties governing alternately, under the conservative constitution of 1876. As suggested above, the years of Restoration Spain benefited from previously created conditions for an incipient process of modernisation. During the late nineteenth century the political panorama was modified and economic change occurred on a noticeable scale. Nevertheless, the weight of tradition and the extreme rigidity of the institutional framework held back the evolution of the country and, compared with progress in other European countries, the total change was small. Spain on the whole remained largely agrarian, traditional, and backward.[30] Changes in the social and institutional structure, such as fiscal and administrative reorganisation, liberalisation of the economy, and educational reform, were achieved, but the process was not consolidated. Economic historians have identified the reason for the failure of the process as the lack of preparation in the country for the assimilation of such a process of modernisation.

The slow and unbalanced process of modernisation was one of the many factors that contributed to the Spanish end-of-century crisis. The historian Jover Zamora has identified the years 1885–1887 as the starting point for this critical period.[31] The beginnings of the crisis have been linked to the death of Alfonso XII in 1885 and the subsequent regency of María Cristina of Habsburg. This period was characterised by a sense of pessimism, largely a consequence of ongoing calamities such as epidemics, floods, wars, mutinies, civil disorders, etc. It was also related to factors present in other European countries, such as the agrarian crisis and the insecurity of the middle classes brought on by the emergence of an organised working class. The economic crisis of the late 1880s in Europe

---

[30] For an introductory and general overview of the different processes of industrialisation all over Europe, see: Sydney Pollard, *La conquista pacífica. La industrialización de Europa 1760–1970* (Zaragoza: Prensas Universitarias de Zaragoza, 1991).

[31] José María Jover Zamora, 'Aspectos de la civilización española en la crisis de fin de siglo', in Jover Zamora (dir.), *Historia de España Menéndez Pidal. Tomo XXXVI. Volumen II* (Madrid, 2002), pp. 747–784.

meant that confidence in the development of science and technology faltered somewhat. It was clear by then that the rise of capitalism had created dreadful problems and exacerbated the misery in which working classes lived.

The crucial point of this period in Spain was the year 1898. The military defeat of that year, in which the United States of America shamed the Spanish Navy and the last overseas colonies (Cuba, Puerto Rico, and the Philippine Islands) were lost was seen as evidence that the country was in terminal decline. If the declaration of war against Spain by the United States engendered a wave of patriotic feeling,[32] the *Disaster* of 1898 troubled the national conscience.[33] Many writers focused on the situation of the country and its position in history. Criticism from the *regenerationist* movement, a legacy of nineteenth-century Spanish liberal ideas, was particularly strong. The aims of regeneration spread throughout various sectors of Spanish society, through the press, universities, and the Parliament.

The turn of the century saw the spread throughout Spain of a very heterogeneous *regenerationist* movement. It developed many different branches, representing particular interests and philosophical theories.[34] There were also *pseudo-regenerationists* within the existing political system, interested in exploiting the popularity of the movement for political ends, and other sectors, such as the socialist party and the anarchists, who called for a complete rupture from the political system. The latter groups, however, were relatively weak and their influence was restricted to local, rather than national, spheres.

---

[32] Sebastian Balfour, 'War, nationalism and the masses in Spain, 1898–1936', in Edward Acton and Ismael Saz (eds.), *La transición a la política de masas* (Valencia: Universitat de València, 2001), pp. 75–91.

[33] Juan Pablo Fusi and Jordi Palafox, *España: 1808–1996. El desafío de la modernidad* (Madrid: Espasa Calpe, 1997).

[34] See: Pedro Cerezo Galán, 'El pensamiento filosófico—El 98: la generación trágica', in José María Jover Zamora (dir.), *Historia de España Menéndez Pidal. Tomo XXXIX, La edad de plata de la cultura española (1898–1936). Volumen I, Identidad, Pensamiento y Vida. Hispanidad* (Madrid: Espasa Calpe, 1993); pp. 133–152; and Pedro Cerezo Galán, *El mal del siglo. El conflicto entre Ilustración y Romanticismo en la crisis finisecular del siglo XIX* (Madrid: Biblioteca Nueva, 2003).

During the first years of the twentieth century almost all sectors of Spanish society adopted a *regenerationist* mentality which involved rejecting the situation of contemporary Spain and putting faith in the future of the country.[35] In general terms, *regenerationists* wished for a nationwide programme of public works to create the infrastructure needed for a modern economy. The programme should include measures to promote agricultural productivity, decentralise the State, and invest in education, and proposals to increase the networks of communication and distribution (more railways) and to reform the judicial and electoral systems.[36] The modernisation of Spain was understood in terms of the French, British, and German models, and the intervention of the State in the economy was seen as crucial to revitalise the country.[37] This initiative, known in Spain as *europeisation*, sprang from the new social and cultural framework developed in Spain during the last quarter of the century.[38]

The fact that the country did not collapse under the strain of economic problems during the difficult years after 1898 allowed the regime of the Restoration to survive.[39] *Regenerationism* strongly influenced the political life of the country during these years, but the concept was finally emptied of meaning, and the word was absorbed by and included as part of the established political system. Almost every critic of the corrupt constitutional monarchy, unrepresentative of the great interests of the nation, assumed the title of *regenerator*. But the protest movements failed

[35] Carlos Seco Serrano, 'Las ideologías políticas', in Jover Zamora (dir.), *Historia de España Menéndez Pidal. Tomo XXXIX. Volumen I* (Madrid, 1993); pp. 319–467.

[36] Sebastian Balfour, *El fin del imperio español (1898–1923)* (Barcelona: Crítica, 1997).

[37] The importance of the State in the process of industrialisation has been studied in: A. Gerschenkron, *El atraso económico en su perspectiva histórica* (Ariel, Barcelona, 1968).

[38] Juan Pablo Fusi, *Un siglo de España: La cultura* (Madrid: Marcial Pons, 1999).

[39] Nonetheless, Spain found itself immersed in a slow and progressive process of weakening. This probably had more to do with the imbalance of the process of modernisation in Spain rather than with the *Disaster* itself. See for instance: Jordi Nadal and Albert Carreras (dirs. and coords.), *Pautas regionales de la industrialización española (siglos XIX y XX)* (Barcelona: Ariel, 1990). This lack of cohesion in late nineteenth- and early twentieth-century Spanish society could be understood as a particular and characteristic stage of an early economic and industrial development. See: Carr, *Modern Spain 1875–1980* (Oxford, 1980). For a study of the *centrifugal* forces of the period 1898–1909 which caused this weakness, see: Balfour, *El fin del imperio español (1898–1923)* (Barcelona, 1997).

either to overthrow, or to radically reform, the discredited political system. As Carr explained:

> The protest movements were fissiparous, divided in tactics and as much rent by personal feuds as the "corrupt" parties they attacked. Secondly, the political system was more open and flexible than its critics admitted. Finally, apathy and the regionalist issue inhibited the creation of mass *nationwide* parties of protest.[40]

Economic and regional interests on the one hand, and the rejection of an alliance with non-dynastic movements on the other, inhibited any alternative to the established political system. Traditional networks of social and ideological control continued in a country that was yet to be modernised, with high levels of illiteracy and underdeveloped transport systems.[41] The political system was now more and more isolated from Spanish society, and its structural characteristics frustrated all efforts to renew it.

By the turn of the century, Spain was still not a modernised country. From the economic point of view, there was in Spain, from 1827 to 1902, an average increase over 215% in the cost of articles of prime necessity, while the augment in wages was 15%. Since 1902 the prices of bread and wine slightly decreased, but in the cost of other necessities was an average increase over 19%. Rents also raised enormously in most industrial centers.[42] In contrast to most advanced European states, during the whole of the nineteenth century Spain's population had grown very slowly, due to a relatively moderate birth rate and high mortality. Several factors contributed to this, among them poor sanitation, low agricultural productivity, chronic housing shortages, and an expensive and inadequate transportation network. One of the most interesting elements was widespread ignorance, in particular about the causes and transmission of illness. There is no doubting the relationship between literacy and economic development and productivity. The investment in human

---

[40] Carr, *Modern Spain 1875–1980* (Oxford, 1980).

[41] Balfour, *El fin del imperio español (1898–1923)* (Barcelona, 1997).

[42] By the early twentieth century, the average salary of industrial workmen, miners, and artisans varied from 3 to 4 pesetas for a working day, while women got from 1 to 2.5 pesetas and children 1 peseta. Therefore, families spent from 75% to 90% of their incomes for necessary food. The cost of the absolute necessities of life was such that an unmarried man could not live on less than an average of 2.75 pesetas a day. See: Clarence Perkins, 'The Social and Economic Problems of Modern Spain', *Political Science Quarterly*, 27 (1912), 92-108.

resources was a great stimulus to economic innovation, both in the sense of inventiveness itself and other more subtle aspects related to a society's capacity to absorb and adopt appropriate technologies. In Tortella's words:

> Poverty limits the potential for investment in education, and in turn the lack of education makes economic growth difficult, thereby perpetuating poverty. On the other hand … a population that is poor and ill-informed in a country that is technologically backward has difficulty appreciating the importance of education, beginning with the ability to read.[43]

In the Spanish case, poverty resulted not only from insufficient economic resources, but also from a historical inertia and lack of political pressure. Spain entered the modern era with an educational system insufficiently developed despite the late but serious efforts of the mid-nineteenth century. In other words, although the restoration of the Spanish monarchy meant political stability, it also reinforced the strong social power of the Catholic Church through the control of education.[44] Spain, by the beginning of the twentieth century, was still mainly an agrarian country where the Catholic Church had strong influence through its control of primary and secondary education. Despite a period during the last decades of the nineteenth century in which the Church lost some economic and political power, there emerged in the late nineteenth and early twentieth centuries what Pérez Díaz has described as

> a strategy for recovering the lost ground and for a partial comeback to a position of influence. It [the Catholic Church] made a concerted effort to extend a network of primary schools through Castile; to revive the missions of the Old Regime with an emphasis on a sort of sentimental religion targeted on women; and to build up a network of secondary schools, Catholic newspapers and Catholic agrarian institutions and political parties.[45]

---

[43] For a more detailed explanation, see: Tortella, *The Development of Modern Spain* (Cambridge, 2000).

[44] See: Antonio Viñao Frago, *Política y educación en los orígenes de la España contemporánea* (Madrid: Siglo XXI, 1982); and Manuel De Puentes Benítez, *Educación e ideología en la España contemporánea* (Madrid: Tecnos, 1999).

[45] Víctor Pérez Díaz, *Peasant Ethics and Religion: Towards a Theory of the 'Three Cultures'* (Madrid: Instituto Juan March, 1991); p. 15.

Publications such as *Razón y Fe* and *El Debate* are good examples of the influence that reactionary, fundamentalist, and anti-liberal positions exerted.[46] As Carr described it:

> Every tension in Spanish society was refracted through the prism of the religious issue ... To liberals, the attempt to re-Catholicize Spain meant handing over the nation to "the irreconcilable enemy of modern institutions, the recognition of the forces of the past as the directing element in society". To Catholics the Institute of Free Education was a step towards godless Spain; to anti-clericals Spain was already in danger of falling into the hands of Jesuits.[47]

Nevertheless, liberal culture largely dominated the press, literature, and the arts during the second half of the nineteenth century in Spain.[48] The main philosophical schools were krausist and positivist,[49] and private, liberal, and secular institutions such as the krausist Institución Libre de Enseñanza, founded in 1876, spread throughout the country. As the century advanced, a public debate developed regarding the state of public education in Spain, including that of universities, and scientific research. Particularly interesting was the krausist approach, which conceived of a higher education system in which scientific research played a crucial role.[50] This rhetoric was echoed in the different Spanish governments of the period. The State became more and more interested in creating an appropriate educational system, and the different reforms promoted took

---

[46] Fusi, *Un siglo de España* (Madrid, 1999).
[47] Carr, *Modern Spain 1875–1980* (Oxford, 1980).
[48] Fusi and Palafox, *España: 1808–1996* (Madrid, 1997).
[49] Krausism was one of the most important intellectual movements in nineteenth-century Spain. Based on the ideas of the German philosopher Karl Christian Friedrich Krause (1781–1832), it was introduced into Spain by the jurist and philosopher Julián Sanz del Río (1814–1869). In their search for a more ethical and just social system, krausists played a crucial role in the social movement which led to the Revolution of 1868. See, for example: Juan José Gil Cremades, *Krausistas y liberales* (Madrid: Seminarios y Ediciones, 1975).
[50] However, some authors have argued that the pragmatic character of the philosophy of science derived from Krause did not support the pursuit of scientific research. This has been presented for the case of medicine by Montiel Llorente in 'Las consecuencias de una elección para la Filosofía de la Medicina Española: Krause frente a Schelling', in Elvira Arquiola and José Martínez Pérez (cords.), *Ciencia en Expansión: estudios sobre la difusión de las ideas científicas y médicas en España (siglos XVIII-XX)* (Madrid: Complutense, 1995).

into account the reformist ideas of previous years.[51] The foundation in 1900 of the Ministry of Public Instruction is a good illustration of this. Working in this new Ministry, Antonio García Alix (1852–1911) and Álvaro de Figueroa y Torres (1863–1950) benefited from the anticlerical atmosphere of those years to re-institute the State's role in educational matters. Education was conceived of as a public good for the instruction of responsible citizens playing their roles in the social and political life of a free country.[52]

In the scientific context, such reformist approaches had already been encouraged during the last decades of the nineteenth century: for example, faculties of medicine promoted a new clinical education, and universities demanded more public funding and support for scientific education.[53] Although Spanish governments were reluctant to accept some of these proposals, universities in some regions benefited from an increase in the number of scientific chairs. This was the case in Barcelona, for example, where progress in natural sciences, physics, and chemistry was seen as essential for the development of agriculture. These new attitudes and interests boosted future reforms during the following century. In fact, the twentieth century saw the strongest attempt to institute a modern and well-constructed system of higher education which promoted science and its applications.

2.4. Science and industry in Restoration Spain

The political instabilities of the first half of the nineteenth century were disastrous to the development of science in Spain. However, the last third of the century brought some progress in several scientific fields, such as medicine and the natural sciences—particularly in botany, zoology,

---

[51] See: Alfredo Baratas Díaz, 'La cultura científica en la Restauración', in Manuel Suárez Cortina (ed.), *La cultura española en la Restauración* (Santander: Sociedad Menéndez Pelayo, 1999), pp. 279–295; and José Luis Peset and Elena Hernández-Sandoica, 'La recepción de la cultura científica en la España del siglo XX: la Universidad', in Antonio Morales Moya (coord.), *Las Claves de la España del siglo XX. La cultura* (Madrid: España Nuevo Milenio, 2001), pp. 127–151.

[52] Estíbaliz Ruiz de Azúa y Martínez de Ezquerecocha, 'El bachillerato y la Universidad', in Jover Zamora (dir.), *Historia de España Menéndez Pidal. Tomo XXXVI. Volumen II* (Madrid, 2002), pp. 337–359.

[53] See: Mariano Peset and José Luis Peset, 'Las universidades españolas del siglo XIX y las ciencias', in José M. López Piñero (ed.) *La ciencia en la España del siglo XIX* [Serie *Ayer*, 7] (Madrid, Marcial Pons, 1992).

geology, and mineralogy.[54] The freedom projected by the *Sexenio Revolucionario* can be identified with the beginnings of this small recovery. A good example of this revitalisation is the aforementioned scientific—but also political and religious—debate set up in the country from the late 1860s onwards around the evolutionary ideas of Charles Darwin (1809–1882).[55] Many of the scientific societies in which this debate took place were founded after 1868. Prominent amongst these new institutions were the Spanish Society of Natural History, created in 1871, and the Sociedad Libre de Histología (Free Histological Society), founded in 1874. This new attitude towards science in Spain was also detected abroad, with journals such as *Nature* writing that:

> A movement has been commenced in Spain for the formation of an association similar to the British Association. The Madrid Societies of Natural History, Anthropology, and Geography have appointed a joint commission to consider how best to organise an annual meeting in different parts of the kingdom for the purpose of investigating matters of scientific interest within the domain of these societies, and also to arrange for the publication of the results that may thus be obtained. A movement like this leads to hope that a fair future is yet in store for Spain.[56]

There are many examples of the international community's interest in a revival of science in Spain, such as the reviews that appeared in *Nature* about the creation in Spain of the Institución Libre de Enseñanza[57] and the works of Salvador Calderón on fossils and petrography.[58] But the beginnings of modern Spanish science have been identified by some historians with a new generation of scientists, active in the last years of the

---

[54] For a general overview of the situation in Spain during the nineteenth century, see for instance: José Manuel Sánchez Ron, *Cincel, martillo y piedra. Historia de la ciencia en España (siglos XIX y XX)* (Madrid: Taurus, 1999). See in particular the second chapter: 'Naturaleza y biología en la España del siglo XIX: de González de Linares a Cajal', pp. 47–83. A regional approach which confirms and completes this description can be found in: José M. López Piñero and Víctor Navarro Brotóns, *Història de la Ciència al País Valencià* (Valencia: Alfons el Magnànim, 1995); Chapter 5: 'El segle XIX', pp. 411–558.

[55] See: *Glick, Darwin en España* (Barcelona, 1982).

[56] See: *Nature*, 16 (1877), 130–131.

[57] On the public support to the 'Institute for Free Education', see: *Nature*, 14 (1876), 132.

[58] A small note on his work *Enumeración de los Vertebrados Fósiles de España*, also published in 1877 in the *Quarterly Journal of the Geological Society of London*, appears in *Nature*, 16 (1877), 140. On the evolution of volcanic rocks, see: 'Petrography in Spain', *Nature*, 21 (1879–1880), 165.

nineteenth century and the early decades of the twentieth, who began to *do science* in stead of *talk about science*.[59] In fact, the paradigmatic case of Spanish scientific revival is that of physiology and histology. The most notable figure in these fields was Santiago Ramón y Cajal (1852–1934), whose international prestige grew progressively during the 1890s, leading him to be awarded a Nobel prize in 1906.[60]

Popularisation of science was a relatively common activity in Spain during the late nineteenth century. The various articles by Echegaray published up to 1905 in journals such as *España*, *El Imparcial*, *Revista Hispano-Americana*, *El Liberal*, *Diario de la Marina* of La Habana, *Ilustración Artística* and *Revista de Obras Públicas* are good examples of these popular writings.[61] The pedagogical and popular works by the chemists José Casares Gil (1866–1961) and José Rodríguez Carracido (1856–1928) are further examples. These and other figures of the second half of nineteenth-century Spain began a process which, by the turn of the century, had evolved into a *civil discourse* for science.[62] This *civil discourse* was defined by Glick as the "process wherein a divided elite agrees to place in abeyance in mutually agreed upon areas the habit of making all ideas serve the ends of ideology". It has been claimed that a scientific middle class that resulted from the slow process of industrialisation of the country probably supported this new attitude.

---

[59] This generation, the *generation of 1880*, together with that of the *Sexenio Revolucionario* and the *generation of 1898*, took an active part in the regenerationist movement which followed the *Disaster* of 1898. However, Laín Entralgo has claimed that, while the regenerationist proposals of the older generation were essentially economic and political and that of the younger mainly literary, the *generation of 1880* played a leading role in the scientific and educational regenerationism. See: Laín Entralgo, Pedro, 'Prólogo', in Jover Zamora (dir.), *Historia de España Menéndez Pidal. Tomo XXXIX. Volumen I* (Madrid, 1993); pp. 11–52.

[60] On Cajal, see for instance the recent publications: Antonio Calvo Roy, *Cajal: triunfar a toda costa* (Madrid: Alianza, 1999); and José María López Piñero, *Cajal* (Madrid: Debate, 2000).

[61] See: José Manuel Sánchez Ron (ed.), *José Echegaray* (Madrid: Biblioteca de la ciencia española, 1990); and Sánchez Ron, *Cincel, martillo y piedra* (Madrid, 1999), Chapter 4: 'Echegaray, entre la ciencia, la literatura y la política', pp. 123–151.

[62] See: Thomas F. Glick, *Einstein in Spain: Relativity and the Recovery of Science* (Princeton: Princeton University Press, 1988); Thomas F. Glick, *Einstein y los españoles: Ciencia y sociedad en la España de entreguerras* (Madrid: Alianza, 1986).

Slowly but increasingly, scientific and technological issues began to be approached separately from political and ideological debates.

Medicine and the natural sciences are the subjects claimed by traditional Spanish historiography to have developed some original activity in mid- to late nineteenth-century Spain. Several authors have argued that the slow process of industrialisation in Spain made the development of other subjects such as physics and chemistry very difficult during these years.[63] Their argument is based on the premise that the development of these subjects required the stimulus of industry and of appropriate institutions, which Spain lacked. Popular scientific works, rather than original research results, are generally pointed out as the most significant contributions to physics and chemistry during this period. That the very limited industrialisation of the country was responsible for the lack of original science in these specific fields seems to be confirmed by the considerable development of science in Barcelona during the second half of the nineteenth century. It was only in Catalonia and the Basque Country that the process of industrialisation took off, and, at least in Catalonia, this has generally been linked to the development of both science and scientific institutions.

Spanish governments of the nineteenth century adopted a similar approach, assuming that there existed a close relationship between the development of science and of industry. Aware of the importance of promoting some particular scientific branches of knowledge, the Ministry of Commerce, Instruction, and Public Works pointed out in 1847 the strong influence of these subjects on the industry and prosperity of nations:

> One of the branches of human knowledge that the Ministry … must preferentially promote is that of the Physical and Natural Sciences, which so powerfully influence the Industry and prosperity of Nations, but which, unfortunately, did not occupy in our former system of education the pre-eminent place that rightly corresponds to them. Shortly … Universities will be provided with the means necessary to cultivate them … but nonetheless, the minister who signs below believes that it is indispensable to come to other no less effective means, which in foreign countries have powerfully

---

[63] See for instance: Sánchez Ron, *Cincel, martillo y piedra* (Madrid, 1999).

contributed to the aggrandizement of those sciences and to the importance of scientific applications of all kinds.[64]

These lines were written to justify the foundation of the Academy of Exact, Physical, and Natural Sciences in Madrid. It is interesting that such ideas encouraging utilitarian science meant the rejection of scientific research with no explicit links to practical applications. This is illustrated by the astronomical work pursued at the Naval Observatory of San Fernando. During the first half of the nineteenth century, the observatory was devoted to meridian and other observations, such as the occultation of stars by the Moon, eclipses of Jupiter's satellites, and circumpolar stars. Furthermore, from 1856 and under the direction of José Sánchez Cerquero (1784–1859), the Observatory delivered an advanced scientific course for officers. This reactivated a range of diverse scientific fields, such as hydrography. The Comisión de la Península, created in 1860 and responsible for cartographical studies of the coasts of Spain and of several islands, was largely staffed by officers who had studied this course. The activities of the Observatory were expanded further later in the century. In 1869, when Cecilio Pujazón (1833–1891) was appointed director of the Observatory, the main activities of the institution were astronomical observations, calculating ephemerides, and storage and conservation of *Armada* nautical instruments. More applied and utilitarian scientific activities were added to these in the late 1870s, with the creation of the Centro Meteorológico Marítimo (Maritime Meteorological Centre) in 1877 and the foundation of the Centro de Agujas Magnéticas (Centre of Magnetic Needles) in 1878.

Modernisation during the nineteenth century has generally been associated with industrialisation and economic progress. In Spanish historiography this association has been seen as crucial in understanding the progress—or stagnation—of both scientific research and education in Spain. One of the aims of this book is, however, to review the validity of this argument in the particular case of astronomical and astrophysical research and education during late nineteenth- and early twentieth-century Spain. Particularly interesting is the study of the progress of astrophysics in the country, a subject strongly linked to the development of physical and chemical research. These fields experienced gradual changes during the late nineteenth century. In fact, different European countries followed

---

[64] See: 'Real Decreto del Ministerio de Comercio, Instrucción y Obras Públicas, 25 de Febrero de 1847', cited in: Sánchez Ron, *Cincel, martillo y piedra* (Madrid, 1999), pp. 70–71.

the development of disciplines such as physics in Spain with great interest, as the following extract of a note published in 1880 demonstrates:

> Don Eduardo Lozano of Teruel, Spain, has lately published a modest little volume of ninety pages, entitled "Estudios Fisicos", in which some of the more recent advances in physical science are explained in an easy and popular form … It is interesting to observe such signs of a revival of interest in the physical sciences in Spain … a country where science has unfortunately been so long at a low ebb.[65]

Any historical approach to the evolution of science in late-modern Spain should seek to identify the specific links to industrial and economic progress claimed by the traditional historiography. The analysis of their importance and the way they work provide an interesting challenge for the historian. In Spain, it was not until the turn of the century that the results of the slow process of industrialisation became noticeable. The legislative programme promoted by Raimundo Fernández Villaverde (1848–1905) at the Ministry of Economy in 1899–1900 balanced the budget and produced a series of surpluses for a whole decade. The considerable economic and industrial growth of the first third of the twentieth century brought serious social tensions, which were ultimately fatal to the political system. However, authors such as Carreras have argued that Spanish process of industrialisation was independent of political changes.[66] As Tortella has explained:

> It is very doubtful that any particular policy aimed at enhancing industrial potential would have played an important role in this process … The growing integration of Spain into the European economy had more influence on its economic development than policies that tended to isolate the country and stand in the way of integration.[67]

There is no doubting the increasing intervention of the State in scientific policy issues during the early twentieth century. Could a similar argument, however, frame the problem of the development of science itself during the late nineteenth and early twentieth centuries in Spain? Was the growing integration of Spain into Europe's scientific network

---

[65] See: *Nature*, 22 (1880), 302.

[66] Albert Carreras, 'An annual index of Spanish industrial output', in Sánchez Albornoz (ed.), *La modernización económica de España* (Madrid: Alianza, 1985), pp. 75–89.

[67] Tortella, *The Development of Modern Spain* (Cambridge, 2000).

more relevant than any State policy? And if so, what exactly does *integrating* Spain into Europe's scientific network mean?

This book seeks to address these questions. It takes into account the way state policies affected the development of science in the country and the evolution of astronomy and astrophysics from the mid-nineteenth century onwards, before any explicit policy on science was adopted. Different Spanish moderate governments attempted to catch up with Europe during the mid-nineteenth century. The new image of the country projected abroad, mainly in pursuit of capital for the modernisation of the country, inevitably affected science. For some scientists this new context brought the opportunity to set up a process which finally evolved, by the end of the century, into a favourable attitude towards science. If this process, rather than any State policy, was what largely generated the growth of scientific activity in Spain during the first decades of the twentieth century, it is worth studying in greater detail. This particular study focuses on the Spanish astronomical community, formed during the late nineteenth and early twentieth centuries.

2.5. Astronomy in Spain by the turn of the century

The turn of the nineteenth century was a period of significant change in the institutionalisation of science in Spain, and the case of astronomy is particularly interesting. From the few private observatories of the last decades of the nineteenth century, such as that of the amateur Rafael Patxot i Jubert (1872–1964), near Barcelona, Spanish astronomers moved into a new context.[68] New institutions sponsored and supported different astronomical and meteorological observatories such as the Fabra (built from 1902 to 1904 in Barcelona), the Cartuja (founded in 1903 in Granada), and the Ebro (finished in 1904 in Roquetes). This new context offered fresh possibilities to the Spanish astronomical community. Spanish clergymen also played a crucial role in the development of astronomy and astrophysics in the country during these years. Members of the Society of Jesus founded both the Cartuja and the Ebro observatories. A few years later, in 1912, the clergyman Ramón María Aller Ulloa (1878–1966) built Galicia's first astronomical observatory, at Lalín. Finally, higher education made a significant contribution to the development of astronomy in Spain during the first decades of the twentieth century. The poor standard of

[68] On the figure of Patxot, see for example: J. Maluquer, *Rafael Patxot i Jubert: Mecens i científic* (Barcelona: Pòrtic, 1994).

lectures on astronomy offered by Spanish universities during the previous century had in many cases been due to a lack of facilities for practical training and education. Academics' frustration with this state of affairs is clearly illustrated by the following text:

> The Faculty has not even a tower in which to place the most modest meteorological instruments and the few that it has for Astronomy and Geodesy cannot be installed due to the absolute lack of suitable conditions … Instruction has to be reduced to oral explanations, which … are far from what is required in the experimental sciences, of observation and application, since the most illuminating descriptions can never replace the experiment, the examination of the objects, or the practical exercise.[69]

The situation was to alter slightly in the new century, with the construction of university observatories in Barcelona and Valencia, both promoted by Ignacio Tarazona y Blanch (1859–1924).[70]

The Spanish astronomical community of the late nineteenth and early twentieth centuries was complex. One of the aims of this book is to show how such an astronomical community was formed and developed. The analysis of these astronomers' methodology, research, results, and publications relating to the total solar eclipses that crossed Spain during these years provides important clues to how social and scientific networks were constructed and developed. The study of these issues not only explains the interest in astronomy in Spain during these years, but also answers questions such as how astronomy was supported and how Spanish astronomers felt about their place as part of a local and/or international community.

This brief introduction has described the general context in which Spanish scientists, particularly astronomers, found themselves by the 1890s. This book also focuses on the total solar eclipses of 1900 and 1905 as catalysts for the modernisation of Spanish scientific activity, in particular astronomy and astrophysics. If *regenerationists* wished to

---

[69] *Memoria del Claustro de Profesores de la Facultad de Ciencias de la Universidad de Madrid, 25 de Febrero de 1898* [Archivo General de la Administración, Sección Educación. Caja 8102. Legajo 8874–1]. Cited in: Baratas Díaz, 'La cultura científica en la Restauración', in Suárez Cortina (ed.), *La cultura española en la Restauración* (Santander, 1999), pp. 279–295.

[70] Enrique de Benito and Vicente Martí y Ortells, 'Don Ignacio Tarazona y el Observatorio Astronómico de Valencia', *Anales de la Universidad de Valencia*, 34 [Año 5] (1924–25).

promote scientific research and dialogue with foreign scientists,[71] then foreign total solar eclipse expeditions to Spain provided extremely good opportunities to develop these networks.

As regards to the institutionalisation of knowledge, it required the building of both a cognitive and a professional identity.[72] The former involves tools, techniques, methodologies, and orientations. The latter, largely studied by scholars, embraces prizes, research grants, positions, appointments, and career-related opportunities. One identifies the structure of a discipline's cognitive identity through journals, reference works, teaching texts, and a sense of common orientation and agreement on central problems and on the relevant techniques of conceptualisation. At the same time, the professional identity of a discipline is discerned in the formation of learned societies, the production of publicity, the recruitment of followers and students, and the creation of satisfactory career structures. Some of these factors can be detected in the discipline building of astronomy during the late nineteenth and early twentieth centuries in the international community, mainly linked to the development of new astrophysical techniques. Indeed, the tensions and cooperation between amateur and institutionalised astronomers show the building of an agreement on central problems regarding astrophysical techniques. Astronomers could also present their results in specialist journals that began to be published during the first half of the nineteenth century, such as *Astronomische Nachrichten* (founded in 1823), the *Monthly Notices of the Royal Astronomical Society* (created in the early 1830s, although published as the *Monthly Notices of the Astronomical Society of London* from 1827 to 1831), and the *Astronomical Journal* (first published in 1849).

However, it was during the last third of the nineteenth and the first third of the twentieth century that many important astronomical and astrophysical journals were begun. *The Observatory*, addressed to a large community of amateur astronomers, was launched in the late 1870s. The

---

[71] This is what the monthly publication *Archivo de Matemáticas* aimed for. Founded in 1893 by Luis Gonzaga Gascó, it had amongst its objectives spreading the works of Valencian and other Spanish mathematicians, establishing personal links with the international scientific community, and translating different works published in foreign journals. See: López Piñero and Navarro Brotóns, *Història de la Ciència al País Valencià* (Valencia, 1995); pp. 411–558.

[72] See for instance: Arnold Thackray and Robert K. Merton, 'On discipline building: The paradoxes of George Sarton', *Isis*, 63 (1972), 437–495.

first *Publications of the Astronomical Society of the Pacific* are from the late 1880s. In France, *L'Astronomie* was published for the first time in 1882—it was absorbed in 1895 by the *Bulletin de la Société Astronomique de France*, created in 1887, and again separately published in 1911. The first issues of the *Journal of the British Astronomical Association*, *Popular Astronomy*, and the *Astrophysical Journal* are from 1890, 1893, and 1895, respectively. Other astronomical journals, such as the Russian *Astronomicheskii Zhurnal*, did not appear until the early 1920s (the English version, *Soviet Astronomy, A.J.*, was first published in the late 1950s). The German *Zeitchrift für Astrophysik* was launched in 1930 and the French *Annales d'Astrophysique* in 1938. Some of these journals were attached to astronomical societies or associations. The foundation of these institutions, with their publicity and recruitment of members, could be seen as the roots of a professional identity. In some countries, the establishment of career structures completed this process of creating a professional identity that, together with a cognitive identity, completed the institutionalisation of astronomical and astrophysical knowledge.

The institutionalisation of astronomy and astrophysics in Spain during these years is presented in this book along the following lines. During the late nineteenth century Spanish astronomers had made use of the institutional structure created abroad, particularly in France, but in the first decades of the twentieth century Spain undertook a discipline-building process in astronomy and astrophysics similar to that of other countries. During these years, Spanish astronomers introduced themselves to the international scientific community through the already established foreign structure, as a consequence both of their aims to achieve status and of the lack of resources in their own country for the development and institutionalisation of astronomy and astrophysics. When these resources became available, Spanish astronomers took similar initiatives to establish themselves in the international context as a respected national community. These ideas are presented here in several sections exploring the development of a Spanish community of professional and amateur astronomers, the foundation of new astronomical societies and their published literature, and the development of astronomy and astrophysics linked to the higher education system, illustrated in the creation of astronomical observatories at universities.

To better understand this process, one has to bear in mind the crucial stimulus provided by the series of total solar eclipses whose path of totality crossed Spain during the second half of the nineteenth and early twentieth centuries. They enabled the Spanish to compare themselves, both socially and politically, with those who visited the country from abroad to observe these events. The facilities and good image that Spain could present to foreign visitors was of course in the minds of many Spaniards, particularly after the military defeat of 1898, in which Spain lost a good part of its credibility as a developed country.

CHAPTER THREE

THE OBSERVATORY OF MADRID

3.1. Astronomy in the early nineteenth century

There were two main branches of astronomy under development by the mid-nineteenth century. The first was related to practical applications, such as navigation and cartography. The second, more theoretical and speculative, was mainly devoted to the observational confirmation of celestial mechanics. During the second half of the nineteenth century some Spaniards became very interested in this kind of astronomy. An observatory had already been created in Madrid for this purpose by the end of the eighteenth century. King Carlos IV founded the Observatory of Madrid in 1790 to deal explicitly with astronomical research. Nonetheless, the political instabilities of the early nineteenth century hugely disrupted the project,[73] and in the French invasion of 1808 the observatory building was almost totally destroyed. All attempts to revitalise the institution during the first half of the nineteenth century were unsuccessful, and very little astronomy was done at the observatory during these years.

The political instabilities and wars of these years had similar consequences for many other European astronomical observatories. For example, the thirty French astronomical observatories that were active during the late eighteenth century were reduced to only two: those of Paris and Marseille. The State funded only the Observatory of the Académie des Sciences. Religious orders were dispersed and provincial universities closed. In Germany, however, individual states of the confederation and

---

[73] See: Manuel López Arroyo, *El Real Observatorio Astronómico de Madrid (1785–1975)* (Madrid: Dirección General del Instituto Geográfico Nacional, 2004). Particularly relevant to the early years of the Observatory are the first two chapters: 'Primera etapa: de la fundación a la destrucción por los franceses', pp. 23–58; and 'Difícil supervivencia. Dependencia de la Junta de Protección del Museo de Ciencias', pp. 59–85.

the universities re-established the observatories and provided them with excellent equipment. In Britain, donations by wealthy people revitalised astronomical observatories, although some universities and religious bodies also took part in the process. In France, it was not until the second half of the nineteenth century that new observatories like those of Longchamps, Toulouse, and Alger were founded.[74]

Similarly to what happened in France, astronomical activity in Spain was revitalised by the re-foundation of the Observatory of Madrid. The difficulties in staffing and equipping the institution, as Ten and Moya Carcel have shown,[75] are evidence of the limitations of the Spanish scientific community at the time. Astronomy in Spain was then monopolised by the Navy, and nobody from the Spanish naval forces felt like relinquishing their military career for the position of Director of the Observatory, a post with an uncertain future.[76] Civilian astronomers were needed to run the observatory, but there were none. Two mathematicians, Antonio Aguilar y Vela (1820–1882), Professor of Mathematics at the University of Santiago de Compostela, and Eduardo Novella y Contreras (1819–1865), a secondary school teacher in Cádiz, were chosen to fulfil this role.

3.2. Reorganisation of the Observatory

Aguilar and Novella were sent to Cádiz in 1847, where the director of the Observatory of San Fernando, Francisco de Paula Márquez, began their astronomical instruction.[77] After a one-year stay, both Aguilar and Novella left on a tour of Europe. The Royal Order of 24 September 1851,

---

[74] C. André and G. Rayet, *L'Astronomie practique et les observatoires en Europe et en Amérique, despuis le milieu du XVIIe siècle jusqu'à nos jours. Vol. 1–2* (Paris: Gauthier-Villars, 1874).

[75] Antonio E. Ten and Teodoro Moya Carcel, 'La formación de un astrónomo en la España del siglo XIX', in Mariano Esteban Piñeiro *et al.* (coords.), *Estudios sobre la Historia de la Ciencia y de la Técnica. Actas del IV Congreso de la Sociedad Española de Historia de las Ciencias y de las Técnicas* (Valladolid: Junta de Castilla y León, 1988), pp. 469–484.

[76] Antonio Gil y Zárate, *De la instrucción pública en España. Tomo III* (Madrid: Imprenta del Colegio de Sordo-Mudos, 1855), p. 371.

[77] Brigadier Francisco de Paula was director of the Observatory from 1856 to 1869. He was succeeded by two prolific astronomers: Captain Cecilio Pujazón y García (1833–1891), in charge from the observatory form 1869 to 1891, and Rear Admiral Juan Bautista Viniegra y Mendoza (1842–1918), director from 1891 to 1903 and Chevalier de la Toison d'Or (Toison Espagnole) in 1914.

which officially refounded the Observatory of Madrid, clearly presents the main reasons for this trip. In part, it was related to governmental instructions to buy the best available astronomical instruments for the observatory, in particular a meridian circle for measuring celestial coordinates and an equatorial telescope for astronomical observations. The trip also provided an opportunity to learn from the most qualified astronomers and, most importantly, allowed the establishment of professional and personal relationships with prestigious scientists. In the words of the Royal Order:

> The creation of an astronomical observatory in this court has been one of the most constant objects of Her Majesty's request, aware as she is of the importance of such a useful establishment in the scientific world … To this effect … her good will arranged that two of the most outstanding young men in the exact sciences passed first to the Observatory of San Fernando and later abroad, in order to acquire the necessary knowledge to put themselves at the head of the planned establishment, to contract for the best instruments, and to know personally the best astronomers of Europe, establishing with them the scientific relations that are indispensable if they are to undertake and realise works of any consideration and importance.
>
> Four years have the *pensionados* spent on these studies and preliminary trips, returning to their mother-land after visiting the principal observatories of Italy, France, Belgium, England, and Germany, and everywhere they have been welcomed not only in a most satisfactory way, but even with joy by all those who are interested in the progress of the sciences and want to see Spain take an active part in them.
>
> The time has come to realise an idea that has to redound to the glory of our country, since, though the astronomical observatory of Madrid does certainly not yet have all the importance that is due to it, it will acquire it in a few brief years, there being perseverance to supply it with all that it needs until it should attain its complete development …[78]

This leads into a current historiographical debate. Twenty years ago, Collins wrote the following words in the postscript to his well-known book *Changing Order*: "individuals' knowledge must be acquired by contact with the relevant community rather than by transferring programmes of instruction".[79] The idea was soon developed by

[78] Cited in: Antonio E. Ten Ros, 'Ver el cielo desde Madrid. La invención de un observatorio astronómico (1846–1860)' (http://www.uv.es/~ten/umlv/oamil.html, 2003). All translations from Spanish sources into English have been made by the author.

[79] Harry M. Collins, *Changing Order: Replication and Induction in Scientific Practice* (Beverly Hills and London: SAGE, 1985), p 159.

constructivists in terms of situating scientists in *networks of trust*.[80] The construction of such relationships meant that not only scientists but also the knowledge they produced became of public interest, mainly through their integration into the international community. The realm of personal experience and connections was as essential to success as the formal constitution of any institution. Indeed, the building up, from the mid-nineteenth century onwards, of appropriate networks by means of which Spanish scientists established personal relationships with prestigious foreign scholars was crucial and boosted the development of astronomy in Spain. This process later influenced the State policies of the early twentieth century which, impregnated of by *regenerationism*, supported scientific research.

Turning back to the historical issue under discussion, the European tour of the two Spanish astronomers can be understood as part of a slow process of establishing networks of trust and credit amongst a broader international community of scientists, particularly in the field of astronomy. Table 3-1 provides a summary of their tour.[81] The trip was concluded in 1851, the same year that the observatory was officially re-founded.[82] The buildings were completed three years later, and the observatory was divided into two main sections: astronomy and meteorology. It was initially staffed by Aguilera and Novella as Director and First Astronomer, respectively. Manuel Rico y Sinobas (1821–1898), lecturer on physics at the Central University of Madrid, was in charge of the meteorological section. The international contacts made by Aguilar and Novella were reinforced through the exchange of scientific papers at an institutional level. There were new civilian astronomers, a new observatory, and a network of trust in its early stages of development.

---

[80] For a brief historiographical introduction to constructivism and its uses in the history of science, see: J. Golinski, *Making Natural Knowledge. Constructivism and the History of Science* (Cambridge: Cambridge University Press, 1998), Chapter 1, 'An outline of cosntructivism', pp. 13–46. See also: Steven Shapin, *A Social History of Truth: Civility and Science in Seventeenth-Century England* (Chicago: University of Chicago Press, 1994).

[81] From: Ten and Moya Carcel, 'La formación de un astrónomo en la España del siglo XIX', in Esteban Piñeiro *et al.* (coords.), *Estudios sobre la Historia de la Ciencia y de la Técnica* (Valladolid, 1988), pp. 469–484.

[82] This section is almost entirely based on: Antonio Aguilar, 'Informe del Director del Observatorio de Madrid al Comisario Regio …', *Anuario del Real Observatorio de Madrid. Año VIII.-1868* (Madrid: Imprenta Nacional, 1867), pp. 235–304.

Table 3-1: Aguilar and Novella's European tour

City	Contacts	Remarks
Turin (Italy)	Giovanni Plana (1781–1864)	Letters of reference to observatories and instrument-makers
Milan (Italy)	Francesco Carlini (1783–1862), Paolo Frisiani (1797–1880)	
Padua (Italy)	Giovanni Santini (1787–1877)	Aguilar and Novella mentioned in *Astronomische Nachrichten* [720 & 723]—discovery of new asteroid and new comet
Geneva (Switzerland)		
Paris (France)	François Arago (1786–1853)	Negotiations with instrument suppliers
Brussels (Belgium)	Adolphe Quetelet (1796–1874)	Close friendship
Greenwich (England)	George Airy (1801–1892)	Letters of reference. Introduction to international scientific community
	Dent (instrument-maker)	Chronometers and pendulums ordered
Bonn (Germany)	Friedrich Argelander (1799–1875)	Decision to buy meridian circle
Hamburg (Germany)	Repsold (instrument makers)	Theodolite and meridian circle bought
Altona (Denmark)		
Berlin (Germany)	Johann Encke (1791–1865)	Close friendship. Optical and meteorological instruments purchased
Strasbourg (France)		
Marseilles (France)		

In 1853, two assistants were added to each section. One assistant of the meteorological section resigned in 1855, as did another from the astronomical section in 1857. As Aguilar pointed out about the latter, he resigned as soon as he had acquired the necessary instruction for the good performance of his post. The new assistants appointed followed a similar path. The main reason was that they were at the observatory while they finished their scientific education at the Central University, but once they were in a position to aim for a better paid post with better prospects, they left the observatory.

During these first years, new instruments were acquired, in particular two chronometers made by Dent (acquired in 1851 and 1853, respectively), a sextant by Oerling (1853), and a theodolite by Repsold (1853). Later arrived a pendulum made by Dent (1854), a meridian circle (objective of 15 cm diameter and 2 metres focal length) by Repsold (1854), and an equatorial (objective of 27 cm diameter and about 5½ metres focal length) by Merz (1858). For the upkeep of the meridian circle, the pendulum, and the chronometers, the Observatory had Augusto Schwerd as instrument-maker, but also asked the clockmaker Villanueva to repair some of these instruments. In order to study how chronometers worked, a programme of meridian observations was initiated in which the altitude of the Sun was measured systematically with the sextant acquired in 1853. It was with the Repsold theodolite that Aguilar in 1853 accurately determined the geographical latitude of the observatory. In November 1854 the meridian circle was installed, and from 1 January 1855 systematic observations were made, including those of the Sun, the Moon, planets, fundamental stars, and circumpolar stars. The zenith distances of these objects on the meridian were also determined. In general terms, this was the scientific programme pursued by the Director and First Astronomer of the Observatory until 1858.

In 1858 the observatory was reorganised. Both sections were combined under the direction of Aguilar. The post held by Rico y Sinobas, who had been in charge of systematic meteorological observations since 1854, was abolished. A new position of Second Astronomer was created, together with four posts for assistants. However, from 1858 to 1859 only three members staffed the observatory: the Director, the First Astronomer, and an assistant. In fact, the new equatorial telescope, acquired in 1858, was used only to observe solar and lunar eclipses, occultations, comets, and other transient events. The main reason was that it was cumbersome to operate, requiring two staff members and an office boy to move the rotating dome. A new assistant was appointed on 1 January 1859. Eleven days later, on 12 January 1859, Miguel Merino y Melchor (1831–1905) was appointed Second Astronomer, also in charge of the meteorological observations. With the addition of these new members of staff, the number of meridian observations increased considerably, although several problems reduced these observations during the following years.

In 1860, both applicants for the post of assistant were accepted—one of them was Eulogio Jiménez Sánchez (1834–1884). The Observatory, now fully staffed, seemed to be in perfect health, as the various astronomical programmes of these years suggest. The most important one was the total solar eclipse of 18 July 1860. Expeditions to observe this eclipse were sent by the Observatory to Moncayo (where the party from the Observatory of Paris was also established), under the direction of the First Astronomer, and to the Desierto de las Palmas, directed by Aguilar. For this solar eclipse new instruments were acquired: two small portable equatorials (objective 12 cm in diameter and 1.85 metres focal length) made by Steinheil and a theodolite by Utzschneider and Fraunhofer, a small comet searcher, and a chronometer capable of measuring time in seconds. The last three were acquired after the death of the owner of a private observatory at Seftenberg. Finally, two (Pistor) reflecting circles were acquired in 1863 for the determination of local time and latitude in special events or circumstances. A geographical programme was also pursued by the Observatory during these years. From 1860 Aguilar worked on the determination of geographical positions with a portable meridian telescope made by Repsold and a pendulum and chronometers made by Dent—they were contracted by Aguilar in 1857 and received in 1860. Although these instruments belonged to the Depósito de Guerra (War Depository), they were kept in the Observatory. This work had almost ceased by 1865, but it did lead to a later collaboration with the Observatory of Paris to find the difference in longitude between the two institutions. It was also from 1860 that the Observatory published its yearbook. The aim of the *Anuario* was inform the people of Spain about the principles and results of astronomy and other sciences related to it, and to wake up and to warm the taste towards similar kinds of studies. Amongst the most outstanding contributions of the early years was the article "Figura de la Tierra", written by Miguel Merino and published in 1863. It was translated into English and published a year later in the Yearbook of the Smithsonian Institution of Washington.

However, the start of 1864 saw not only one assistant resign, after obtaining a chair at the University of Valencia, but also the First Astronomer. After two public examinations, the first of which was unsuccessful because of the lack of preparation of the applicants to the post, Vicente Ventosa y Martínez de Velasco (1837–1919) was appointed as the new assistant in May of that year. But in December another assistant resigned. The subsequent reorganisation of the Observatory led to a new structure. The following members staffed the Observatory: Antonio Aguilar as Director, Miguel Merino y Melchor (1831–1905) as First Astronomer, and Cayetano Aguilar y Vela and Eulogio Jiménez y Sánchez, the older previous assistants, upgraded to Second Astronomers.

There should have been two assistants, but after Vicente Ventosa was appointed, a vacant post remained after public examinations in May 1865 and January 1866 produced no suitable candidate. The two auxiliaries were Eduardo Sánchez Pardo y Ovallo, previously scribe at the observatory since 1857, and, from 1865, Jose Bonet y Alabern.

The Observatory staff were organised into two groups. The First Astronomer, a Second Astronomer and an auxiliary formed the first, while the other Second Astronomer, the assistant, and another auxiliary formed the second. The Director occasionally joined one group or the other as necessary. With this arrangement, each group made systematic observations for one week. In principle, the auxiliary would make the meteorological observations, the astronomer would make the celestial observations during the day and the first hours of night, and the remaining member of the group would make the observations from nine in the evening until midnight or later. Most of them had to work at their offices from nine in the morning until two in the afternoon on calculations based on their observations. The exceptions were the auxiliaries, who as students at the Central University had to attend to lectures. This work of reducing observations and calculation, largely delayed during the first years due to the shortage of staff, had to be done in the mornings by all members of the observatory, even if they had been observing the previous night. The programme of meridian observations was also expanded from 1864 to the observation of asteroids and fundamental stars, and two years later to the revision of the positions of the stars in the British Association Catalogue. The observatory also provided public services such as the daily measurement and telegraphic transmission of the time, and the comparison of the pendulum with chronometers from the observatory and from both other institutions and private individuals. They also wrote daily weather forecasts, analysed and studied different meteorological apparatus, etc.

The good health of the institution during these years allowed the Director of the Observatory to set up new projects, including the building of an annexe to install the two Steinheil portable equatorials. As Aguilar wrote in 1867:

> In completing this project there will be a triple advantage: the instruments will be sheltered and will be preserved in good condition, for any unforeseen occasion on which they will be of absolute need; the Auxiliaries of the Observatory will be able to practise observations that some day they will have to carry out with major and more delicate apparatus; and the simple amateurs of astronomy, who are found in a

capital like Madrid, and to whom it is not just to close the doors of the Establishment, will equally be able to satisfy their natural and legitimate curiosity, contemplating the principal celestial bodies, without bothering the professional observers and inadvertently spoiling any other instrument of major importance.[83]

One year after the publication of these lines, the political situation changed dramatically. Isabel II, the re-founder of the Observatory, was dethroned by the Glorious Revolution of 1868. The Comisario Regio (Royal Delegate) to the Observatory was removed after the Glorious Revolution, and the Observatory came under the control of the Central University. But the University wanted to play an active role in the organisation of the Observatory's astronomical activity, not like the Royal Delegate, who simply helped with the administration. Subsequently, the Director, Antonio Aguilar, was removed from office in 1871, although he was reinstated a year later.[84] There is a record of him representing Spain, as Director of the Observatory of Madrid, at the International Meteorological Conference in Rome in 1879, three years before his death.

The decades of the 1860s, 1870s, and 1880s were years in which the Observatory of Madrid began to be involved in some astrophysical tasks. First, the Observatory began a programme of systematic observations of sunspots during the 1860s with the 27 cm aperture equatorial by Merz. The direct observation of sunspots changed in January 1870, after the visit of the British astronomer Richard Carrington (1826–1875) to the Observatory. It was during this visit that Carrington suggested the use of the projection method that he had developed. A systematic programme based on this new method was fully established in 1875 and lasted until 1897. As a result, more than four thousand drawings were obtained, as well as large quantities of data.[85] The Observatory also tried to photograph sunspots, although without success due to the lack of appropriate instruments. Finally, in 1887 and 1888 the Observatory attempted a spectroscopic study of solar prominences. However, because of the

---

[83] *Ibid.*

[84] López Arroyo has suggested that political and ideological motives were behind the 'persecution' of Aguilar, as he was a Carlist. See: M. López Arroyo, 'Los dos siglos de existencia del Observatorio Astronómico de Madrid', in *Doscientos años del Observatorio Astronómico de Madrid* (Madrid: Asociación de Amigos del Observatorio Astronómico, 1992), pp. 1–26.

[85] 'El eclipse total de Sol de este mes', *El Mundo Científico*, 7 (1905), 527–534, here pp. 529–530.

difficulties of this task, the observations with the Merz direct vision spectroscope were quickly abandoned.[86]

The last third of the nineteenth century saw a progressive decline of the Observatory of Madrid. On the one hand, there were significant problems of getting and keeping staff. Young students had not the training and formation needed to work at the Observatory and it seemed that the only way to recruit staff was to convince some well-established Spanish scholars to join in. However, the personal problems of the First Astronomer with the Director and the rest of the staff of the Observatory, which resulted in the resignation of Novella in 1864, may have made it difficult to work in the institution. It is not surprising, then, that some of the new astronomers and assistants moved into new positions in the academic world as soon as they had an opportunity. On the other hand, the political instabilities of the period meant constant changes and restructurings in the organisational chart. These issues, together with the several public works projected by the City Council of Madrid around the Observatory, were translated into uncertainties as regards the future of the institution. During these years the Observatory took on almost no new staff, renewed none of its astronomical equipment after the acquisition of two Steinheil telescopes for the eclipse of 1860, saw the meteorological studies gradually become independent of the Observatory, and cancelled the publication of its yearbook (from 1881 until 1907).[87]

3.3. The beginnings of the twentieth century

This decline of the Observatory of Madrid has to be viewed against the background of the state of Spanish science at the time. As the astronomer Miguel Merino pointed out in 1893, in describing the last gasps of the *polémica* about Spanish science:

> Let us concentrate our thoughts and eyes in the wretched current state of the sciences in Spain and form brave resolutions to contribute to their return from their lethargy and to come to fertile and vigorous life.[88]

---

[86] *Ibid.*, p. 531.

[87] López Arroyo, *El Real Observatorio Astronómico de Madrid (1785–1975)* (Madrid, 2004).

[88] Miguel Merino, 'Discurso de contestación de D. Miguel Merino' in Acisclo Fernández Vallín, *Cultura científica en España en el siglo XVI* (Sevilla: Padilla Libros, 1989) [*Discursos leidos ante la Real Academia de Ciencias Exactas,*

It seems that Ventosa, as First Astronomer, was in charge of the Observatory of Madrid when the Englishwoman Elizabeth Brown (1830–1899) visited it on 6 May 1894. A sense of the state of the observatory can be got from her description of the visit.[89] The Observatory had a Transit House where a very fine transit instrument was kept. Close by there was a small room which contained a great variety of instruments, "many of which were preserved as venerable relics in glass cases, curious through their age".[90] A separate building with a revolving dome contained a Repsold refractor of 11 inches aperture. As mentioned above, the observation of sunspots was made by projection, and for ascertaining heliographic latitude and longitude, the method developed by Carrington was employed.[91] Although sunspots had been observed systematically at the Madrid Observatory since 1868, the lack of staff members had made it impossible to keep a daily record in the form of drawings. As Brown wrote, the Observatory had "only quite recently begun to photograph the spots".[92] Finally, the lack of funds and the small number of staff meant that none of the observations were published during these years. As Brown concluded, "it does seem a cause for regret that the splendid climate of Spain cannot be better utilised".[93]

More discouraging was the ministerial inspection of the Observatory carried out by Juan de la Cierva in 1900:

> The most important telescope was not working because the rails on which it turned had to be fixed, and the money had been spent on decorating the offices of the astronomers in the same building; that two other magnificent telescopes acquired for the last solar eclipse visible in Spain were lacking electrical connections of scant cost. They wanted me to see the stars with those telescopes and naturally I could see none! Then the business about

---

Físicas y Naturales en la recepción pública del Excmo. Sr. D. Acisclo Fernández Vallín (Madrid: Sucesores de Rivadeneyra, 1893)], p. 335.

[89] E. Brown, 'A visit to the Madrid Observatory', *Journal of the British Astronomical Association*, 4 (1893–1894), 361–362.

[90] *Ibid.*

[91] See for instance: E. H. T. Teague, 'Carrington's method of determining sunspot positions', *Journal of the British Astronomical Association*, 106 (1996), 82–85.

[92] Brown, 'A visit to the Madrid Observatory', *Journal of the British Astronomical Association*, 4 (1893–1894), 361–362.

[93] *Ibid.*

> the missing light bulb came out. The result was that the three telescopes I mentioned were impossible to use.[94]

But the series of total solar eclipses of 1900, 1905, and 1912, together with the observations of Halley's Comet in 1910, would be crucial in revitalising the interest in astronomy in Spain.

After Miguel Merino left the management of the Observatory in 1898, Ventosa would have been appointed Director if the tradition had been followed. But the University, under the auspices of the Royal Academy of Sciences, appointed Professor Francisco Iñiguez Iñiguez (1853–1922) on 24 August 1899 as the new Director of the Observatory of Madrid. His relationship with some members of the staff at the Observatory was a very difficult one, particularly so with Vicente Ventosa, First Astronomer and Interim Director of the Observatory during the first eight months of 1899. The origins of their conflicts may have been the unwillingness expressed by Ventosa about a director who, contrary to himself, was not an experienced professional astronomer.[95] The tensions were explicit from the first year of Iñiguez's directorship. The lack of information received and the cooperation demanded by Ventosa from the new director is a good example.[96] For instance, the programme of meridian observations distributed amongst the Second Astronomers, Antonio Tarazona y Blanch (1843–1906), Antonio Vela Herranz (1865–1927), Carlos Puente Úbeda and Ramón Escandón Piñeiro, and the assistants Francisco Cos Mermería (1864–1943) and Miguel Aguilar Cuadrado (1869–1925), was never delivered to the First Astronomer.[97]

---

[94] Juan De la Cierva, *Notas de mi vida* (Madrid: Instituto Editorial Reus, 1955), pp. 65–66; cited in Juan Vernet, *Historia de la ciencia Española* (Madrid: Instituto de España, 1975), p. 223; translated into English by Glick, *Einstein in Spain* (Princeton, 1988), pp. 48–49.

[95] As an example, here are the words that Ventosa addressed to the director after his appointment: 'I do not believe myself authorised to judge the meridian observations made by Your Excellency from the end of 1899 until mid-April 1900, with the enthusiasm of those who want to gain experience in this kind of study.' See: [Vicente Ventosa], *Memoria del primer astrónomo del Observatorio de Madrid al director del mismo establecimiento. 1899–1900* (Madrid, 1902), p. 43.

[96] *Ibid.*, p. 7.

[97] *Ibid.*, p. 11.

Fig. 3-1: Francisco Iñiguez. Director of the Observatory of Madrid from 1899 to 1919. Published in the Spanish periodical *El Universo* on 29 August 1905.

Meridian observations were one of the main activities of any official observatory. In January 1900 the new director established a programme of observations for the Repsold meridian circle that the observatory owned.[98] There were two daily activities: on the one hand, time determinations and the supervision of the proper working of the astronomical instruments; on the other, the observation of specific stars to study atmospheric refraction. Divided into groups, the astronomers and assistants spent a week at a time on these observations. One astronomer and an assistant observed circumpolar stars to study atmospheric refraction. Another astronomer was in charge of time determination and other regular activities. An astronomer who one week observed the circumpolar stars had to devote himself the following week to routine tasks. While the astronomer in charge of the routine services had to observe six reference stars (two circumpolar and four equatorial), the other astronomer and the assistant had to observe between them at least twenty circumpolar stars and reduce their observations. The programme was based on a selection of 292 stars from

[98] *Ibid.*, pp. 10–11.

the Greenwich Catalogue of 1880, the positions of which were reduced to 1900 by Vela with the help of the assistant Cos.[99]

This programme of meridian observations was seriously criticised by Ventosa.[100] First, he considered that the number of observers was too large. There were three observers each day who observed in total about thirty stars daily. Due to the magnificent weather in Spain, there were on average two hundred clear nights each year, which equated to about six thousand observations annually. In other words, the Observatory of Madrid could make more meridian observations per year than many prestigious international observatories could manage. Second, the orders of the new director implied that astronomers had to reduce their observations the morning after making them. The Observatory did not have people to specifically carry out this task, as other observatories did. This was particularly inconvenient during the summer, as the meridian observations were extended until late in the morning. Third, all this work produced an enormous amount of results which, due to the lack of resources, could not be published. Only a few of these results were published in international journals or sent to foreign scholars who demanded them for their work.

Other astronomical observations, of for example the opposition of Mars, and the meteorological and magnetic observations pursued at the Observatory, are also worth mentioning. Astronomers also had to do some work in their offices, both administrative and scientific. As has been mentioned, the lack of staff and economic resources meant that astronomers had to copy and reduce their own observations. This was in contrast with what happened at other national observatories. For example, in 1879 the Observatory of Paris already had twenty-six employees. During the following years, the number decreased gradually, to twenty by 1926, but there were still six assistants, six calculators, and one person working exclusively on administrative matters.[101] In general terms, and as Ventosa argued, the model of work pursued at the Observatory of Madrid was in many ways inefficient and a waste of time.[102]

---

[99] *Ibid.*, p. 117.

[100] *Ibid.*, pp. 12–13.

[101] See: S. Débarbat, S. Grillot and J. Lévy, *Observatoire de Paris: son histoire 1667–1963* (Paris: Observatoire de Paris, 1984).

[102] *Memoria del primer astrónomo del Observatorio de Madrid ... 1899–1900* (Madrid, 1902), pp. 113–116.

Ventosa was not a man who kept criticism to himself. His own published description of the astronomers of the Observatory of Madrid shows what he thought of them:

> It is evident that Mr. Puente is of all the observers the most constant and assiduous, as might be expected from his calm, methodical, and committed approach to work. Given these conditions of his character, his uncommon erudition, and his experience, acquired during many years of unsurpassable services, it is not necessary to say that the observations and calculations made by Mr. Puente will have all the guarantees of precision and seriousness.
>
> Mr. Tarazona probably follows Mr. Puente, in the relative quantity of work done; but this observer is, on the contrary, rather passionate and inconstant: when he tackles the resolution of any problem, and he does not lack intelligence and knowledge for this, he usually begins his task with singular eagerness, employing many consecutive hours in this without rest; sometimes he finishes it, but every so often … he needs from further stimulus to focus his attention. This volubility of Mr. Tarazona's character makes him a bit untidy and not as methodical as he should be. One notices also that he worries too much about what his colleagues do to classify his own work, when this should matter little to someone who feels real love for science.[[103]] As regards the quality of his observations, they don't seem to be unworthy of those of Mr. Puente, although I do not think of him as having as much aptitude as observer than as calculator …
>
> Mr. Vela is a very erudite and enthusiastic person, of vivacious and resolved character, and, although with less experience than the two astronomers above mentioned, having not exercised his profession for so long, he probably overcomes them in shrewdness, and he is not unworthy of them in the care and attention in his job, or the sincerity with which he does it, without looking for ostentation …
>
> From the Observatory of San Fernando, where he worked as astronomer many years ago, Mr. Escandón had great practice in managing instruments and in calculations, and this was undoubtedly in another time; but at present, probably not being that young and having some physical weariness, or weak sight, or for any reasons that I do not know, his work, I find hard to say it, corresponds neither in quantity nor in quality to that of his colleagues, nor to what from his precedents and his extensive and varied knowledge should be expected.[104]

---

[103] Ventosa later described him as a "tireless advocator of the *equal distribution* of work". See: *Ibid.*, p. 114.
[104] *Ibid.*, pp. 40–42.

Fig. 3-2: Antonio Tarazona. Astronomer of the Observatory of Madrid. Published in the Spanish periodical *El Universo* on 29 August 1905.

With regard to the links between the Observatory and the University of Madrid during these years, illustrated in the figure of the new director of the Observatory, it has also to be mentioned that the Royal Decree of 4 August 1900 meant that Tarazona and Cos had to lecture for the last year of undergraduate studies at the Faculty of Sciences of the Central University. The former lectured on physical astronomy; the latter on meteorology. These lectures were delivered daily in the Observatory's library, leading to complaints from the staff, unable to use the library while the lectures were in progress.[105]

During the years that followed, new scientific instruments were purchased to expand and improve the range of observations made at the Observatory. One of the most important was the spectroheliograph made by Grubb, an instrument for forming a photographic image of the sun in a

[105] *Ibid.*, p. 122.

narrow wavelength, of special design.[106] In 1919 Antonio Vela Herranz (1865–1927) was appointed Director of the Observatory of Madrid, a position he held to the time of his death. The life of Vela illustrates the difficult path that those interested in developing a professional career in astronomy had to follow during the late nineteenth and early twentieth centuries. After graduating at the Central University of Madrid as a Doctor of Science, Vela became attached to the staff of the Observatory of Madrid as assistant in 1884. Five years later he was promoted to Astronomer, his fields of study being meteorology and positional astronomy. In 1894 he also became Assistant Professor in the Faculty of Sciences and in 1908 he was elected Professor of Astrophysics, another position he held until his death. His observations for determining latitude and longitude at different localities in Spain and at the Observatory entailed a very large number of meridian observations of stars, ending in the compilation of a catalogue. During this last period of his life, though weakened in health, he secured the publication of all the work on astrophysics carried out at the Observatory, and was contemplating the publication of photometric observations of Nova Cygni 1920. He was elected a Fellow of the Royal Astronomical Society on 10 March 1922.[107]

The figure of Vela typifies, in certain ways, the work of many astronomers who worked at the astronomical Observatory of Madrid during the second half of the nineteenth and the early twentieth century. A glimpse through the most relevant publications of these years reveals several notable names such as Miguel Aguilar Cuadrado, Cayetano Aguilar Vela, Pedro Carrasco Garrorena, Honorato Castro y Bonet, Francisco Cos y Mermería, Ramón Escandón y Piñeiro, Victoriano Fernández Ascarza, Enrique Gastardi Peón, Francisco Íñiguez e Iñiguez, Pedro Jiménez Landi, Eulogio Jiménez Sánchez, Miguel Merino y Melchor, Eduardo Novella, Carlos Puente y Ubeda, Gonzalo Reig y Soler, Manuel Rico y Sinobas, Antonio Tarazona y Blanch, José Tinoco Acero, Eduardo Torroja y Caballé, Antonio Vela y Herranz, and Vicente Ventosa y Martínez de Velasco. The evolution of Spanish official astronomy is illustrated by these names, working in an institution that during the first decades of the twentieth century not only employed similar instruments to

[106] Howard Grubb, 'The new spectroheliogaph for the Madrid Observatory', *Report of the British Association for the Advancement of Science* (Dublin, 1908), 603–604.

[107] See: *Monthly Notices of the Royal Astronomical Society*, 88 (1928), 258–259.

those of the most prestigious observatories in the world, but also had similar observational programmes.[108]

[108] Ignacio Tarazona y Blanch, *La fotografía solar. Discurso leído en la Universidad Literaria de Valencia en la solemne apertura del curso 1909–1910* (Valencia: Miguel Gimeno, 1909), p. 60.

CHAPTER FOUR

PRIVATE ASTRONOMERS AND THE ORIGINS OF NEW ASTRONOMICAL INSTITUTIONS

4.1. Introduction

Some years ago, the Spanish historian of science López Piñero pointed out the importance of the generation of Spanish scholars born by the mid-nineteenth century and active during the 1880s for the development of medicine in Spain.[109] These scholars, building on the achievements of a few isolated figures in previous decades, were in general self-taught people open to new doctrines developed in the rest of Europe. The development of astronomy in Spain followed similar lines. The solar eclipses of 1860 and 1870 illustrate the attempts of some Spanish scientists to prepare the ground for the development of astronomy and astrophysics in their country. They also established international cooperation which helped the progress of astronomy in Spain. That was the case with the official parties sent by the Observatory of San Fernando to Oropesa (Castellón), together with the Portuguese Commission from the Observatory of Coimbra, and the expedition from the Observatory of Madrid to the Desierto de las Palmas (Castellón), organised jointly with the Italian party led by the Jesuit Father Pietro Angelo Secchi (1818–1878) from the Observatory of the Gregorian University of Rome.

These collaborations also led to the first recognised publications by Spanish astronomers in international scientific journals. For instance, the new Director of the Observatory of Madrid, Antonio Aguilar y Vela (1820–1882), had his observations published in the *Comptes Rendus* of the

---

[109] See for instance: José María López Piñero, Luis García Ballester, and Pilar Faus Sevilla, *Medicina y sociedad en la España del siglo XIX* (Madrid: Sociedad de Estudios y Publicaciones, 1964).

French Academy of Sciences.[110] Another example is the note on the results of the observations of the eclipse of 1860 by the Spanish Naval Commission published in the *Monthly Notices of the Royal Astronomical Society*. These results, noticed (but not printed) in the *Monthly Notices*, were published by the Director of the Observatory of San Fernando, Francisco de Paula Márquez de Prado y Roco (1816–1886), in the *Memoria sobre el Eclipse de Sol de 18 de Julio de 1860*. This work presented an elaborate optical theory for some eclipse phenomena, including the corona and the red prominences, "considered as arising from the diffraction and reflection in our atmosphere of the light refracted at the border of the Moon's disk, and from the motion of the Moon".[111]

For the total solar eclipses of 1860 and 1870, Aguilar, as the Director of the Observatory of Madrid, invited the international astronomical community to visit Spain. Information was provided and facilities were offered to expeditions interested in going. All this was warmly welcomed by the international scientific community. Several official expeditions were sent to Spain to observe these eclipses, including those from Lisbon (Portugal), Rome (Italy), Pulkovo (Russia), Greenwich (England), Paris (France), Lund (Sweden), Copenhagen (Denmark), Munich (Bavaria), Leipzig (Germany), and Geneva (Switzerland), with their respective directors travelling ahead. The eclipses of 1860 and 1870 were therefore not only considered of general interest, but also proved very helpful in projecting an international image of Spain, largely through the new Observatory of Madrid. The scientific activity at the Observatory of Madrid and the visits by foreign astronomers were in perfect accord with the prestige policy of mid-nineteenth-century Spanish governments.

Visits by foreign astronomers enriched the Observatory of Madrid through the establishment of institutional and personal relationships. Foreign astronomers had to write to the Observatory to take advantage of the facilities offered by the Spanish government. This correspondence was not only useful for foreign astronomers, who received special treatment at custom houses when entering the country, but also for Spanish astronomers.

---

[110] Antonio Aguilar, 'Mesures prises en Espagne pour l'observation de la prochaine éclipse totale de soleil' [Lettre de M. Aguilar à M. Le Verrier], *Comptes Rendus*, 50 (1860), 483–484; for correspondence with the Secretary of the French Academy regarding the results of these observations see: *Comptes Rendus*, 51 (1860), 412, and 889–890.

[111] 'Recent publications', *Monthly Notices of the Royal Astronomical Society*, 22 (1861–1862), 22.

Foreign parties were encouraged to read the publications of the Observatory of Madrid and to provide details of the instruments they had brought with them.[112] The Director of the Observatory of Madrid would thus have been well informed about the most recent developments in astronomical research, particularly about solar physics. This opened new doors to the practice of astronomy in Spain, which until then had relied very much on the Navy. Not only the Observatory of San Fernando, but even amateur astronomers had links with the military. For example, Salvador Clavijo y Clavijo, one of the few amateur astronomers who published on astronomical issues in the *Comptes Rendus* before 1860, was a military engineer from Tenerife amongst whose works were a *Mémoire sur la vitesse de rotation des planètes* and *Réflexions sur le système planétaire*.[113]

The eclipses of 1860 and 1870 also gave Spanish scientists the opportunity to pursue some original astronomical research. For example, copies of photographs of the corona during the 1860 eclipse taken by José Monserrat, a member of the expedition co-organised by the Observatories of Madrid and Rome, were sent all over Europe for further study. As the *Monthly Notices of the Royal Astronomical Association* recorded:

> There were exhibited at the Meeting [14 December 1860] four photographs of the Sun during the Eclipse of the 18th July, 1860, obtained at Desierto de las Palmas in Spain, by M. Monserrat, Professor of Chemistry at the University of Valentia [sic], with the telescope by Cauchoix of the Observatory of the Collegio Romano, brought by the Director, P. Secchi. The photographs were communicated to the Astronomer Royal by Signor [sic] Aguilar, the Director of the Madrid Observatory.[114]

At the same time, the strong links between astronomy and mathematics encouraged the pursuit of astronomical research by figures such as Eduardo León y Ortiz (1846–1914). León y Ortiz secured a position in the Astronomical and Meteorological Observatory of Madrid after graduating

---

[112] See for instance the translation of a letter from Antonio Aguilar to the *Astronomische Nachrichten* that appeared in *Nature*, 3 (1870–1871), 69.

[113] References to Calvijo are found in *Comptes Rendus*, 47 (1858), 258; 48 (1859), 414; and 71 (1870), 387. A copy of his *Reflexiones sobre el sistema planetario* (Santa Cruz de Tenerife: Imp. Isleña … de J.C. Hernandez, 1870) is kept in the Biblioteca Nacional de España (Signatura: 1/1229).

[114] See: *Monthly Notices of the Royal Astronomical Society*, 21 (1860–1861), 51. The only other photographs of the solar corona obtained at this eclipse were the two taken by the expedition of Warren de la Rue to Spain.

in sciences in 1870. However, he soon left to be appointed Professor of Superior Algebra and Analytical Geometry, firstly at the University of Granada and later at University of Valencia. He moved back to Madrid in 1882 to occupy the Chair of Geodesy (renamed Spherical Astronomy and Geodesy in 1900) at the Central University. Another example is Lorenzo Presas y Puig, Professor of Differential Calculus at the University of Barcelona, who had written three short works on astronomical and physical subjects by 1862.[115]

4.2. The new generation of Spanish astronomers

These individuals were the predecessors of the Spanish astronomers who were born in the middle decades of the nineteenth century, those who gradually applied to their observations the new astrophysical techniques developed in Europe and latterly in America. This new generation of Spanish astronomers, most of them self-taught people, produced some important original research which was duly recognised by the international scientific community. This was particularly so for planetary, lunar, and solar observations, at a time when government support for astronomical research was still very poor. In fact, any attempt to present a reasonably complete picture of science in Restoration Spain must look beyond the institutions and take into account the activities of such individuals.[116] As in other European countries, amateurs played a crucial role in the development of astronomy during the second half of the nineteenth century. By the last third of the century, the progress of astronomy and astrophysics in Spain was linked not only to institutions such as the Observatories of Madrid and San Fernando, but also to the personal work of individuals such as José Joaquín Lànderer Climent (1841–1922), Josep Comas i Solà (1868–1937), and Eduard Fontserè Riba (1870–1970). By the 1890s, the importance of their work had already been acknowledged in various French scientific journals.[117]

---

[115] These short works were presented to the *Comptes Rendus* of the French Academy by the French Minister of Public Instruction. See: *Comptes Rendus*, 60 (1862), 835.

[116] See: Antoni Roca Rosell and Víctor Navarro Brotóns, 'Reflexions sobre la ciència i la tècnica a la Restauració', in G. Blanes, L. Garrigós, A. Roca and J. Assizabalaga (coords.) *Actes de les IV Trobades d'Història de la Ciència i de la Tècnica* (Alcoi and Barcelona: Societat Catalana d'Història de la Ciència i de la Tècnica, 1997), pp. 181–190.

[117] See references to Lànderer, Comas i Solà, and Fontserè in: Bouquet de la Grye, 'Les Progrès de l'Astronomie en 1891', *Bulletin de la Société Astronomique de*

The astronomical activity of José Joaquín Lànderer[118] began in the 1870s with his systematic observations of Jupiter's satellites. For this purpose he acquired a 95 mm aperture reflector made by Secrétan. Lànderer also made, from Tortosa, some astronomical observations with a small telescope of 57 mm aperture.[119] He later purchased two more reflectors, of 108 and 135 mm aperture, and a refractor made by Bardou of 75 mm aperture. As Navarro Brotóns and Gozalo Gutiérrez have shown, his results, published in several Spanish and French scientific journals, led to the confirmation of the analytical theory of the movement of Jupiter's satellites developed by Cyrille-Joseph Souillart (1828–1898).[120] His observations covered the study of not only the period of rotation of Jupiter, but also Jupiter's spots and the diameters of its satellites.[121] His broad knowledge of geology led him to his work on the composition of the

---

France, 6 (1892), 151–153, 155; 'Annuaire Astronomique pour 1892', *L'Astronomie*, 11 (1892), 2; and Camille Flammarion, 'Les Progrès de la Société Astronomique de France', *Bulletin de la Société Astronomique de France*, 6 (1892), 159–163, here p. 161. See also references to Lànderer and Comas i Solà in: Camille Flammarion, 'Nouvelles observations sur la planète Mars', *L'Astronomie*, 11 (1892), 36–382, here pp. 374, 377.

[118] On the figure of Lànderer and his important role in the development of astronomy and astrophysics in Spain during the late nineteenth and early twentieth centuries, see: R. Gozalo Gutiérrez and V. Navarro Brotóns, 'Josep Joaquim Lànderer i Climent (València 1841—Tortosa 1922). La recerca fora del món acadèmic: astronomia i geologia', in J. M. Camarassa and A. Roca (dirs.), *Ciència i Tècnica als Països Catalans. Una aproximació biogràfica* (Barcelona: Fundació Catalana per a la recerca, 1995), pp. 457–493; V. Navarro Brotóns and R. Gozalo Gutiérrez, 'Entre "amateurs" i professionals: l'obra astronòmica de Josep Joaquim Lànderer i Climent (1841–1922)', in Carles Puig-Pla, Agustí Camós, Jon Arrizabalaga, and Pasqual Bernat (coords.), *Actes de les III trobades d'història de la ciència i de la tècnica als Països Catalans* (Barcelona: SCHCT, 1996), pp. 303–312, here pp. 305–306; and López Piñero and Navarro Brotóns, *Història de la Ciència al País Valencià* (Valencia, 1995); pp. 411–558.

[119] J. J. Lànderer, 'La Scintillation et les climats', *L'Astronomie*, 13 (1894), 341–343.

[120] Navarro Brotóns and Gozalo Gutiérrez, 'Entre "amateurs" i professionals ...' in Puig-Pla *et al.* (coords.), *Actes de les III Trobades d'Història de la Ciència i de la Tècnica ...* (Barcelona, 1996), pp. 303–312, here pp. 305–306. See also for instance: Lànderer, 'La Tache rouge de Jupiter', *L'Astronomie*, 11 (1892), 394–395; and Lànderer, 'Sur la théorie des satellites de Jupiter', *L'Astronomie*, 13 (1894), 256–257.

[121] Lànderer, 'Diamètre des satellites de Jupiter', *L'Astronomie*, 12 (1893), 255–256; 'Sur un passage de l'ombre du quatrième satellite de Jupiter', *Comptes Rendus*, 120 (1895), 248–250.

Moon. His most outstanding contribution on the lunar composition was his method based on the polarisation of light reflected by the Moon. From these studies, made with a Cornu photopolarimeter, Lànderer concluded that the surface of the Moon was mainly formed of basaltic rocks.[122] His studies of the polarisation of light during the eclipse of May 1900, his observations of Jupiter and its satellites, and his work on sunspots and lunar eclipses were recognised when the Société Astronomique de France awarded him the Janssen Prize at the beginning of 1901.[123]

Another example of this new generation of amateur astronomers is Augusto T. Arcimis Werle (1844–1910). Elected a Fellow of the Royal Astronomical Society on 10 December 1875,[124] he made several observations during the 1870s with a 100 mm (4-inch) Secrétan refractor with a 92-power eyepiece from his private observatory in Cadiz.[125] He

---

[122] Julio Garrido, 'Un precursor español de las investigaciones petrográficas de la superficie lunar: José Joaquin Landerer', *Revista de la Real Academia de Ciencias Exactas, Físicas y Naturales de Madrid*, 67 (1973), 285–287; and Navarro Brotóns and Gozalo Gutiérrez, 'Entre "amateurs" i professionals ...' in Puig-Pla *et al.* (coords.), *Actes de les III Trobades d'Història de la Ciència i de la Tècnica ...* (Barcelona, 1996), pp. 303–312, here p. 307. See the following works by Landerer: 'Essai de sélénologie', *Les Mondes*, 51 (1880), 854–856; 'Histoire ancienne de la lune', *Les Mondes*, 54 (1881), 184–192; 'Las revoluciones del globo lunar', *Anales de la Sociedad Española de Historia Natural*, 11 (1882), 153–191; 'Sur l'angle de polarisation de lune', *Comptes Rendus*, 109 (1889), 360–362; 'Sur l'angle de polarisation des roches ignées et sur les premières déductions sélénologiques qui s'y rapportent', *Comptes Rendus*, 111 (1890), 210–212; 'Le Problème cosmologique', *Bulletin de la Société Astronomique de France*, 16 (1902), 270–274; 'Sur la polarisation de la lumière lunaire', *Comptes Rendus*, 150 (1910), 1164–1165; and his books *Principios de Geología y Paleontología* (Barcelona: Librería Religiosa, 1878) and *La evolución del globo lunar* (Barcelona: Herederos de Juan Gili, 1910). He also submitted to the French Académie des Sciences in 1880 a work entitled *Géologie lunaire* in which he determined the lithographic nature of the Moon. See: *Comptes Rendus*, 90 (1880), 1018.

[123] Cammille Flammarion, 'La Société Astronomique de France', *Bulletin de la Société Astronomique de France*, 15 (1901), 205–217, here p. 216.

[124] See: *Monthly Notices of the Royal Astronomical Society*, 36 (1875–1876), 53.

[125] See: 'Observations of the zodiacal light at Cadiz', *Monthly Notices of the Royal Astronomical Society*, 36 (1875–1876), 48; 'Observations of the lunar eclipse, 1876 September 3, made at Cadiz', *Monthly Notices of the Royal Astronomical Society*, 37 (1876–1877), 12–13; 'On the visibility of the unilluminated portion of the disc of Venus', *Monthly Notices of the Royal Astronomical Society*, 37 (1876–1877), 259; 'Phenomena of Jupiter's satellites observed at Cadiz', *Monthly Notices of the Royal Astronomical Society*, 37 (1876–1877), 259–260; 'Observations of

Fig. 4-1: Josep Joaquim Lànderer. Published in *La Ilustración Española y Americana*, 15 June 1900 [Año 44, nº 22].

made several astronomical observations in the following decades,[126] although he later specialised in meteorology, a subject much more affordable and accessible in Spain during the second half of the nineteenth century. His interest in both astronomy and meteorology led him to create an observatory at the Institución Libre de Enseñanza, where he lectured:

> To the *Boletín de la Institución Libre de Enseñanza* for January 15, D. Augusto Arcimis sends an account of the meteorological branch of the observatory recently attached to the institution … This meteorological station is thus one of the best equipped in Europe, and in fitting it up advantage has been taken of the experience already acquired from the working of similar establishments elsewhere.[127]

occultations of stars by the Moon, made at Cadiz in the year 1876', *Monthly Notices of the Royal Astronomical Society*, 37 (1876–1877), 261; 'Lunar eclipse of 1877 February 27, observed at Cadiz', *Monthly Notices of the Royal Astronomical Society*, 37 (1876–1877), 400.

[126] See: *Nature*, 30 (1884), p. 324.

[127] See: *Nature*, 31 (1884–1885), p. 349.

4.3. The influence of French astronomy

Arcimis probably knew a small group of amateur astronomers from Cadiz under the leadership of Manuel S. Navarro. Some of their observations with a 94 mm (3¾-inch) telescope made by Negretti and Zambra at magnifications of 142 and 200 were published in *L'Astronomie*.[128] This French journal became one of the main ways for Spanish amateur astronomers to present their observations. There was no similar periodical in Spain. At the same time, the lack of local astronomical instrument-makers (Óptica Roca, the first one, did not start until 1910) forced many Spanish amateur astronomers to buy French telescopes from firms such as Bardou, Mailhat, Secrétan and Vion.[129] All this explains why, even when both Great Britain and, later in the century, Germany were seen by the Spanish elite as the most prominent examples of material progress, France remained during the nineteenth century the reference point for many Spanish astronomers. Indeed, French astronomy was very influential in Spain, particularly amongst amateur astronomers.

A glance at the articles in *L'Astronomie* during the early 1890s shows the success of amateur astronomy in Spain. Amateur observers were spread all over the country: there were for example Pedro de A. Peña, from Palma in Majorca,[130] Enrique López Morales, Director of the Meteorological Observatory of Almeria,[131] Juan Valderrama, from Santa

---

[128] Philippe Gérigny, 'Société Astronomique de France. Séance du 8 janvier 1890', *L'Astronomie*, 9 (1890), 69–73, here p. 71; Manuel S. Navarro, 'Nouvelles de la science. Variétés. Le Cirque Archimède', *L'Astronomie*, 9 (1890), 74; Philippe Gérigny, 'Société Astronomique de France. Séance du 5 février 1890', *L'Astronomie*, 9 (1890), 103–105; Philippe Gérigny, 'Société Astronomique de France. Séance du 5 mars 1890', *L'Astronomie*, 9 (1890), 148–150; Manuel S. Navarro, 'Un nouveau cratère dans l'arène de Plinius', *L'Astronomie*, 9 (1890), 187; Manuel S. Navarro, 'Gutenberg et Godenius', *L'Astronomie*, 9 (1890), 187–188.

[129] Josep M. Oliver, *Historia de la astronomía amateur en España* (Madrid: Sirius, 1997).

[130] Philippe Gérigny, 'Société Astronomique de France. Séance du 8 janvier 1890', *L'Astronomie*, 9 (1890), 69–73, here p. 71.

[131] 'L'Eclipse de Soleil du 17 Juin', *L'Astronomie*, 9 (1890), 241–257, here p. 256; and Gaston Armelin, 'Société Astronomique de France. Séance du 1er octobre 1890', *L'Astronomie*, 9 (1890), 424–428. Given the conjunction of astronomy and meteorology during the nineteenth century, it might seem odd to classify López Morales as an amateur astronomer. However, he has left no other trace. In fact, he was probably no more than a meteorological observer.

Cruz, Tenerife,[132] and Idelfonso González, secretary of the Sociedad Científica Flammarion of Jaén, who observed with a telescope of 108 mm aperture and 160 power enlargement.[133]

Camille Flammarion (1842–1925) played a particularly significant role during these years. Flammarion became the most prestigious astronomer and populariser of astronomy of the late nineteenth and early twentieth centuries after the publication of his book *La Pluralité des mondes habités* (1862; published in Spain by 1866). His popular books, such as *Astronomie populaire* (1880), together with his work in journals such as *Cosmos*, *Magasin Pittoresque*, and *Le Siècle*, and the foundation of the journal *L'Astronomie* (1882) and of the Société Astronomique de France (1887), made of him one of the great names of this period. His influence may partially explain the increasing interest in astronomy in Spain, especially amongst Spanish amateur astronomers. The following lines illustrate the esteem in which he was held in Spain:

> The famous astronomer Camille Flammarion, when delivering an enthusiastic speech on the marvels of the Jovian world in the city hall of some small Spanish town, was welcomed by hysterical crowds, and his path was covered with thousands of flamboyant roses.[134]

As has been mentioned, scientific societies named after him were founded in Spain during these years. The Sociedad Científica Flammarion of Jaén was founded in 1881/1882 under the presidency of José M. Folaché. The Flammarion Scientific Society of Madrid, promoted by Ernesto de la Guardia, was created in 1899.[135] Francisco Cantillo, José Guillén de la Cadena, Ramiro Blanco, and Leonardo Ortega Andrés were also members of this society, also known as the Astronomical Society of Madrid. Many of the new Spanish amateur astronomers subscribed to

---

[132] 'L'eclipse de Soleil du 17 Juin', *L'Astronomie*, 9 (1890), 241–257, here p. 256; 'Nouvelles de la Science.–Variétés. L'éclipse totale de Soleil du 16 avril', *L'Astronomie*, 12 (1893), 230–233, here p. 232; 'Nouvelles de la Science. Variétés. Conjonction de Saturne avec l'étoile double gamma de la Vierge', *L'Astronomie*, 12 (1893), 233–235; and Juan Valderrama, 'Nouvelles de la Science. Variétés. Conjonction d'Uranus avec α de la Balance', *L'Astronomie*, 13 (1894), 234–235.

[133] 'Nouvelles de la science. Variétés. Une épée dans la Lune', *L'Astronomie*, 9 (1890), 75.

[134] Jean-Claude Pecker and Gisèle Ringeard, 'Astronomy, the people and the governments', in *Vistas in Astronomy*, ed. Arthur Beer, 15 (1973), 1–12.

[135] See: Oliver, *Historia de la astronomía amateur en España* (Madrid, 1997).

L'Astronomie and were members of Société Astronomique de France. The increasing number of Spanish amateur astronomers during the second half of the nineteenth century is also illustrated by the award received by Ernesto de la Guardia from the Société Astronomique de France: the "médaille frappée à la Monnaie de Paris pour récompenser le zèle du sociétaire qui aura amené le plus grand nombre de nouveaux adhérents".[136] He was awarded this medal after introducing fourteen new members to the Society in 1900. In fact, this award illustrates how rapidly the number of Spanish amateur astronomers increased during the first decade of the twentieth century. This was possible, at least in part, because of the development of amateur astronomy in the neighbouring country of France. In the absence of any Spanish equivalents, the French amateur astronomical institutions brought together this important group of Spaniards interested in astronomy and astrophysics.[137]

4.4. From amateurs to professionals

During the second half of the nineteenth century, Spain experienced a rapid increase in the number of amateur astronomers. The theosophist Mario Roso de Luna (1872–1931), from Logrosán (Extremadura), is another example. His name became familiar to amateur and professional astronomers after his discovery of a comet in July 1893. This discovery was, in fact, the observation of an object which he supposed to be a new star. He made a sketch of its position and sent it with a letter to the Director of the Observatory of Madrid.[138] Similar observations had been made independently by other observers such as William E. Sperra, and Alfred Rordame, who were also in the running for the Comet Medal to be awarded by the Astronomical Society of the Pacific that year for the

---

[136] Cammille Flammarion, 'La Société Astronomique de France', *Bulletin de la Société Astronomique de France*, 15 (1901), 205–217, here p. 210.

[137] The election of Luis de Ocharán and Ramón Sovaluce as members of the Société Astronomique de France in 1902 is a good example. Ramón Sovaluce was presented to the Société by Flammarion and Janssen as the 'Director of the Observatory of San Sebastian'. My efforts to find more information about this individual have been unsuccessful. However, the lack of information about the institution he was in charge of suggests that it was a meteorological rather than an astronomical observatory. Luis de Ocharán, from Bilbao, was proposed to the Société by Flammarion and Bertaux. He was elected *perpetual member* of the society. See: 'Admissions et présentations', *Bulletin de la Société Astronomique de France*, 16 (1902), 528.

[138] See: *Publications of the Astronomical Society of the Pacific*, 6 (1894), 155.

discovery of the comet. The difficulty of ascertaining the first discoverer of the comet meant that the Comet Medal Committee withheld the medal, and no award was made that year.[139] However, the Comet 1893 II was finally named in honour of Rordame, an amateur astronomer residing in Salt Lake City.[140]

Younger generations of Spanish amateur scientists played their part in the professionalisation of science in Spain. As Glick pointed out, Spanish scientists born around 1880, the so-called *generation of 1914*, "attempted to institutionalise the regenerationist goals".[141] One of them was Eduard Fontserè, who began as an amateur scientist and later became one of the most prominent Spanish meteorologists of his time. His early meteorological and astronomical studies[142] were considered by many scientists, as recorded in *L'Astronomie*:

> M. Gérigny, secrétaire, lit un rapport sur deux mémoires qui ont été soumis à son examen : la premier, par M. Ed. Fontséré, sur la rotation du Soleil, travail que le rapporteur estime très sérieux, tout en faisant ses réserves sur la conclusion.[143]

But perhaps the figure that stands out amongst the other astronomers of this period is that of Comas i Solà. In writing about Comas i Solà, mention must be made of Rafael Patxot Jubert (1872–1964), who in 1894 founded a private observatory at San Feliu de Guíxols known as the Catalan Observatory. The 5 m diameter dome of this observatory housed a refractor of 22 cm aperture made by Mailhat. The visual and photographic equatorial had attached a micrometer and a spectrograph. After the

---

[139] 'Discovery of comet B 1893', *Publications of the Astronomical Society of the Pacific*, 6 (1894), 146.

[140] In the new system of notation this comet is C/1893 N1. Known as Rordame–Quénniset, this comet was named also after Ferdinand Jules Quénisset (1872–1951).

[141] Thomas F. Glick, 'Emilio Herrera and Spanish technology', in Emilio Herrera, *Flying: The Memoirs of a Spanish Aeronaut* (Albuquerque: University of New Mexico Press, 1984), pp. 173–215, here p. 173. There is a Spanish version: Emilio Herrera, *Memorias*, ed. Thomas F. Glick and José M. Sánchez Ron (Madrid: Universidad autónoma de Madrid, 1988), pp. 161–195.

[142] See for example: 'L'eclipse de Soleil du 17 Juin', *L'Astronomie*, 9 (1890), 241–257, here p. 250; and Gaston Armelin, 'Société Astronomique de France. Séance du 3 Janvier 1894', *L'Astronomie*, 13 (1894), 70–73, here p. 71.

[143] Gaston Armelin, 'Société Astronomique de France. Séance du 3 Mai 1893', *L'Astronomie*, 12 (1893), 226–229.

inauguration of this private observatory, Patxot contracted Comas i Solà as astronomer. Notable amongst their work at the Catalan Observatory are several planetary observations, mainly of Mars, Jupiter, and Saturn, and a catalogue of 156 double stars.

The popularity achieved by Comas i Solà during the early 1890s was probably one of the main reasons why Patxot chose him to work at the Catalan Observatory. Indeed, Comas i Solà had become a recognised astronomer due to his original contributions to journals such as *L'Astronomie*. In his late teens, when he began his university degree in the physical and mathematical sciences at the University of Barcelona, he bought a refractor made by Bardou of 108 mm aperture which he installed on the flat roof of his house. With it he made not only many solar, lunar, and planetary observations, but also observations of stars and nebulae that were published in Flammarion's *L'Astronomie*.[144] He was also well known in Barcelona for his regular popular-astronomy contributions from 1893 to the Catalan newspaper *La Vanguardia*.

Comas i Solà worked at the Catalan Observatory on a regular basis from 1897 until 1899, when he finally resigned from his position and set up a private observatory at his house. Here he installed a telescope of 156 mm (6 inches) aperture made by Grubb.[145] With this telescope he made

---

[144] See for instance: 'Nouvelles de la science.- Variétés : Une épée dans la Lune', *L'Astronomie*, 9 (1890), 75; J. Comas, 'Nouvelles de la Science.- Variétés : L'étoile multiple σ d'Orion', *L'Astronomie*, 9 (1890), 154; 'L'eclipse de Soleil du 17 Juin', *L'Astronomie*, 9 (1890), 241–257; Comas, 'Nouvelles de la Science.- Variétés : Antarès', *L'Astronomie*, 9 (1890), 311–312; Comas, 'Nouvelles de la Science.- Variétés : Saturne, ζ du Cancer, Antarès', *L'Astronomie*, 9 (1890), 353–354; Gaston Armelin, 'Société Astronomique de France. Séance du 1er octobre 1890', *L'Astronomie*, 9 (1890), 424–428; Philippe Gérigny, 'Société Astronomique de France. Séance du 3 décembre 1890', *L'Astronomie*, 10 (1891), 70–72; 'La Tache Rouge de Jupiter', *L'Astronomie*, 10 (1891), 321–322; 'L'éclipse totale de Lune du 4 Novembre dernier', *L'Astronomie*, 12 (1893), 13–18; 'Nouvelles de la Science.–Variétés. Conjonction de Saturne avec l'étoile double gamma de la Vierge', *L'Astronomie*, 12 (1893), 233–235 ; 'Observation de Jupiter faites à l'Observatoire de Juvisy, à Bruxelles et en Espagne. Observations de M. José Comas, à Barcelone', *L'Astronomie*, 12 (1893), 81–90; Gaston Armelin, 'Société Astronomique de France. Séance du 3 Janvier 1894', *L'Astronomie*, 13 (1894), 70–73.

[145] The following lines, which appeared in the *Journal of the British Astronomical Association*, show the exact location of the new observatory: 'Our Fellow Member, Señor J. Comas Solà, has just inaugurated his private observatory, Sant Felip, 29,

many interesting observations of Mars and Jupiter, and in the late 1910s and the 1920s discovered eleven asteroids and a couple of comets.[146] For his observations at Villa Urania, as he renamed his house, he also purchased a micrometer, a spectroscope to study solar prominences, and a camera. Comas i Solà became one of the most prestigious Spanish astronomers of the first decades of the twentieth century. His work and original contributions to astronomy were recognised early in his career, as his election as Fellow of the Royal Astronomical Society on 14 March 1902 shows.[147] He also wrote several books, amongst which stand out *Astronomía y Ciencia General* (1907), *Astronomia* (1919), and *El Cielo* (1927).

As the work of these Spanish astronomers became more accepted and respected abroad, their observations and results were taken more seriously. In some cases, their claims for priority show not only their confidence in their observations and work but also the progress of Spanish astronomy. A good example is that of Lànderer as regards the exponential form of the Titius–Bode Law. The idea of regularity in the mean distances of the planets from the Sun was formalised through an arithmetic series in the eighteenth century by Johann Daniel Titius (1729–1796). The law $r_n = 0.4 + 0.3 \times 2^n$ (with $n = -\infty, 1, 2, \ldots$) was popularised by Johann Elert Bode (1747–1826) and hence it became known as *Bode's law*. However, the discovery of Neptune divided astronomers into two camps: the opponents, such as Urbain Jean Joseph LeVerrier (1811–1877), John Frederick William Herschel (1792–1871), Elias Loomis (1811–1889), and Simon

---

Sant Gervasi, Barcelone [sic]; N. lat. 41°24′2″, E. long. 8^m35^s2'' '. See: 'Notes: A new observatory', *Journal of the British Astronomical Association*, 10 (1899–1900), 79.

[146] For his observations of Jupiter, see: 'Observaciones de Júpiter durante la Oposición de 1900', *Astronomische Nachrichten*, 153 (1900), 417–430. These observations attracted the attention of some other astronomers, as one can see in: Touchet, 'Société Astronomique de France', *Bulletin de la Société Astronomique de France*, 16 (1902), 518–528, here p. 520. On his observations of Mars, see for instance: 'Observaciones de Marte', *Astronomische Nachrichten*, 191 (1912), 5–8. The asteroids Hispania, Alphonsina, Barcelona, Amelia, Sadeya, Reginita, Pepita, Mercedes, Comas Solá, Pólit, and Gothlandia were discovered in March 1915, January 1920, February 1921, October 1922, January 1927, May 1927, November 1928, October 1929, November 1929, December 1929, and September 1930, respectively. His first periodic comet (originally designated 1926f and now P/32 Comas-Sola) was discovered in November 1926.

[147] See: *Monthly Notices of the Royal Astronomical Society*, 62 (1901–1902), 345. Comas i Solà was proposed by A. Stanley Williams.

Newcomb (1835–1909), versus the proponents, who included John Pringle Nichol (1804–1859), Daniel Kirkwood (1814–1895), Stephen Alexander (1806–1883), and Edouard Roche (1820–1883). Furthermore, during the nineteenth century alternative laws were proposed. The first one to state the exponential form $r_n = r_0 \times k^n$ was Lànderer, choosing the value 1.7 for k.[148] Lànderer published his formula first in Spain, in 1866, and then in France,[149] but Louis Gaussin was credited in 1880 as the author of the exponential law, having in his formula a slightly different k.[150] In the *Comptes Rendus* in 1880, Lànderer claimed "priorité des idées émisses par M. Gaussin, concernant l'arrangement des planètes", soon after the publication of Gaussin's results.[151]

Another example is that of Comas i Solà, who had an interesting public discussion in the *Journal of the British Astronomical Association* in the 1910s with Antoniadi about some of Solà's astronomical observations:

> Après la galante critique de M. Antoniadi par rapport à mes observations de Mars faites en 1894 (avec une lunette de 108^{mm}), critique publiée dans une note du "Sixth Report of the Section for the Observation of Mars", page 75, et après ma protestation contre cette critique, protestation publiée dans le "Journal of the British Astronomical Association", page 438 du n° 8 de l'année courante, M. Antoniadi explique la raison de sa critique à la page 47 du "Seventh Report of the Section".
>
> M. Antoniadi dit, simplement, qu'avec un 4½ pouces il était impossible de voir, en 1894, le Lac du Phoenix, obscur, défini et rougeâtre, puisque le grand Dawes ne put jamais l'apercevoir auparavant avec un 8 pouces.[152]

---

148 See: S. R. Ignjatovlae, 'History of the exponential form of the Titius–Bode Law', in S. Jokić, I. Milošević, A. Balaž and Z. Nikolić (eds.), *Proceedings of the Fifth General Conference of the Balkan Physical Union. Vrnjačka Banja, Serbia and Montenegro, August 25–29, 2003* (http://www.phy.bg.ac.yu/~bpu5/proceedings/Papers/SP18%20-%20003.pdf, 2003).

149 See: J. J. Lànderer, 'Rapport entre les distances mutuelles des planètes et des satellites', *Les Mondes*, 15 (1867), 61–62; and 'Notes sur les planètes intra-mercurielles. Vulcain et Pluton', *Les Mondes*, 48 (1879), 420–423.

150 Louis Gaussin, 'Lois concernant la distribution des astres du système solaire', *Comptes Rendus*, 90 (1880), 518–520; and Louis Gaussin, 'Lois concernant la distribution des astres du système solaire', *Comptes Rendus*, 90 (1880), 593–596.

151 See: *Comptes Rendus*, 90 (1880), 717.

152 José Comas i Solà, 'Quelques considérations sur mes observations de Mars', *Journal of the British Astronomical Association*, 21 (1910–1911), 199–201, here p. 199.

His defence was based on the observations of Mars made by Stanley Williams in 1894, in his "sincérité scientifique, jamais déméritée formellement", and in his "practique d'observateur, déjà reconnue".[153] His rivalry with Antoniadi was long-lasting, as the following lines, written in 1913 about his observations of Jupiter's Great Red Spot, prove:

> Dans le No. 4668 des A[stronomische].N[achrichten]., M. *Antoniadi* dit que le 1 octobre 1902, dans une carte postale adressée au Dr. *Crommelin*, il avait attiré son attention sur la probabilité que l'accélération de la Tache Rouge, lors de sa conjonction avec la tache grise tempérée australe, pourrait être due à une poussée de cette dernière, et que Mr. *Denning*, dans "Nature" (18 décembre 1902), énonça la même hypothèse en toute indépendance, hypothèse qui a été confirmée par les faits. Je dois ajouter que j'avais déjà publié, dans le No. de septembre 1902 du Bulletin de la Soc. Astr. de France l'hypothèse de la "poussée". Voyez ce No., page 389, où je dise, en comparant les rotations de la pointe suivante de la Tache Rouge en 1901 et 1902 : "La poussée du courant de protubérances (actuallement, la zone grise tempérée australe) sur la Tache Rouge est donc bien sensible".[154]

4.5. The evolution of astronomy in Catalonia and Spain

As has already been pointed out, Catalonia was the Spanish region where the process of industrialisation experienced a major boost during the nineteenth century. Together with this industrial and economic progress, Barcelona became one of the most productive centres of scientific research in the country during these years. The development of scientific activity in Catalonia during the nineteenth century was strongly linked with the Acadèmia de Ciències i Arts of Barcelona.[155] To advance the study of meteorology and astronomy, this Academy promoted the foundation of an observatory in the early 1880s. Lànderer, a member of this Academy, was asked to provide a budget to purchase astronomical and meteorological

---

[153] *Ibid.*, p. 200.

[154] Comas i Solà, 'Sur l'accélération de la Tache Rouge de Jupiter pendant les conjonctions avec la tache grise tempérée australe', *Astronomische Nachrichten*, 195 (1913), 413–414.

[155] This institution was founded in 1764. Originally named the Conferencia Physico-matematica Experimental, it changed its name four years later to become the Real Academia de Ciencias Naturales y Artes. The current name, Real Academia de Ciencias y Artes, was finally adopted in 1892. See: A. Nieto-Galán and A. Roca Rosell (dirs.), *La Reial Acadèmia de Ciències i Arts als segles xviii i xix. Història, ciència i societat* (Barcelona: RACAB/IEC, 2000).

instruments. In 1884 he was appointed director of the (non-existent) observatory, but the lack of economic resources finally led to the cancellation of the project.[156]

Ten years later, in 1894, this Academy supported a similar project proposed by Eduard Fontserè: the building of an observatory next to the city, on top of Tibidabo Hill. The proposal was originally rejected but was resurrected later, in 1902, under the directorship of Comas i Solà. The observatory was finally inaugurated in the presence of King Alfonso XIII on 7 April 1904. Comas i Solà was appointed Director of the Observatory that same year, a position that he kept until his death. This appointment precipitated a conflict between Comas i Solà and Fontserè which the Academy tried to solve in 1912 by making the two recognised scientists co-directors. However, the enmity between them remained and led in the 1910s to the creation of two different astronomical societies in the country.

Fig. 4-2: The Fabra Observatory. From: *El Mundo Científico*, 7 (1905).

[156] Gozalo Gutiérrez and Navarro Brotóns, 'Josep Joaquim Landerer i Climent …' in Camarassa and Roca (dirs.), *Ciència i Tècnica als Països Catalans* (Barcelona, 1995), pp. 457–493.

The new observatory was named in honour to the nobleman and industrialist Camil Fabra, who since 1900 had provided crucial funding for the project. In general terms, non-religious bodies had a lack of economic resources in Spain, and the wealthy classes did not give financial support to any scientific projects. But the economic development of Catalonia during the late nineteenth century made this region a special case in Spain. As Tarazona wrote in 1909:

> There are not in our land any Carnegie, Yerkes, or Lick, American multimillionaires; but believe me, it is not easy to find people here in Europe who, as the Marquis of Alella, funds Observatories such as the Fabra, of Barcelona.[157]

All in all, it was clear that progress in astronomy was being made in Spain during the 1900s, even when it seemed that at times it was more a matter of chance rather than a consequence of a carefully planned strategy. This is illustrated by the introduction of new techniques such as photography to astronomical research. For example, the success of the photography by the official Spanish expedition sent by the Observatory of Madrid to observe the eclipse of 1900 depended on the comments and suggestions from the Irish astronomers next to the Spanish official party in Plasencia. The director of the Observatory of Dublin, Charles Jasper Joly (1864–1906), helped with focusing the equatorial telescope; and Arthur Alcock Rambaut (1859–1923) and William Edward Wilson (1851–1908) provided the Spanish astronomers with the appropriate formula to obtain good pictures with long exposures. As Ventosa wrote:

> Mr. Vela has confessed to me that … [he] experienced disagreeable surprise when … he was ordered, together with Mr. Aguilar, to obtain the photographs of the solar corona, without any preliminary tests … not knowing what kind of plates were more suitable to use in this case; which were the best developers and toners; nor how to prepare the plates to ensure as far as possible a flattering success.[158]

---

[157] Tarazona y Blanch, *La fotografía solar* (Valencia, 1909), p. 60.
[158] *Memoria del primer astrónomo del Observatorio de Madrid ... 1899–1900* (Madrid, 1902), p. 78.

Similarly, Comas i Solà, observing from Elche, obtained one photograph of the coronal spectrum, but the lack of apparatus to measure the lines prevented him from publishing any results.[159]

These shortcomings contrast with one of the most important events organised after the eclipse of 1912 by the Astronomical Society of Barcelona: the General Exhibition of Lunar Study held at the University of Barcelona from 15 May to 15 June 1912. The exhibition had the object of "grouping in one harmonious whole the discoveries of human intelligence" relating to the Moon, and consisted of the following sections: lunar cartography; representations of the Moon by drawings and models; photographs of the Moon; lunar physics; the study of the tides; observatories and instruments; "apparatus for cosmographical demonstration"; history and biography; and lunar astrology.[160] As Glick has noted, this event was a celebration of Spanish astronomy and meteorology:

> Spanish exhibitors gave a strong impression of the maturity of Spanish astronomy and the ability of Spanish astronomers to occupy a place in the forefront of research, particularly in the area of photography.[161]

In sum, it was for disciplines in which amateurs had more opportunity to develop their own work, such as natural history and astronomy, that important scientific societies were created in Spain during the last third of the century. This is very well illustrated by the creation of the Spanish Royal Society of Natural History in 1871. However, the creation of the most relevant societies had to wait until the last years of the nineteenth century and the early twentieth century. As regards the physical and mathematical sciences, the nineteenth century was characterised in Spain by a lack of institutions. The only one was the Real Academia de Ciencias Exactas, Físicas y Naturales, created in 1847. However, it was not until the twentieth century that institutions of experts and specialists in scientific subjects were set up, exemplified by the creation of new associations and

---

[159] Victoriano Fernández Ascarza, 'Eclipse total de sol de 1900. Nota sobre la longitud de onda de la raya verde (1474 K), del espectro de la corona solar', *Revista General de Marina* (April 1901), 1–15, here p. 14.
[160] S. Raurich, 'Correspondence: The Astronomical Society of Barcelona', *Journal of the British Astronomical Association*, 22 (1912), 239–240.
[161] Glick, *Einstein in Spain* (Princeton, 1988), pp. 50–51. See also: *Exposición General de Estudios Lunares: celebrada en mayo y junio de 1912 en la Universidad de Barcelona ... Catálogo* (Barcelona: Sociedad Astronómica de Barcelona, 1912).

societies such as the Royal Society of Physics and Chemistry (1903) and the Spanish Association for the Advancement of Science (1908). At the same time, there were initiatives to bring science to the Spanish public. Particularly significant was the foundation of learned amateur societies in Catalonia. For example, the Catalan Institution of Natural History was created in 1899, the Astronomical Society of Barcelona was founded in 1910, and the Astronomical Society of Spain and America was established in 1911. Some of these new scientific societies were set up largely for political reasons, with the aim of changing Catalonia economically and socially. Science was considered by many Catalan nationalists as an element of their specific identity. As stated in the charter of the Institute of Catalan Studies (1907), Catalans required new scientific institutions and research centres that reflected the cultural identity of Catalonia.

But this new context also meant a conflict for astronomy in Spain. The Astronomical Society of Barcelona was founded in January 1910. After its first year of existence, the Society had about four hundred members and was expected to be soon equipped with an observatory for the use of its members. The new observatory had to house instruments donated by Patxot, whose observatory at San Feliu de Guíxols was to be re-erected in Barcelona: a double equatorial (visual and photographic) of 230 mm (9 inches) aperture, and several accessories for spectroscopic, photographic, and micrometric work.[162] Under the auspices of Salvador Raurich i Ferriol (1869–1945) and Fontserè, the Astronomical Society of Barcelona was the meeting point and institution in charge of coordinating Spanish, and particularly Catalan, astronomical and meteorological amateurs. On the other hand, the Astronomical Society of Spain, an offshoot of the Society of Barcelona promoted and presided over by Comas i Solà, aimed to encourage cooperation between amateur and the few local professional astronomers.[163] The Society, founded in February 1911, had Mariano Rubio and Manuel Olió as Vice-President and Treasurer, respectively. The subscription to the Society included receiving its journal, the first issue of which was published in April 1911, and its governing body was formed of twelve people, ten of them living in Barcelona.

Even if one might be tempted to explain the creation of these two different astronomical societies in terms of political agendas (Fontserè was an important protector of the cultural identity of Catalonia), the foundation

---

[162] 'Notes: The Astronomical Society of Barcelona', *Journal of the British Astronomical Association*, 22 (1911–1912), 111–112.
[163] 'Notiz', *Astronomische Nachrichten*, 188 (1911), 15–16.

of these associations has also to be understood in terms of personal animosities. A few months after the foundation of the Astronomical Society of Barcelona, Comas i Solà accused Raurich of campaigning against him and his observations on Mars in the Society's journal. As the references were not explicit, the Governing Body of the Astronomical Society requested Comas i Solà not to feel insulted by Raurich's comments. However, Comas i Solà did not accept this decision and announced his rupture with the Society, which he thought had been set up specifically to antagonise him.[164] One year later the Astronomical Society of Spain was founded, promoted by Comas i Solà.

This situation proved dysfunctional. There was not enough activity in astronomy for both of them to survive. The Astronomical Society of Barcelona soon made meteorology its main activity, partly encouraged by Fontserè's own interests and personal relationships with other important Spanish meteorologists. The Astronomical Society of Spain, on the other hand, had a higher level of participation and proved more committed to astronomy. While the Astronomical Society of Barcelona was extinguished in the 1920s, after the instauration of the Catalan Meteorological Service, the Astronomical Society of Spain has survived until today.

164 Antoni Roca Rosell, 'Una vida consagrada a la ciència' in A. Roca Rosell (coord.) *Josep Comas i Solà, astrònom i divulgador* (Barcelona: Ajuntament de Barcelona, 2004), pp. 13-42.

CHAPTER FIVE

ASTRONOMY AND RELIGION

5.1. Jesuits in the history of astronomy

Almost since Ignacio de Loyola (1491–1556) founded the Order in 1540, Jesuits have had a permanent interest in the study of geophysical phenomena. Amongst the most outstanding work are the several astronomical, meteorological, and seismological observations made by members of the Order during the seventeenth and eighteenth centuries. Such activity has generally been explained in terms of their educational work in colleges and universities; their missionary endeavours to remote lands, where they observed interesting and often as yet undocumented natural phenomena; and the network of communication that brought research by other Jesuits to their attention.[165] This scientific tradition led to the establishment of different Jesuit geophysical research centres in the nineteenth century. By the first third of the twentieth century there were more than thirty Jesuit geophysical observatories around the world.[166] Some of them were mainly astronomical institutions, but carried out geophysical observations as well. Others, however, were created specifically as meteorological, magnetic, or seismological observatories. The foundation of Jesuit observatories dedicated to meteorology had the aim of preventing damage from natural catastrophes such as hurricanes and severe weather. These studies were particularly important in the Caribbean and the Pacific. Similar motives led to the creation of seismological observatories for the study of earthquakes, volcanic eruptions, and so on.

---

[165] Agustín Udias and William Stauder, 'Jesuit geophysical observatories', *EOS*, 72 (1991), 185–189.

[166] *Ibid.* See also: Agustín Udias, *Searching the Heavens and the Earth: The History of Jesuit Observatories* (Dordrecht: Kluwer Academic Publishers, 2003).

The first modern Jesuit observatories were those at Georgetown (United States of America) and Stonyhurst (England). The latter was perhaps one of the most famous astronomical observatories of the nineteenth century. Founded in 1838, the Observatory of the Jesuit Stonyhurst College gained great fame with the geomagnetic work initiated by Stephen J. Perry (1833–1889) and later continued by Walter Sidegreaves (1887–1919). Amongst their most relevant contributions were their research on the relation between magnetic variations and solar activity and their magnetic surveys in many areas of Europe and Asia. This observatory became a research centre for talented astronomers, many of whom were later selected by the British Government to be in charge of and take part in important astronomical expeditions. The last British solar eclipse expedition led by an astronomer from the Stonyhurst Observatory was the one to the Tonga Islands in 1911, which was led by the Director of the Observatory, Father Aloysius Laurence Cortie (1859–1925).

5.2. The Ebro Observatory

By the early twentieth century, the Province of Aragon of the Society of Jesus had established in Tortosa (Tarragona) a philosophical faculty, a chemical laboratory, and a biological station. Moreover, the first years of that century saw the foundation of a new Jesuit astrophysical observatory in Roquetes, a neighbouring village of Tortosa. This initiative had a great impact on foreign scientists all over the world. As recorded in *Knowledge & Scientific News*:

> Many circumstances have combined to render the foundation of a new Spanish observatory of great interest and importance. There is first the feeling of not unnatural surprise at such a happening in a country which had for many years been regarded as non-progressive, and by some writers freely labelled decadent, dying, effete, or some other equally uncomplimentary epithet. Signs of revival are now more numerous, and a King on his own feet with an English Queen to sympathise with his modern ideas bids fair to modify many of the harsh criticisms passed of late years upon his historic heritage. Yet the observatory is only in a sense of Spanish growth. It is the latest branch of the scientific activity of the powerful Society of Jesus, whose labours at Kalocsa, at Herény, at Stonyhurst, at Georgetown, D.C., at Havana, at Manila, and many other posts are landmarks in the domains of astronomy and cosmological physics.[167]

[167] 'Observatory of the Ebro. Notes of the 1905 eclipse', *Knowledge & Scientific News*, 4 (1907), 3–5.

Vue des pavillons de l'Observatoire
1) Pavillon électro-météorologique.—*2)* Pavillon sismique.—*3)* Pavillon astrophysique.—*4)* Kiosque néphoscopique de réfraction.—*5)* Kiosque pour les appareils magnétiques absolus.—*6)* Pavillon pour les appareils magnétiques de variation.

Fig. 5-1: The Ebro Observatory. From: R. Cirera, *Notice sur l'observatorie et sur quelques observations de l'éclipse du 30 Août 1905* (Barcelona: Gustavo Gili, 1906).

Indeed, the observatory was designed to be one of the most complete for the study of solar-terrestrial relationships and astrophysical observations, as the French astronomer Charles Nordmann (1881–1940) wrote:

> L'Espagne sera bientôt dotée d'un Observatoire d'Astronomie physique qui n'aura rien à envier aux plus perfectionnés de ceux qui se trouvent en Angleterre et en Amérique, et qui même réalisera en quelque sorte un établissement unique par l'idée générale qui a présidé à sa fondation et à l'organisation de ses divers services, à savoir: l'étude des relations, de plus en plus évidentes, qui existent entre l'activité du Soleil et les divers phénomènes électriques et magnétiques de notre globe.[168]

The idea of a cosmic-physics research centre was conceived by Father Ricardo Cirera Salse (1864–1932) at the magnetic section of the

[168] C. Nordmann, 'Fondation d'un observatoire astrophysique en Espagne', *Revue générale des sciences pures et appliquées*, 15 (1904), 474–475.

Observatory of Manila (Philippine Islands).[169] In order to give life to the project, Cirera spent a whole year in Paris, where he picked up several ideas from the International Exposition of 1900, and he also spent ten months at Stonyhurst, with "shorter visits being paid to Kew and Greenwich".[170] After visiting several British and French observatories, which were not very well equipped with modern electrical and seismological equipment, he proceeded to Belgium, Germany, Austria-Hungary, and Italy. He visited, amongst others, the observatories of Jersey (United Kingdom), Strasbourg (France), Meudon (France), Montsouris (France), Uccle (Belgium), Potsdam (Germany), Kalocsa (Hungary), Ogyalla (Hungary), Vienna (Austria), the Vatican, and Rocca di Papa (Italy). He spent a long time at these observatories, usually choosing and testing the instruments he purchased in the country of origin: the Dover magnetometer at Kew, the dip instrument at Potsdam, the spectroscopic apparatus at Meudon, and so on.[171] The definitive location of the future observatory was chosen by early 1902. Several prestigious astronomers and scientists were involved in the realisation of the project: they included Henri Alexandre Deslandres (1853–1948), John Evershed (1864–1956), Jean Mascart (1872–1935), Maurice Loewy (1833–1907), Karl Schwarzschild (1873–1916), Giulio Grablowitz (1846–1928), Charles Chree (1860–1928), Hermann Carl Vogel (1841–1907), Joseph Maria Pernter (1848–1908), Charles Nordmann, and Lànderer.

Philanthropy, a characteristic of nineteenth- and twentieth-century British and American societies, was not very common in Spain, and the few universities, colleges, societies, associations, and other scholastic institutions were generally not founded and funded by donations from wealthy Spaniards. Everything was expected to be provided by a State engaged in a complex process of modernisation. Huge investments were made by Spanish governments to achieve in the country the modernisation which had been characteristic of more developed European countries: to create new industries and develop the country's transport systems. But the promotion of basic science was left behind in favour of more utilitarian science, such as engineering and other applications of scientific knowledge. (In Catalonia the situation was slightly different because there the process of industrialisation really took off in the late nineteenth

---

[169] Ignacio Puig, *El Observatorio del Ebro* (Tortosa: Imprenta Moderna del Ebro de Algueró y Baiges, 1927); Chapter 1, 'Labor de Organización', pp. 3–16.

[170] 'Observatory of the Ebro …', *Knowledge & Scientific News*, 4 (1907), 3–5.

[171] 'Astronomical publications: A new Spanish observatory', *Journal of the British Astronomical Association*, 17 (1906–1907), 152–153.

century.) Nevertheless, Spanish science was slowly but gradually developing.

In some cases, projects such as the creation of the Ebro Observatory clashed with the desire to modernise the country. As Nordmann explained:

> Nous venons d'apprendre qu'il est question de construire, au voisinage du nouvel observatoire, un tramway électrique d'intérêt local.—Ce serait rendre impossible toute observation magnétique, et ruiner, dès son début, l'idée féconde qui a présidé à la fondation de ce bel établissement.
>
> Il faut espérer que le Gouvernement espagnol saura imposer aux constructeurs de ce malencontreux tramway un mode de traction qui ne soit pas funeste à l'Observatoire de Tortosa; il ne voudra pas permettre qu'on détruise une œuvre qui promet d'être à la fois si fructueuse pour la science et si honorable pour l'Espagne.[172]

In the end no electric tramway was constructed. Indeed, a few minutes' walk from the Observatory was a horse-tramway which connected Tortosa with Roquetas, which had a station on the line to Barcelona.

Despite the push for modernisation, Spain remained a traditional society. Religious orders were largely influential in the upper echelons of Spanish society, and their requests were generally listened to and complied with. Indeed, religious orders were perhaps the only institutions capable of attracting, in general terms, important private donors for projects, some of which were related to the progress of science. As Selga wrote with regard to the foundation of the new Ebro Observatory promoted by Cirera:

> Prominent business men and highly cultured citizens gave their support to the observatory: the Ebro spectroheliograph, the first of the kind to be used in Spain, is a munificent gift of Mr. Girona; but it is to Mr. Pedro Gil de Mora that the observatory looks as its greatest benefactor and quasi-founder.[173]

Pedro Gil Moreno de Mora was a "wealthy Spaniard residing in Paris" whose powerful aid was acknowledged by Cirera before the plans for the buildings to be constructed in Tortosa were drawn up.[174] As Cirera himself

---

[172] Nordmann, 'Fondation d'un observatoire astrophysique en Espagne', *Revue générale des sciences pures et appliquées*, 15 (1904), 474–475.
[173] M. Selga, 'The Ebro Observatory, Tortosa, Spain', *Publications of the Astronomical Society of the Pacific*, 27 (1915), 21–27.
[174] 'Observatory of the Ebro …', *Knowledge & Scientific News*, 4 (1907), 3–5.

recognised, it was Gil Moreno de Mora who paid for the offices of the Observatory.[175] Whenever resources ran short, “Providence”, in Father Cirera’s pious narrative, found new patrons both in Barcelona and in the Spanish community in Paris.

At the same time, the extensive network developed by the Jesuits made possible the communication with and contributions of many scholars to such a project. When equipment began to arrive at the Ebro Observatory, many distinguished Jesuit scientists came forward to lend their valuable assistance at the new installation. Amongst the Spanish Jesuit Fathers who took an active part in the project there stand out not only Ricardo Cirera, Director of the Observatory from 1904 to 1919, but also Mariano Balcells (1864–1911), who was in charge of the construction of the buildings and cooperated with the solar section, from 1904 to 1908, and Juan Comellas, who after finishing his astronomical studies in Paris took charge of installing the meridian circle.[176] The experience of Comellas in France was very valuable indeed. It was the Ambassador of Spain in Paris, Fernando de León y Castillo (1842–1918), who wrote to the French Ministry of Public Instruction on 10 November 1903 to obtain permission for Comellas to assist with the practical and theoretical courses at the Observatory of Montsouris. The request was forwarded to the Bureau des Longitudes on 18 November 1903. Ten days later, the President of the Bureau des Longitudes answered the Ministry:

> Les cours et les exercices d’astronomie pratique de l’Observatoire du Bureau des Longitudes au parc de Montsouris ayant précisément pour objet l’instruction des voyageurs et des géographes, il y a lieu d’accorder à M.M. René Michel et Jean Comellas les autorisations qu’ils sollicitent … des ordres on été donnés à l’Observatoire afin qu’ils y soient admis tous deux dès qu’ils se présenteront.[177]

On 30 November 1903 the consul of Spain in Paris was informed that Comellas was now authorised to attend all the lectures delivered by Guillaume Bigourdan (1851–1932) at the Observatory of Montsouris.

---

[175] Ricardo Cirera, ‘Prólogo. El Observatorio del Ebro’, *Boletín del Observatorio del Ebro*, 1 (1910), 7–14, here p. 9.

[176] Puig, *El Observatorio del Ebro* (Tortosa, 1927), pp. 3–16.

[177] Letter signed on 28 November 1903. For this and other related documents, see: Archives Nationales de France—F^{17} I [F17/01(B)]), F^{17} 13571: *Bureau des longitudes. Affaires scientifiques diverses 1881–1922*. Folder: *Bureau des Longitudes. Affaires diverses (1901–1914)*. Folder: ‘Observatoire de Montsouris. Affaires diverses (Visites, admission aux cours) 1901–1923’.

With regard to the creation of the Ebro Observatory, it is worth naming some of the foreign Jesuits who were involved in the early stages of the project: the Frenchmen A. Willot and E. Merveille, who installed the magnetic instruments and later became, from 1905 to 1906, in charge of the magnetic section of the Observatory; the German L. Dressel, in charge of the electrical observations; the French B. Beloty, in charge of the time service and of calculating the times of contacts as seen from the Observatory at the eclipse of 1905; and the Dutchman Johan Stein (1871–1951), doctor in astronomy at the University of Leiden, who determined the coordinates of the Observatory.

The project also received contributions from many other Jesuit Fathers and Brothers, who were always willing to respond to their Society's requests: Bienvenido Clusella, who worked at the meteorological section from 1905 to 1915; Juan Ortega, who was at the magnetic section from 1913 to 1914; Enrique de Rafael Verhulst (1885–1955), member of the magnetic section from 1920 to 1921; Joan Rosanas, who joined the electrical section from 1909 to 1911; Pere Trullás, in the seismological section from 1917 to 1926; Josep Ubach, in the magnetic section from 1910 to 1911; Joan Forcadas, who was in the photographical section from 1904 to 1925; Francisco Rubio, member of the solar section from 1916 to 1931; and Josep Albiñana (1875–1922), who worked at the seismological section. Joan García Mollá worked at the geophysical, seismological, and magnetic sections from 1906 to 1908 and from 1915 to 1921, and Joaquín Pericas was in the solar and seismological sections from 1910 to 1919. The Observatory also had amongst its staff other important members such as Luis Rodés i Campderà (1881–1939), who after studying astronomy in Holland and the United States of America cooperated with the observatory from 1914 and was appointed its director in 1920; and Antonio Romañá Pujó (1900–1981), responsible for the geomagnetic section and later appointed Director of the Observatory in 1939.

The Observatory was inaugurated and blessed on 8 September 1904, in the presence of about a thousand people. The Spanish Government saw such a centre as important in the development of science in Spain and declared the Observatory an Institution of Public Utility just one month after its inauguration, on 18 October 1904. One year later, on the occasion of the total solar eclipse of 30 August 1905, it was visited by about seventy scientists, almost all of whom were members of the order of Jesuits. Although not everything worked as expected during the event, the observation of the eclipse was a complete success from the scientific point of view. Amongst the problems was the second coelostat purchased by the observatory, acquired in time for the eclipse but not in time to alter the arrangement by which the first coelostat had to do duty for both instruments. The spectroheliograph had to be returned to its makers, even though it was used during the eclipse. All in all, the success of the observations of the eclipse of 1905 was far beyond scientific interests.

The aim of the observatory was to provide an explanation for the relationships between solar activity and terrestrial electric and magnetic phenomena.[178] The scientific activity of the Observatory was mainly devoted to heliophysics, geophysics, and meteorology.[179] The observation of the Sun was confined to the astronomical department, which occupied a whole building. Photographs of the Sun were taken on every available day with an equatorial telescope. The equatorial, made by Mailhat (Paris), had two objectives of 162 mm (6½ inches) aperture and about 2 m focal length (one objective corrected for photographic use and the other for visual use). An enlarging system brought the focal image of the sun, about 20 mm in diameter, to the standard size of 200 mm on the photographic plate. The equipment also included an Evershed two-slit spectroheliograph, a spectrogoniometer, and a prominence spectroscope. These instruments, together with a Grubb coelostat and a Rowland grating, almost completed the department's inventory. A summary of the activity of the astronomical, magnetic, electrical, and meteorological departments, and of their instruments, is found in the article by Selga. Specifically dealing with the astronomical section, he wrote:

> The heliographic positions of the sun-spots were deduced from the plates either by the method used at Greenwich, by means of the position micrometer or by different graphs. The measured surface of the sun-spots is read in square millimeters and tenths of a square millimeter, and the reduced surface is computed to millionths of the solar hemisphere. Flocculi or calcium vapors in the Sun's chromosphere are studied by means of an Evershed [photographic] spectroheliograph. An excellent [photographic] spectrogoniometer of four prisms, in connection with a coelostat, affords an opportunity of determining the radial velocities of different vapors of the solar atmosphere. The importance of solar radiation work to determine

[178] Cirera, 'Prólogo …', *Boletín del Observatorio del Ebro*, 1 (1910), 7–14, here p. 7.

[179] Some authors, however, assured that this institution was organized into four different sections: astronomy, terrestrial magnetism, atmospheric electricity, and meteorology. See: Selga, 'The Ebro Observatory …', *Publications of the Astronomical Society of the Pacific*, 27 (1915), 21–27.

the solar constant was early recognized and fully realized by the director. Observations have been carried on daily with an Ångstrom compensating pyrheliometer.[180]

Fig. 5-2: View of the building for astrophysical studies and the photographic laboratory at the Ebro Observatory. From: R. Cirera, *Notice sur l'observatorie et sur quelques observations de l'éclipse du 30 Août 1905* (Barcelona: Gustavo Gili, 1906).

The results of the observations were published regularly from 1910 thanks to a government subsidy.[181] This same year, the Ebro Observatory became the first observatory in the world to systematically publish observations on the solar chromosphere.[182] The significance of the

[180] *Ibid.*

[181] This subsidy was for the publication of the *Memorias del Observatorio del Ebro*. This publication, whose first issue was paid for by Pedro Gil Moreno de Mora, was intended to popularise the scientific activity pursued at the Ebro Observatory. In contrast, the *Boletín del Observatorio del Ebro* was a more 'complete' publication addressed to specialists. See: Cirera, 'Prólogo …', *Boletín del Observatorio del Ebro*, 1 (1910), 7–14, here p. 9.

[182] It seems that the Ebro Observatory was also the fifth in the world to perform systematic spectroscopic observations of the Sun. This information has been

Observatory in the context of Spanish science is well shown by the high-profile visit to the Observatory by King Alfonso XIII on 5 May 1912. Two years later he awarded the Director of the Observatory, Ricardo Cirera, the Gran Cruz de Alfonso XII.

The importance of the Ebro Observatory was not just national. At the international level, the Observatory was seen as a reference point for the scientific activity developing in Spain. Its foundation led not only to the establishment of formal institutional relationships with other observatories, such as the French Observatory of Lyon and the Spanish observatories of San Fernando and Fabra, but also to visits by prominent astronomers such as the French Comte Aymar de la Baume Pluvinel (1860–1938), the American Frank Hagar Bigelow (1851–1924), the Russian Aleksey Pavlovich Hanski (1870–1908), and the Italian Annibale Riccò (1844–1919).[183] These links proved very important, and cementing them became one of the first main objectives of the Observatory. As Cirera wrote in the *Memorias del Observatorio del Ebro* when he presented the first astronomical results obtained at this institution: "Ce présent écrit nous servira, en attendant, à établir des relations avec les Observatoires et centres scientifiques".[184]

Immediately after the Ebro Observatory made its impact on the scientific community through the observations of the eclipse of 1905, its director received two invitations to attend international meetings. He declined the invitation to the conference organised by the International Meteorological Committee in Innsbruck, with an agenda that included topics such as terrestrial magnetism and atmospheric electricity, but accepted the one to the meeting of the International Union for Cooperation in Solar Research at Oxford.[185] Cirera was the only Spaniard and

obtained from a poster presentation, 'L'observatori de l'Ebre i la seva contribució a la física còsmica', by Josep Batlló and Antoni Roca Rosell, at the *VIII Trobada d'Història de la Ciència i de la Tècnica* in 2004.

[183] Ricardo Cirera, *Notice sur l'observatorie et sur quelques observations de l'éclipse du 30 Août 1905* (Barcelona: Gustavo Gili, 1906), pp. 11–15.

[184] *Ibid.* [Édition Française; Traduction par E. Merveille; *Mémoires de l'Observatoire de l'Ebre*, 1 (1906)]. French was one of the languages most employed for international scientific communication during the nineteenth and early twentieth centuries. It seems clear that these lines were deliberately published both in Spanish and French in order to reach a broader audience and to better engage with the scientific community of the period.

[185] See the note published in October 1905, 'International Unit for Co-operation in Solar Research: Meeting at Oxford, 1905 September 27–29', *The Observatory*, 28

representative of a Spanish institution at the meeting. As he wrote, it provided an excellent opportunity to start taking part in regular international collaboration in solar studies.[186]

5.3. The Cartuja Observatory

The origins of the Cartuja Observatory are linked with another Jesuit institution: the Jesuit Colegio Noviciado del Sagrado Corazón, built in Granada in 1894.[187] The college students, future members of the order of Jesuits, had to acquire a broad and current knowledge for when they became teachers, and studied not only philosophy and theology but also scientific subjects. A few years after the establishment of the institution, the college built for its students one of the most important observatories in the country, devoted to meteorological, seismological, and astronomical work.

The Cartuja Observatory was founded after the total solar eclipse expedition organised by the college in 1900.[188] The expedition was formed by the Rector of the College, Father Juan de la Cruz Granero, reader in physics, Father Ramón Martínez, and an assistant, Brother Luis Hurtado. They arrived at Totana on 26 May 1900, after a two-day trip. As they recorded, the only rail link between the two localities involved a journey of more than 600 km, as opposed to the geographical separation of 230 km.[189] This expedition took with them various instruments from the

---

(1905), 382–387. For primary sources on the International Union for Cooperation in Solar Research, strongly promoted by the American astronomer George Ellery Hale, see for instance: G. E. Hale and C. D. Perrine, 'International Co-operation in Solar Research', *Astrophysical Journal*, 20 (1904), 301–305; and 'International Union for Cooperation in Solar Research', *Nature*, 72 (1905), 563.

[186] Cirera, *Notice sur l'observatorie ...* (Barcelona, 1906), p. 15.

[187] Most of the information presented in this section has been obtained from: Manuel Espinar, José Antonio Esquivel, and José Antonio Peña (eds.), *Historia del observatorio de Cartuja*, 1902–2002. *Nuevas investigaciones* [CD-ROM] (Granada: Ayuntamiento de Granada, 2002).

[188] Antonio Due Rojo, 'El Cincuentenario del Observatorio de Cartuja', *Urania*, 38 (1953), 67–80.

[189] There is confusion in the literature about the exact destination of the expedition. In some documents the party was said to have visited Tobarra (Albacete); and in others, Totana (Murcia). The distance between Granada and Totana is about 225 km, which fits better with how the members of the expedition described their trip than the 360 km between Granada and Tobarra. However, it seems that the expedition did stay in Tobarra, in a property owned by Javier Rodríguez, as

college's physics cabinet, including some photographic cameras and a telescope made by Secrétan, of 60 mm aperture and about 2 m focal length, for the observation of the solar event.

Fig. 5-3: The Cartuja Observatory in 1902. From: M. Espinar, J. A. Esquivel and J. A. Peña, *Historia del observatorio de Cartuja, 1902-2002. Nuevas investigaciones* (2002). Courtesy of the University of Granada.

The results of the observations were presented in a public lecture illustrated with slides on 10 June 1900. It was during these events that the idea of founding an observatory attached to the college took shape. One of the students, the future Father Antonio Osborne, son of the Count of Osborne, offered a sizeable part of his family legacy to fund the project. In order to properly prepare and supervise it, and with the intention of purchasing the necessary equipment for the new observatory, Father Granero visited some of the most important European observatories during that year. One was to the Observatory of Paris, where the French astronomer Bigourdan received and helped him. After his return to Granada, the construction of the observatory began in June 1901. The architect, Enrique Fort, had provided the plans of the building for free in April 1901.[190] The project was finally finished thanks to the help of

recorded in: 'El eclipse de Sol', *El Heraldo de Madrid*, 21 May 1900 [Año 11, nº 3478], 1.

190 Manuel Espinar Moreno and Juan de Dios Morcillo Puga, 'Nacimiento y evolución del edificio del Observatorio de Cartuja desde 1902 a 2002', in Espinar,

several benefactors. People from a variety of religious communities were involved, including Fathers Antonio Osborne, Francisco Cuenca, Ricardo Garrido, Juan Leal, Francisco Maldonado and his family, Antonio Revuelto, Antonio Schormarrandi, Antonio de Viu, and the families of Father José Mier y Terán and Brother José Ridruejo. Other individuals also contributed, such as José Artega, Marcelo Blanca, Enrique Mendoza Calvo Flores, Antonio Aranda Casanova, José Criado Tejada, Abelardo Fajardo Aguilar, Mercedes García Verde, Fermín Garrido, Carmen Godoy Fonseca, Josefa Gordo de Guerrero, Antonio Knörr, Rosa Emertina Lastra, Soledad Lobatón, José Manuel Morales Belmonte, and Fernando Nuñez Estremera. The benefactors of the observatory were not only Spaniards. In fact, amongst the first donors were the Argentineans Rosa and María Beltrán, the German Pedro Breuel, and the Englishman W. A. S. Davenhill. The project also had institutional support from the City Hall of Granada and the Provincial *Diputación* of Granada.

The building was finished in March 1902 and inaugurated in June 1902. The first director of the Observatory was Father Granero. He organised the Observatory, with the inestimable help of the Jesuit Brothers Luis Hurtado and Luis Camarero, into three main sections: astronomical, meteorological, and seismological.[191] Amongst his most relevant contributions were not only the determination of the geographical coordinates of the observatory, but also some observations of nebulae and double stars. His activity at the forefront of the Observatory's work finished in 1904, when he was appointed Visitor (and later General Superior) of the Ecuador, Peru, and Bolivian Missions.

His successor was the Mexican Father José Mier y Terán. The strong interest of Mier y Terán in solar physics led the Observatory to focus on this subject, and it began to publish statistics on solar activity in the *Boletín Mensual del Observatorio de Cartuja*. At least two leaflets on solar observations and one on lunar eclipses were published during these years.[192] Amongst his achievements at the Cartuja Observatory was also the expedition sent to Carrión de los Condes (Palencia) to observe the total solar eclipse of 30 August 1905. This expedition included Fathers Julio

---

Esquivel, and Peña (eds.), *Historia del observatorio de Cartuja* ... (Granada, 2002).

[191] Manuel Merlo Vallejo, 'Observatorio Universitario de Cartuja. Prefacio' in Espinar, Esquivel and Peña (eds.) *Historia del observatorio de Cartuja* ... (Granada, 2002).

[192] *OAUV*, 11/16—1.

Fenyi and Teodoro Angern, respectively Director and Sub-director of the Kalocsa Observatory (Hungary), with whom the Spanish Jesuits later cooperated and published the results of their observations. The publication of the results of this expedition, in which around ninety photographs of the eclipse were obtained,[193] firmly established the Cartuja Observatory amongst the small number of Spanish astronomical research centres. The solar studies pursued at the Observatory were made by Mier y Terán, a well known astronomer in the international scientific community. In fact, his reputation was such that the Belgian Astronomical Society chose him to be in charge of the official Belgian expedition sent to Rance (near Chimay) to observe the eclipse of 17 April 1912, once he had moved to that country having been appointed Professor at Enghien College.[194] The prestige and position that the former astronomer of the Cartuja Observatory achieved in Belgium may have been crucial to the publication of astronomical work by this observatory in *Ciel et Terre*, the journal of the Belgian Astronomical Society.

The first years of the Cartuja Observatory were devoted not only to meteorological observations, but also to the study of solar activity, as its first publications show.[195] The astronomical observations were made with a Mailhat equatorial telescope of 330 mm aperture and 5.35 m focal length, acquired during Granero's tour of Europe, before the construction of the Observatory. An equatorial telescope made by Grubb, of 152 mm aperture and 2.3 m focal length, was also purchased thanks to a donation by Soledad Lobatón. The Observatory also acquired a Littrow autocollimator spectrograph of 2 m focal length and diffraction grid of 14,438 lines per inch (567 lines per millimetre), and a Hilger "macro-micrometer" (a measuring machine) for solar physics research. Equipment for meridian observations was also acquired, including a Mailhat meridian

---

[193] José Mier y Terán, *Eclipse total de sol del 30 de agosto de 1905. Observaciones hechas en Carrión de los Condes (Palencia) por la sección astronómica del Observatorio de Cartuja (Granada) dirigido por padres de la Compañía de Jesús* (Granada: L. Guevara, 1905).

[194] G. A. Quignon, 'The eclipse of the Sun, April 17, 1912 (Translation of Letter to the Secretary)', *Journal of the British Astronomical Association*, 22 (1911–1912), 198.

[195] See for example the *Boletín mensual del Observatorio de Granada*. The first and second volumes (1903 and 1904) dealt only with meteorological and seismological observations. However, the third (1905) and subsequent volumes also contained astronomical observations made at the observatory.

circle of 58 mm aperture and 0.62 m focal length, with a micrometric ocular and accurate to 1 second of arc.

The astronomical section later acquired an iris diaphragm, several cameras, two finders (one of them of 109 mm aperture), a photographic objective, a Foucault regulator, two spectroscopes made by Grubb, and a photoheliograph donated by Fathers Manuel María Sánchez-Navarro Neuman (1867–1941) and José Mier y Terán. All these instruments were employed not only to instruct students at the Colegio Noviciado del Sagrado Corazón, but also to pursue research in, for example, stellar astronomy and spectral analysis.

Fig. 5-4: The 330 mm Mailhat equatorial at the Cartuja Observatory. From: M. Espinar, J. A. Esquivel and J. A. Peña, *Historia del observatorio de Cartuja, 1902-2002. Nuevas investigaciones* (2002). Courtesy of the University of Granada.

Many of these instruments, mainly those of the meteorological and seismological sections, were in regular use from 1 January 1903 onwards, although the inauguration of the astronomical section was not until several months later. One of the first records of systematic astronomical observations at the Cartuja Observatory is that of daily visual observations of the solar photosphere, combined with the systematic gathering of sunspot statistics, from January 1905. These results were presented in the early publications of the Observatory. The studies by the astronomical section also included spectrographic work, and observations of eclipses, comets, and planets.

In 1906, Brother Salvador Parra took charge of solar photography and the study of photoheliographic statistics, the main astronomical activity of the observatory during the following years.[196] Some of the results of the observations made at the astronomical section of the Cartuja Observatory were presented in international journals such as *Ciel et Terre*. They dealt mainly with solar observations such as solar activity or photoheliographic statistics, but there were also reports on transient astronomical events such as the lunar eclipse of February 1906 and the transit of Mercury in November 1907.

In 1908, Father Ricardo Garrido was appointed director of the institution, a position he held until 1913. Although the heliographic work continued under his directorship, logistical problems largely limited the activity of the observatory during these years and those of his successors, Fathers Juan Murillo and Cándido Guerrero. Nevertheless, some interesting work was still carried out by the astronomical section. As Garrido pointed out, the most important astronomical activity of the Observatory was the spectral study of sunspots: studies "very advanced in other nations" but "almost abandoned in our Mother Land"—although photoheliographic statistics were collected at the observatories of Ebro and Madrid.[197] Garrido had a very extensive and up-to-date knowledge of solar physics, as he showed in his address to the third conference of the Spanish Association for the Advancement of Science. In it he presented a concise summary not only of the history of solar physics, but also of the most recent discoveries and research in progress, dealing in particular with solar spectral studies and research on sunspots. He was careful enough to recognise what were accepted but not confirmed scientific descriptions of nature. The following words provide an interesting example:

---

[196] This work was relegated to second position in the Observatory's priorities after its confirmation as an important international seismological centre. However, although it was not pursued on a systematic basis, this kind of work was still done sporadically until 1940. See for instance: Ricardo Garrido, 'Estadística solar. Enero-diciembre 1920', in *La Estación Sismológica y el Observatorio Astronómico y Meteorológico de Cartuja (Granada). Memorias y trabajos de vulgarización científica* (Granada: Imprenta Gráfica Granadina, 1921).

[197] Ricardo Garrido, 'El estudio espectral de las manchas del sol' in *Asociación Española para el Progreso de las Ciencias, Tercer congreso (Granada, del 20 al 25 de junio de 1911). Tomo 1–3* (Madrid: Eduardo Arias, 1911), pp. 73–92.

> The wave theory of light, although ingenious and apparently conclusive, is a hypothesis ... Today's knowledge about the intimate nature of light and its way of propagation is a hypothetical knowledge.[198]

Nonetheless, it was the seismological section, under the supervision of Father Sánchez-Navarro Neumann from August 1906, that was most active at this time.[199] In fact, the Observatory became one of the leading centres both in the study of seismic activity in Spain and in the development of seismological instrumentation. The seismological work was transferred from the Observatory to the newly founded Estación Sismológica de Cartuja (Cartuja Seismological Station), which became an institution independent of the Observatory in July 1908. Seismological observations began to be issued by this station in a monthly publication. Several practical seismological problems were solved by original work; the construction of the "Cartuja-bifilar" and the "Cartuja-vertical" (seismological instruments to obtain high-quality traces drawn by pen-recorders) is a good example. This was a period of scientific expansion in Spain, particularly in the field of seismology, where the work undertaken by Sánchez-Navarro Neumann and his collaborators at the Cartuja Observatory was largely copied abroad.[200] Indeed, the importance of the Seismological Station of Cartuja was recognised by the Spanish Government with the declaration of the station as an Institution of Public Utility on 13 October 1920 (noted in the *Gaceta* on 16 October 1920).

An example of how extensive was the network built by the Jesuits during the early twentieth century is provided by the very large number of centres with which the Cartuja Observatory established formal links from 1910 to 1916. These institutions were spread all over the world, in countries such as Australia, Austria, Belgium, Bolivia, Brazil, Britain, Canada, Chile, China, Costa Rica, Denmark, Egypt, France, Germany, Greece, Haiti, Hawaii, Hungary, Italy, Japan, Malta, Mexico, Norway, Peru, the Philippines, Portugal, Romania, Russia, El Salvador, Serbia, Spain, Syria, Turkey, the United States of America, and Uruguay.

---

[198] *Ibid*, p. 90.

[199] Sánchez-Navarro Neumann was told by his superiors in mid-1905 to begin with the study of seismology. This same year he visited the Observatory of San Fernando to gain experience in this field.

[200] Manuel Espinar Moreno, 'Notas sobre la vida y obras del R. P. Manuel María Sánchez-Navarro Neuman, S.J. (1867–1941)', in Espinar, Esquivel, and Peña (eds.), *Historia del observatorio de Cartuja ...* (Granada, 2002).

5.4. Atheism and agnosticism versus religious scientists

It was pointed out in Chapter 2 that conflicts between science and religion affected not only Spain but all of Europe during the second half of the nineteenth century. The strong influence of the Catholic Church in Spain made it a topical issue that persisted during the first decades of the twentieth century. During these years, many scientists expressed their belief in the compatibility of religion with the progress of science. Good examples are some of the first published works by Lànderer on cosmological issues,[201] and his translation of an article published in the French *Revue générale* in which was given a "mathematical demonstration of Christianity".[202] He was a Catholic "with more faith in the dogma than in the formulae of his astronomical calculations".[203]

However, the Glorious Revolution of 1868 had opened the doors to religious scepticism in Spain. The following lines, published in the Bulletin of the Cartuja Observatory a few years after its foundation, perfectly illustrate the situation:

> The uncertain course of Spanish politics at the time, the way the revolution was shaking the faith of the people, and the emerging climate of religious persecution, made many people believe that the project of the Jesuits was crazy and offered little for the future.[204]

Glick has claimed that the years that followed the Spanish military defeat of 1898 ushered in a *civil discourse* for science in Spain, which tried to shift scientific and technological debates away from politics and ideology, in the interests of modernising the country.[205] Scientific inquiry

---

[201] Navarro Brotóns and Gozalo Gutiérrez, 'Entre "amateurs" i professionals ...', in Puig-Pla *et al.* (coords.), *Actes de les III Trobades d'Història de la Ciència i de la Tècnica ...* (Barcelona, 1996), pp. 303–312.

[202] J. J. Lànderer, 'Demonstración matemática del cristianismo', *Crónica Científica*, 14 (1891), 229–230.

[203] See: A. F. Linari, 'Don José Joaquín Landerer y Climent', *Ibérica*, 18 (1922), 178–180, here p. 180; and 'Don José Joaquín Landerer', *Anuario de la Real Academia de Ciencias Exactas, Físicas y Naturales* (1923), 206–211, here p. 210.

[204] Cited in: Manuel Espinar Moreno, 'Fundación del Observatorio de Cartuja. Primeros años de funcionamiento (1902–1906)', in Espinar, Esquivel, and Peña (eds.), *Historia del observatorio de Cartuja ...* (Granada, 2002).

[205] Thomas F. Glick, 'Emilio Herrera and Spanish Technology', in Herrera, *Flying: The Memoirs of a Spanish Aeronaut* (Albuquerque, 1984), pp. 173–215, here p. 174. See also: Glick, *Einstein y los españoles* (Madrid, 1986).

was removed from the arena of political ideology, and new ideas were discussed and evaluated freely and openly. This provided the immediate background for the rapid maturation of fields such as physics and chemistry in the country. In some cases, however, this new attitude towards science contrasted with the values of the conservative elite that had held power in Spain throughout most of the nineteenth century.

One has to bear in mind this general context to better understand the role played by Jesuits in the development of astronomy and astrophysics in Spain during the late nineteenth and early twentieth centuries. To answer the question of why the Jesuits set up such high-quality astronomical and astrophysical observatories in Spain, one has to take into account several factors. On the one hand, as pointed out above, Jesuits were the owners of an important scientific tradition strongly linked to the spiritual character of the Order. As was engraved above the main entrance of the Cartuja Observatory, the heavens were supposed to declare the glory of God (*coeli enarrant gloriam Dei*[206]). On the other hand, the Jesuits are well known for their missionary spirit. Wherever they thought people needed their help, they were prepared to go. This explains their interest in setting up meteorological and seismological observatories. Through the study and analysis of their observations at these research centres, they provided a service to the society to which they were attached. The knowledge of when natural disasters are likely to occur and their characteristics are of great help to a local population. A good example of how useful this information is can be found in the first regulations governing the construction of buildings in the Philippine Islands, encouraged by the seismological observations and results obtained by the Jesuits in that region.[207] The foundation of astronomical observatories, however, is slightly more difficult to explain in these terms. Of course, an important part of the astronomical activity at these observatories, as was explicit in the foundation of the Ebro Observatory, was the attempt to discover the relationships between solar and terrestrial processes. However, the Jesuits already had several astronomical centres devoted to similar studies. Why did they have to set up new observatories in Spain? This question is even

---

[206] From Psalm 19: *coeli enarrant gloriam Dei et opus manus eius adnuntiat firmamentum / dies diei eructat verbum et nox nocti indicat scientiam ...* (translated in the New Revised Standard Versión as 'The heavens are telling the glory of God; and the firmament proclaims his handiwork / Day to day pours forth speech, and night to night declares knowledge …').

[207] I am very grateful to Josep Batlló for this information on an area in which he is currently working and is hoping soon to publish a paper.

more legitimate when one thinks about the poor state of the Spanish economy and the difficulties of founding observatories and purchasing and setting up the necessary instruments in the country.

One is tempted to suggest that behind these projects is the belief of the Society of Jesus that with the foundation of such institutions they were helping Spain and its society. This could be understood from two different points of view. First, it could be that the Jesuits thought that their projects would help Spain to recover its international prestige. After all, Jesuits were very influential in Spain, particularly under conservative governments. A better international image of the country could benefit their interests. As the Jesuit Fathers from Oña pointed out, the study of the eclipses was in the interest of "the glory of God, of the Company, and of our mother country [Spain]".[208] Second, and as has already been pointed out here, the social and political changes in the second half of the nineteenth century had changed Spanish society dramatically. Religious scepticism had spread through much of the population, and in particular to many scientists. At the same time, science had come to be regarded by the ruling classes as a tool for the progress of humanity. Jesuits had to show that such progress was not in opposition to Catholicism.

If religious men could be recognised as prestigious scientists who made original contributions to the progress of scientific knowledge, science would be shown to be compatible with the faith that many Spaniards were gradually losing. The foundation of astronomical observatories could thus be understood as a step forward for the Society of Jesus in this battle. As an example of this conflict, let me reproduce the following lines written in 1910 about the Second Conference of the Spanish Association for the Advancement of Science[209] in the religious and conservative journal *Razón y Fe*:

---

[208] *Observaciones del eclipse total de Sol del 30 de agosto de 1905, hechas por los padres de la Compañía de Jesús en el Colegio de Oña* (Oña: Imprenta privada del Colegio, 1906), p. 6.

[209] The Asociación Española para el Progreso de las Ciencas was founded in 1908 in Zaragoza (Spain) thanks to the efforts of several members of the Real Sociedad Española de Historia Natural, including Bolívar, Simarro, Lázaro, Calderón, and García Mercet. In fact, it was a few days after the Conference of Spanish Naturalists held in Zaragoza in October 1908 that the Association was created. The Association was divided into several sections: Mathematics; Astronomy and Physics of the Globe; Physics and Chemistry; Natural Sciences; Social Sciences; Philosophy, History, and Philology; Medicine; and Applied Sciences. The Jesuit

> What positive results have this second Conference of the Association for the Advancement of Science brought? Has it made science advance? For the impious science, in other words the science that is not a science, one doubts it; for the real science, the Catholic science, one hopes so. This result will not meet the aspirations of some organisers of this Association; but … significant despondency has been observed in some anticlerical elements, which were initially very active in the Association … Summarising, this second Conference of Valencia has been a small victory for Catholic Spanish science. Small, yes, partial, not as universal as we would like, but victory in the end. Let us hope for a complete victory eventually.[210]

Ricardo Cirera founded the section of Astronomy and Solid-Earth Geophysics at the First Conference.

[210] Longino Navás, 'El segundo Congreso de la Asociación Española para el Progreso de las Ciencias celebrado en Valencia los días 15–20 de Mayo de 1910', *Razón y Fe*, 27(1910), 364–368, here p. 368.

CHAPTER SIX

IGNACIO TARAZONA AND THE FOUNDATION OF UNIVERSITY OBSERVATORIES IN THE TWENTIETH CENTURY

6.1. Astronomical university observatories in the international context

During the 1870s, the lack of adaption to a changing science such as astronomy led the few university astronomical observatories existent in the world into crisis. New facilities, staff, and training was required to engage with astrophysics. However, new opportunities arose in the 1880s thanks to some benefactors, who boosted a shift of fundamental research and instruments from amateurs to the universities, particularly in Great Britain and the United States of America.[211] The strength and initiative of amateurs, together with the stimulus to the instrument industry, favoured the application of new techniques and methods to university observatories in these countries, opposed to the very slow and limited response to astrophysics in countries such as France.

In America, dozens of university and college observatories were developed thanks to the availability of funds from multiple sources, which included private donations. The creation of these institutions, a consequence of the decentralisation of American science, was partly motivated by intercity or state rivalry and benefited from the managerial abilities of several individuals. In Europe, it was Charles Pritchard (1808–1893),[212] the first British professor to respond to astrophysics. Being free

---

[211] See Roger Hutchins, *British university observatories, c.1820-1939: ideals and resources* (D.Phil. Thesis, University of Oxford, 1999), from which most of the information in this section has been obtained.

[212] On the figure of Charles Pritchard, see: Ada Pritchard, *The Life and Work of Professor Pritchard* (London: Seeley and Co. Limited, 1897).

of meridian work, he transformed the intended teaching facility at the University of Oxford into a site for research. This was possible, even when he had a very modest quantity was granted for assistance and expenses, thanks to the gifts of several amateur astronomers such as Warren de la Rue (1815–1889), who donated his equatorial photographic reflector. Together with a new Grubb visual refractor, a transit instrument, and a solar spectroscope with six prisms, Pritchard got involved in photometry and photographic parallax, as well as in the Astrographic Catalogue, one of the main international collaborations in the field of astrophysics during these years.

Similarly, the University of Cambridge faced the challenge to engage with astrophysics, and by the turn of the century there had been a complete change in the research activity at this astronomical observatory, with almost no meridian observations and a new photographic refractor as the main instrument. Consequently, by the early twentieth century several British and American university observatories conducted leading astrophysical research projects. The best example is that of American observatories such as those of Harvard and Lick, with observational stations at Arequipa (Peru) and Santiago (Chile) which allowed them to cover the whole sky comprehensively. Moreover, the activities pursued at these university observatories moved also into offering advanced astronomy degrees, which largely contributed to the professionalisation of the discipline.

6.2. Early attempts in Spain

In Spain, like in many other countries, serious attempts were made from the mid-nineteenth century to promote and nationalise education, as the 1857 educational reform promoted by Claudio Moyano y Samaniego (1809–1890) demonstrates. In particular, the law promoted by Moyano (Ley de Instrucción Pública, 9 September 1857) led to the creation of faculties of science; science had previously been studied in faculties of philosophy. However, because of the strong political centralisation of the governments of the time, the State capital, Madrid, was the only place with a complete faculty of sciences in its university. Cities such as Barcelona, Granada, Santiago de Compostela, Sevilla, Valencia, and Valladolid were relegated to a second level, as during these years they could confer no more than *bachiller* degrees. The Glorious Revolution of 1868 returned to provincial universities the ability to provide postgraduate courses. However, this lasted only until the restoration of the monarchy.[213]

Science was seen as the engine of industry and modernisation, but this "science" was mainly applied and utilitarian. It was believed that the modernisation of the country would be easier with the application of new imported techniques and knowledge. The educational reform was based on these ideas, and it placed technical degrees awarded by Professional Schools such as Industrial Engineering on the same level as higher-education degrees conferred at universities. Indeed, the degree of Industrial Engineer could now be obtained not only from the Royal Institute of Madrid, but also from the Professional Schools of Barcelona, Gijón, Sevilla, Valencia, and Vergara. Financial problems soon caused all these courses to disappear, with the exception of that at Barcelona. So although the educational reform helped to spread modern science, it did not encourage original scientific research. The Spanish system of higher education aimed to emulate that of foreign countries, but the system collapsed during the second half of the nineteenth century, mainly due to a lack of economic resources, continual attempts by successive governments to control the universities, and the explicit exclusion of scientific research.[214]

It was after the military disaster of 1898 that the state of the educational system again became a public issue. *Regenerationist* campaigns encouraged educational reform, and the action of the State became progressively more effective, as the creation of a Ministry of Public Instruction and Fine Arts in April 1900 proves. The foundation of new astronomical observatories attached to universities has to be understood in this context. Indeed, they were originally conceived to provide appropriate practical instruction to students. Furthermore, these institutions later provided several important public services and prompted Spanish universities to begin astronomical research.

Physical astronomy had been taught in Spanish universities as part of general physics courses during the first half of the nineteenth century. It continued as an independent subject in the higher education system for the second half of the century after its separation from mechanics in the syllabuses established by the educational reform of 1857. A new subject,

---

[213] See: Sánchez Ron, *Cincel, martillo y piedra* (Madrid, 1999), p. 88.

[214] José Luis Peset, 'Educación y ciencia en el fin del Antiguo Régimen', in Sánchez Ron (ed.), *Ciencia y sociedad en España* (Madrid, 1988), pp. 17–25.

named Astronomy of the Planetary System, was added in 1900.[215] In spite of these efforts to provide a complete astronomical instruction to students in science, nineteenth-century Spanish universities offered poor lectures on astronomy as a consequence of their organisational problems. Academic scholars repeatedly complained about the lack of facilities for practical training and education in the subject.[216]

At the beginning of the twentieth century there were no astronomical observatories attached to Spanish universities. Several initiatives had been taken during previous centuries, but all of them were unsuccessful. One of the best known is the attempt in 1790 by the Chancellor of the University, Vicent Blasco García (1735–1813), to equip the University of Valencia with an astronomical observatory. This observatory was built in the years that followed but was destroyed during the Napoleonic invasion, in 1812.[217] Some universities had a meteorological station where astronomical observations were sporadically made. However, it was not until the first decade of the twentieth century that the universities of Barcelona and Valencia constructed a specific building where instruments for astronomical observations were permanently set up. Both these observatories, conceived almost exclusively for the students' instruction, were promoted by Professor Ignacio Tarazona y Blanch.

Ignacio Tarazona Blanch (1859–1924) obtained his scientific education at the universities of Valencia and Madrid during the late 1870s, obtaining his doctorate at the Central University of Madrid in 1880. In January 1887 he was appointed *profesor auxiliar interino* (interim auxiliary lecturer) at the Faculty of Sciences of the University of Valencia. He won the position of permanent auxiliary lecturer one year later. In October 1893 he took charge of the Meteorological Station of the university. Tarazona moved in 1898 to the University of Barcelona, where he was appointed Professor of

---

215 See: Moreno González, 'De la física como medio a la fin. Un episodio entre la Ilustración y la crisis del 98', in Sánchez Ron (ed.), *Ciencia y sociedad en España* (Madrid, 1988), pp. 27–70, here p. 69.

216 *Memoria del Claustro de Profesores de la Facultad de Ciencias de la Universidad de Madrid, 25 de Febrero de 1898* [Archivo General de la Administración, Sección Educación. Caja 8102. Legajo 8874–1]. Cited in: Baratas Díaz, 'La cultura científica en la Restauración', in Suárez Cortina (ed.), *La cultura española en la Restauración* (Santander, 1999), pp. 279–295.

217 Víctor Navarro Brotóns, 'El conreu de les ciències', in *Cinc Segles i un dia* (València: Universitat de València, 2000), pp. 119–126.

Cosmografía y Física del Globo (Cosmography and Geophysics). He later added to this Chair that of Spherical Astronomy and Geodesy.

Fig. 6-1: Ignacio Tarazona Blanch. Courtesy of the Astronomical Observatory of the University of Valencia.

6.3. The Astronomical Observatory of the University of Barcelona

When Tarazona arrived at Barcelona, the university did not have a building to house the astronomical equipment that belonged to the Faculty of Sciences. Perhaps the most striking of these instruments was a Repsold theodolite that the Dean of the Faculty had obtained from the Instituto Geográfico y Estadístico. The Faculty also owned a topographic theodolite made by Brünner and a sidereal chronometer made by Frodsham, both of them acquired from that same institution in 1899.[218] The permanent installation of the theodolite, together with a chronometer, enabled students to solve a great number of problems related to cosmography and spherical astronomy, particularly those related to the apparent and real movements of stars.[219] Tarazona's interest in the construction of a suitable building led him to seek a draft plan. The first sketch was made before May 1900 by one of Tarazona's students from the Faculty of Architecture, with some data collected by Tarazona himself.[220]

---

218 Tarazona y Blanch, *La fotografía solar* (Valencia, 1909), p. 60.
219 *OAUV*, 11/5—1.
220 *OAUV*, 11/5—6.

The support of the Chancellor of the University of Barcelona and the Dean of the Faculty of Sciences led Tarazona to write to the Spanish Minister of Public Instruction about the project in December of that same year. It was done through the Member of Parliament Pedro Grau Maristany (1863–1926), whose interest in the project ensured a good reception in the Ministry.[221] The installation of the theodolite, accurate to one arc second, was intended to help students to solve problems related to observations in both the meridian plane and the prime vertical (perpendicular to the meridian).[222] The former was linked to solar observations and to the observation of the Polar and circumpolar stars. After considering limitations on the design of the new observatory and the visibility of objects, Tarazona concluded that the project had to be based on the construction of a dome 4.5 m in diameter, similar to the two under construction by 1900 at the Observatory of Madrid by the Irish firm owned by Howard Grubb (1844–1931).

On 22 April 1901 the Ministry of Public Instruction asked the architect Ricardo Magdalena Tabuenca (1899–1910) to estimate the budget needed to accomplish such a project.[223] Ricardo Magdalena was a well-known and prestigious Spanish architect who had been in charge of the construction in 1893 of the Faculty of Sciences and Medicine of the University of Zaragoza.[224] By May 1901 the main aspects of the building's design, which was to be erected in the gardens of the University of Barcelona, had been decided upon.[225] In the following months, Tarazona suggested several improvements to the original design, and he finally contacted Howard Grubb in May and June 1902 to ask for estimates for the different domes that he could construct.[226] The excellent relationship of Grubb with the Observatory of Madrid, where Antonio Tarazona (Ignacio's brother) worked, probably influenced the final decision of Ignacio Tarazona to

---

[221] *OAUV*, 11/5—23. In his letter he requested not only the Minister's approval to start the work needed to permanently install the theodolite and other astronomical apparatus, but also that his students had to pay an extra fee due to the need of using special scientific instruments, as it was done with other experimental subjects. See: *OAUV*, 11/5—2.

[222] This and the following lines are based on: *OAUV*, 11/5—3, 4, 5.

[223] *OAUV*, 11/5—19.

[224] For further information about this architect, see: Ascensión Hernández Martínez, *Vida y obra del arquitecto Ricardo Magdalena (1849–1910)* (Zaragoza: Universidad de Zaragoza, 1999).

[225] *OAUV*, 11/6—3.

[226] *OAUV*, 11/5—17; 11/6—1, 3.

choose the Irish maker—particularly after meeting him during the total solar eclipse of 1900, as it will be shown in the second part of the book. Once Tarazona had chosen the model of dome he wanted for the observatory, Grubb sent to Barcelona the drawings required by the architect.[227] However, bureaucracy delayed progress, and it was not until 25 November 1903 that the Government published in the *Gaceta* an invitation to tender for the construction of the observatory buildings, with a 4 m dome to be provided by Grubb, and construction began in February 1904.[228]

The initial investment by the Spanish Government was later used by Tarazona to further develop his project of building an observatory at the University of Barcelona. He was very interested in the acquisition of new astronomical equipment for the projected observatory, and he wrote to Grubb in 1903 asking him for estimates for several instruments, including three portable equatorials, a wedge photometer, a spectroscope, an enlarger to study sunspots, and a unifilar and a bifilar micrometer.[229] The same year, the observatory acquired a Magnaghi circle made by Salmoiraghi.[230] In May 1904, Tarazona requested from the Ministry of Public Instruction a grant to buy an equatorial.[231] The request was accepted, and the observatory received a grant of 6000 pesetas. The next month, Tarazona agreed to buy for £180 a 5-inch (127 mm) aperture

---

[227] *OAUV*, 11/6—4, 5.

[228] *OAUV*, 11/5—25, 26; 11/6—2, 3. There is more information on the installation of the dome and how it was transported to Barcelona in: *OAUV*, 11/6—33, 50, 53, 54, 67, 109. More details on the evolution of the observatory from February 1904 until the finishing of the building are found in *OAUV*, 11/5.

[229] *OAUV*, 11/6—14, 17, 18, 19, 20, 21.

[230] Tarazona y Blanch, *La fotografía solar* (Valencia, 1909), p. 60. The name of this instrument was given after Giovan Battista Magnaghi, who improved the Amici circle in the second half of the nineteenth century.

[231] *OAUV*, 11/6—40. See also the telegrams at: *OAUV*, 11/6—34, 35.

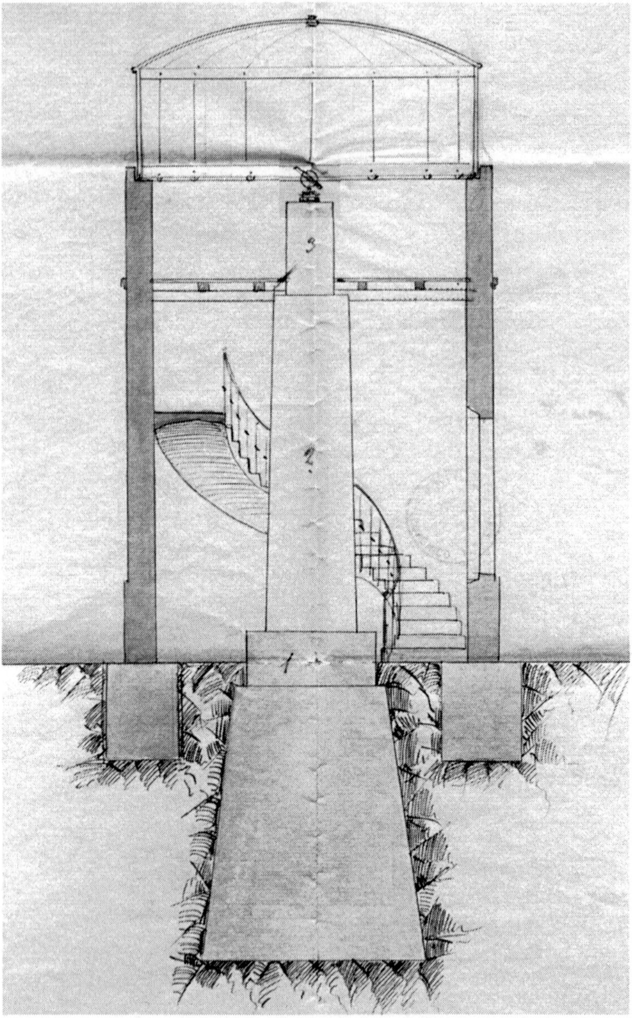

Fig. 6-2: Plans for the building to house the Repsold theodolite in the University of Barcelona. Courtesy of the Astronomical Observatory of the University of Valencia.

equatorial from Grubb.[232] He also asked Grubb for the prices of different micrometers and position circles to be attached to the equatorial.[233] This request for further astronomical instruments seems to be closely linked to Tarazona's original decision to build the new observatory. Indeed, one is tempted to suggest that in proposing to build the observatory he already had in mind the later acquisition of other important astronomical equipment. This would be the main reason why Tarazona wanted to construct such a massive observatory when the only instrument it needed to house was the theodolite.

The dome, with the exception of the papier mâché covering and the copper rivets and washers, was finished by mid-June 1904 and sent to Barcelona in July 1904.[234] By December 1904 the University of Barcelona had a complete observatory to be proud of.[235] As Grubb wrote in a letter to Tarazona on 19 January 1905, the observatory was a striking one, and he wanted to include a view of it in his next illustrated catalogue.[236] The equatorial was despatched in December 1904 and, in spite of problems at

---

[232] *OAUV*, 11/6—41, 50. Grubb estimated the cost of the 5-inch equatorial telescope to be 5500 pesetas. This information, plus the details of the instrument, can be found in: *OAUV*, 11/6—63. In fact, the exchange rate from 1902 to 1904 varied from 1 pound to 31.1 pesetas to 1 pound to 34.7 pesetas, which meant that £180 were about from 5600 to 6250 pesetas –remember that by the early twentieth century, the average salary of men workers in Spain varied from 3 to 4 pesetas for a working day, while women got from 1 to 2.5 pesetas and children 1 peseta. Currency equivalent of the peseta may be found in Maria Christina Marcuzzo, Lawrence H. Officer and Annalisa Rosselli (eds.), *Monetary Standards and Exchange Rates* (London: Routledge, 1997) and Gabriel Tortella (dir.), *La banca española en la Restauración* (Madrid: Banco de España, 1974). Furthermore, Tarazona and Grubb also wrote one to each other about the prices of the different domes available for the new observatory and any reduction that could be made in the final cost. On these issues, see: *OAUV*, 11/6—37, 38, 39. Note here that the budgets of the university observatories of Oxford and Cambridge were approximately £800 per anum, excluding the Professor's salary, while the Radcliffe Observatory at Oxford had a budget of £1700 and that of the Americans Harvard Observatory and Lick Observatory was about £4000 and £5000, respectivley—excluding the cost of their stations at the South emisphere. See: Roger Hutchins, *British university observatories, c.1820-1939: ideals and resources* (D.Phil. Thesis, University of Oxford, 1999)

[233] *OAUV*, 11/6—42.

[234] *OAUV*, 11/6—41, 42, 55.

[235] *OAUV*, 11/5—84.

[236] *OAUV*, 11/6—119.

customs, arrived in Barcelona in January 1905.[237] This same month, the micrometer was said to be finished and the position circle ordered.[238] However, they were not sent until November 1905, when Tarazona received from Grubb the unifilar micrometer with bright and dark field illuminations and the position circle.[239] The items ordered during the first half of the year "had to be laid aside on account of some very pressing orders connected with the Eclipse".[240]

The 5-inch aperture equatorial made by Grubb, with the position circle and the micrometer, took the place of the Repsold theodolite as the new observatory's main instrument.[241] During the following years, Tarazona asked Grubb on several occasions for additional material for the observatory's equipment. He bought from Grubb a so-called prism of total reflection to observe near the zenith, a Dawes helioscope, a spectroscope to study prominences, and apparatus for studying sunspots by projection onto a screen.[242] The contacts made by Tarazona and the experience of setting up a new astronomical observatory proved very useful and encouraged him to start a similar project as soon as he arrived at the University of Valencia.

6.4. The Astronomical Observatory of the University of Valencia

Tarazona was back at the University of Valencia in 1906, appointed Professor of Cosmography and Geophysics.[243] This decision may have been influenced by his strong business interests in Valencia, where he had some country properties whose exploitation provided him significant income. As recorded in the *Anales de la Universidad de Valencia*:

> Once [Tarazona] sat down in the venerable chair of his Professorship, he found himself with the blackboard, the chalk, and a lonely and old

[237] *OAUV*, 11/5—84; 11/6—107, 108, 111, 113, 122 to 125.
[238] *OAUV*, 11/6—110.
[239] *OAUV*, 11/6—131, 133, 136.
[240] Letter from Grubb to Tarazona in: *OAUV*, 11/6—132.
[241] Tarazona y Blanch, *La fotografía solar* (Valencia, 1909); p. 60.
[242] *Ibid.*
[243] He was later also given the Chair of Infinitesimal Calculus.

> theodolite ... with which he could do no more than measure an azimuth or a zenith distance.[244]

During the first Faculty meetings he attended, Tarazona pointed out to his colleagues the urgent need for properly installed equipment for the instruction of the students. It seems that, after the success of his project in Barcelona, Tarazona had little trouble in convincing other lecturers and professors of the advantages of building a new astronomical observatory as soon as possible. By August 1907 Tarazona had sent a formal request to the Dean of the Faculty and to the Chancellor of the University of Valencia in which he proposed, as he had for Barcelona, the construction of an astronomical observatory in the university's botanical garden. This request was immediately forwarded to the Ministry of Public Instruction.[245] In September 1907 the Ministry appointed Alfredo López Alcrudo responsible for the realisation of the project and its budget.[246] The final plans for the new observatory were sent to the Civil Constructions Section of this Ministry by the end of June 1908.[247]

Tarazona's previous experience in Barcelona help to clarify procedures, and the project was slightly speeded up. By December 1907, Tarazona had sent another request to the Ministry asking for some money, between 12,000 and 15,000 pesetas, to buy an equatorial telescope and auxiliary apparatus for the new observatory.[248] The Royal Order of 11 June 1908 (published in the *Gaceta* on 17 June 1908) granted 11,000 pesetas to buy scientific equipment for the University of Valencia.

The University of Valencia delegated to Tarazona the acquisition of an equatorial for a price that was expected to be less than 15,000 pesetas (total sum, including the 11,000 pesetas from the government.) After this, and as he did for Barcelona, Tarazona asked Howard Grubb for a dome which would cost no more than the maximum expense established by the University.[249] In January 1909 the Grubb equatorial was sent to Valencia. The telescope, almost identical to the one ordered previously for

---

[244] See: de Benito and Martí y Ortells, 'Don Ignacio Tarazona y el Observatorio Astronómico de Valencia', *Anales de la Universidad de Valencia* (1924–25).
[245] *OAUV*, 11/8—1, 2.
[246] *OAUV*, 11/8—3, 4.
[247] *OAUV*, 11/8—3, 4.
[248] *OAUV*, 11/8—2.
[249] *OAUV*, 11/10—1.

Barcelona, had been acquired through the firm Hijos de Juan Cuesta.[250] It arrived in December 1908, together with several instruments bought from Grubb, which included a spectroscope (1500 pesetas), a photometer (300 pesetas), and a unifilar micrometer for the telescope (600 pesetas), all for a total cost of £341.[251] The following years tested not only Tarazona's enthusiasm for the project, but also his influence, his personal prestige, and his financial position—he had to use his own savings to pay for the construction of the observatory when part of the credits granted by the State did not arrive in time.[252]

The new observatory was finally built on top of the natural history cabinet of the University of Valencia, at the south-east corner of the university building, instead of in the botanical garden. By March 1909 the construction of the observatory, supported by the Ministry, had already started[253] and during the following months Tarazona had several ideas to improve the new building. Some of these were rejected due to lack of funding,[254] such as his suggestion to Alfredo López Alcrudo in November 1909 that stairs should be constructed to give access to the observatory without having to pass through the cabinet.[255] The observatory was almost completed by the end of 1910 with the installation of a rotating dome, designed by Tarazona himself.

Tarazona had also asked the Government, through the Chancellor of the University, to include the cost of the construction of the dome, 4500 pesetas, in the budget for "scientific material" for 1909 and 1910.[256] During these years the observatory acquired several astronomical instruments. In addition to the topographic theodolite made by Filotécnica

---

[250] This firm probably had something to do with the pharmacy owned by the Hijos de Blas Cuesta, who had also played a crucial role in the acquisition of other scientific instruments such as the topographic theodolite made by Filotécnica of Milan and bought in 1907. See: *OAUV*, 11/9—1 to 7.

[251] *OAUV*, 11/10—3. The exchange rate during these years was fairly stable, 1 pound to 27.2 pesetas, which meant that £341 were about 9275 pesetas.

[252] de Benito and Martí y Ortells, 'Don Ignacio Tarazona y el Observatorio Astronómico de Valencia', *Anales de la Universidad de Valencia* (1924–25). A similar statement is made in a letter from Cordoba (Argentina) on 13 May 1912 signed by his disciple Hermengildo Costa. See: *OAUV*, 12/29—2.

[253] *OAUV*, 11/3—5.

[254] *OAUV*, 11/8—9.

[255] *OAUV*, 11/8—7.

[256] *OAUV*, 11/8—1.

of Milan, acquired at the start of the 1907 academic year,[257] and the scientific material bought in 1909 from Howard Grubb, the observatory also purchased, for example, a stereoscope (1908) by Zeiss of Jena. The following equipment was owned by the observatory by October 1909: a theodolite by Salmoiraghi; a stereoscope and stereomicrometer, both by Zeiss; an equatorial of 6 inches (152 mm) aperture by Grubb, together with a position circle and a filar micrometer; a total reflection prism for observations near the zenith, a photometer, and a spectroscope to study solar prominences, all by the same instrument-maker; and a sidereal chronometer by the British maker Dent.[258]

As Tarazona suggested in 1911, the equatorial combined with the photometer and the spectroscope were enough for students to pursue astrophysical work:

> The Chair of Cosmography and Solid-Earth Geophysics [at the University of Valencia] is relatively well equipped with astrophysical instruments. Nonetheless, when it comes to measuring of azimuths, altitudes, latitudes, times, etc. it is very poorly equipped.[259]

The lack of instruments for spherical astronomy was, in Tarazona's opinion, one of the main weaknesses of the observatory. In fact, time determination was a problem as there were available only the topographic theodolite and the chronometer mentioned above. To obtain some degree of accuracy in determinations based on measurements with these instruments it was necessary to use methods which involved lengthy calculations. Tarazona also considered it crucial that the observatory acquired a camera to attach to the equatorial for photographic observations. All this led him in April 1911 to put forward a request, through the Dean of the Faculty of Sciences, to the President of the Instituto de Material Científico, an institution founded just the previous month by the Ministry of Public Instruction to support the acquisition of scientific instruments for research and educational centres.[260]

---

[257] *OAUV*, 11/9—2 to 7. It seems that by the end of 1909 the instrument had been returned to Milan to be repaired. See: *OAUV*, 11/9—8.

[258] Tarazona y Blanch, *La fotografía solar* (Valencia, 1909), p. 60.

[259] See: *OAUV*, 13/10—3, 7.

[260] For further details of this institution, see: Ana Romero, 'Dos políticas de instrumental científico: el Instituto de Material Científico y el Torres Quevedo', *Arbor*, 160 (1998), 359–386.

Tarazona expected to obtain money from this institution for a portable meridian circle by the firm Filotécnica (A. Salmoiraghi),[261] and a mean-time pendulum made by Dent or by Strasser & Rohde; and for the transformation of the visual equatorial into a photographic and visual telescope by Howard Grubb himself.[262] As Tarazona explained, these instruments were essential to support his cosmography lectures. By the end of the year the Instituto granted the observatory 1500 pesetas to buy an astronomical mean-time clock made by Strasser & Rohde (pendulum by Riefler).[263] Although the sum was insufficient to pay for this instrument (its actual cost is not recorded),[264] the clock was finally acquired in June 1912.[265] The observatory also purchased a sextant with a Pistor system, an artificial horizon,[266] and later an average-time chronometer made by Pérez Shekel. Resolved to improve the work of the observatory in the fields mentioned above, Tarazona again asked the Ministry of Public Instruction, on 21 October 1912, for more money to complete his project.[267]

From 1911 Tarazona had the collaboration of Vicente Martí Ortells as Assistant Professor. Together they carried out the main work of the observatory, as determined by Tarazona. This included not only the practical instruction of students, but also routine work such as providing the official time service and taking daily photographs of the Sun,[268] and processing the information received from meteorological telegrams, as well as making observations of, mainly, eclipses, the planets, and transits of Mercury. Other special services were also provided, mainly testing naval chronometers, obtaining meteorological data, and measuring magnetic declination. At the same time, the observatory paid special attention to the social diffusion of astronomy. In fact, the main objectives of the observatory included not only establishing ongoing relationships

---

[261] This instrument was thought to be like the one dispatched to the Observatory of Madrid in 1900 (70 mm aperture, 0.8 m focal length, with a micrometer attached).
[262] *OAUV*, 13/10—1 to 4, and 13/11—1.
[263] *OAUV*, 11/11.
[264] *OAUV*, 13/11—1.
[265] *OAUV*, 13/12—1.
[266] Also acquired were a dipleidoscope, a so-called prism of total reflection to observe zeniths, and a stereoscope and a stereomicrometer by Zeiss. See: *OAUV*, 13/10—9.
[267] *OAUV*, 13/11—1.
[268] The importance of systematic solar photography, not only at the observatory of the University of Valencia, but also at the observatories of Madrid, San Fernando, Ebro, Cartuja, Fabra, and San Feliu de Guíxols (Girona), was discussed by Tarazona in his *Discurso de apertura de curso sobre la fotografía solar* (1909).

with other scientific institutions, but also supporting the growth of astronomical institutions and contributing to the popularisation of this discipline.[269] The collaborations of Tarazona with different newspapers[270] and the organisation of external visits to the observatory are good examples.

By the end of 1912 the observatory had purchased and installed apparatus for wireless telegraphy made by Ducretet et Roger. Its main function was to help gather time information and meteorological data. Other new instruments followed, such as the Mailhat semi-fixed meridian circle by F. Mouronval (the successor of Mailhat, Paris), obtained by 1914 thanks to the director of the Observatory of Paris, Guillaume Bigourdan. Another acquisition was the chronographic register apparatus bought in 1922, a model developed by Campos Rodrigues, director of the Observatory of Lisbon, and manufactured commercially by Favarger. It was acquired after Tarazona and Martí Ortells visited Lisbon in 1921.[271] After this visit, in October that year, Tarazona and Martí Ortells joined the Astronomical Society of Portugal; the subscription was 20 escudos each.[272]

Tarazona also became a Perpetual Member of the Société Astronomique de France in 1914, which cost him 400 French francs.[273] He was also member of the Astronomical Society of Spain and America, the Spanish Association for the Advancement of Science, and the Spanish Society of Physics and Chemistry.[274] He was awarded the Spanish Royal

---

[269] Domingo Querol Cabot, 'Astronomía en la Valencia de la Restauración: Ignacio Tarazona y el Observatorio Astronómico de Valencia', in G. Blanes *et al.* (coords.), *Actes de les IV Trobades d'Història de la Ciència i de la Tècnica* (Alcoi and Barcelona, 1997), pp. 173–180.

[270] de Benito and Martí y Ortells, 'Don Ignacio Tarazona y el Observatorio Astronómico de Valencia', *Anales de la Universidad de Valencia* (1924–25).

[271] The budget for the new instrument was agreed on 3 November 1921.

[272] *OAUV*, 11/7—5.

[273] *OAUV*, 12/6—1, 3, 6, 7, 9, 18, 19. He later paid another 100 French francs, after the modification of the statutes of the society. Note that by 1914 the exchange rate was 1 peseta to 1.1 French francs.

[274] López Piñero and Navarro Brotóns, *Història de la Ciència al País Valencià* (Valencia, 1995), Chapter 6, 'El segle XX', pp. 559–620; and Víctor Navarro Brotóns and Alvaro López García, 'La colección de instrumentos del Observatorio Astronómico de Valencia', in José Ramón Bertomeu Sánchez and Antonio García Belmar (eds.), *Abriendo las Cajas Negras. Colección de instrumentos científicos*

Order of Alfonso XII and appointed by the French Government an Official of Public Instruction.[275] Both Tarazona and Martí Ortells published some research work, most notably in the *Comptes Rendus des Séances de l'Académie des Sciences*.[276] Tarazona also published two astronomical texts (*1468 estrellas del Preliminary Catalogue de Boss contenidas en los principales efemérides* and *Las estrellas del Preliminary Catalogue de Boss ordenadas según sus declinaciones*) and presented some of his work at the observatory to meetings of the Spanish Association for the Advancement of Science.[277] Tarazona was a very prominent astronomer. He was member of several societies, academies, and other institutions and he had many contacts amongst the international scientific community. However, his prestige was more linked to the success of his projects at the Spanish universities rather than to any striking contribution to the scientific knowledge that he made.

This paucity of original work at the Observatory of the University of Valencia contrasts with that of some comparable observatories abroad. For example, from 1902 to 1914 the University Observatory of Oxford published, on average, thirteen papers a year. The statistics show that of these, nine were published in the *Monthly Notices of the Royal Astronomical Society*. These papers dealt with the most appropriate methods for measuring photographs of stars, meridian observations, the apparent movement of stars, observations of variable stars, new

---

de la Universita tde València, (València: Universitat de València, 2002), pp. 331–336.

[275] de Benito and Martí y Ortells, 'Don Ignacio Tarazona y el Observatorio Astronómico de Valencia', *Anales de la Universidad de Valencia* (1924–25).

[276] I. Tarazona and V. Martí Ortells, 'Observation de l'éclipse de Soleil du 20–21 août 1914, faite à l'Observatoire astronomique de l'Université de Valence (Espagne)', *Comptes Rendus*, 159 (1914), 468–469; and I. Tarazona and V. Martí Ortells, 'Observation de l'éclipse de Soleil du 3 février 1916, faite à Valence (Espagne)', *Comptes Rendus*, 162 (1916), 313. Tarazona also published in this journal the following works: 'Observation de l'éclipse de Soleil du 22 novembre 1919, à l'Observatoire astronomique de l'Université de Valence (Espagne)', *Comptes Rendus*, 169 (1919), 1086–1087; and 'Observation de l'éclipse annulaire du Soleil du 27–28 mars 1922, faite à l'Observatoire astronomique de Valence (Espagne)', *Comptes Rendus*, 174 (1922), 1004–1005.

[277] He presented in Madrid (1913) 'Observaciones en el eclipse de Sol del 16 de abril de 1912', and in Valladolid (1916) 'Las 180 zonas estelares del Catálogo general preliminar de Boss'.

discoveries at the observatory, mathematical issues, and so on.[278] Some important projects were also pursued at these observatories, where original research was an important part of their activity. For example, the Oxford *Astrographic Catalogue* was formally completed in 1911 by the University Observatory with the publication of the seventh volume of this catalogue—which gave the positions of about 200,000 stars to about 12th magnitude from the photographic measurements of about 500,000 star images.

However, once the Observatory of the University of Valencia acquired a photographic camera, the study of the solar surface became one of its main and most interesting activities. This work required systematic high-quality photography of the Sun. Some of the photographic tests of the equatorial, made by Martí Ortells from April 1915, were sent to the Observatory of Madrid. Several comments and suggestions on the quality of these photographs and photographic techniques were sent back to the Observatory of Valencia by Aguilar, the director of the Observatory of Madrid.[279] That was the beginning of a correspondence between Martí Ortells and Aguilar about technical details of the telescope and the photographs obtained.[280] These early photographs, which were merely tests before the decision to undertake systematic work on sunspots was made, had in Aguilar's words a perfect focus and were big enough for good measurements of the heliographic coordinates of sunspots—although a bit small to obtain good sunspot statistics.[281]

Once the technical problems had been overcome, Tarazona and Martí Ortells initiated in January 1916 a programme of systematic photography of the Sun to obtain sunspots statistics. Although the study and analysis of these observations was slow but continuous, publication of results was, however, not forthcoming. There were only two staff members and thus little time to analyse the astronomical observations made at the Observatory, a consequence of their teaching obligations. At the same time, administrative tasks were very tedious and time-consuming, and

[278] See: H. H. Turner (ed.), *Miscellaneous Papers of the University Observatory Oxford. Vol. 2, 1902–1905* (Oxford: University Observatory, 1905); H. H. Turner (ed.), *Miscellaneous Papers of the University Observatory Oxford. Vol. 3, 1905–1909* (Oxford: University Observatory, 1910); and H. H. Turner (ed.), *Miscellaneous Papers of the University Observatory Oxford. Vol 4, 1910–1914* (Oxford: University Observatory, 1914).

[279] Letter signed on 24 October 1915. *OAUV*, 11/8—1.

[280] *OAUV*, 11/8—2, 4, 5.

[281] *OAUV*, 11/8—2.

indeed became obstacles to obtaining results. By 1924, the statistics from more than two thousands plates obtained since 1916 had still not been published.[282] The addition of further staff to the observatory during these years aimed to solve this problem. An assistant, E. Aleixandre y Ferrandis, was appointed in or around 1920[283]; in 1923 Tomás Almer Arnau took over his position and, after a practical training period,[284] became deeply involved in the programme of daily photography of the Sun and the compilation of statistics of sunspots for later publication.[285]

Fig. 6-3: Installation of the meridian circle at the observatory of the University of Valencia. Courtesy of the Astronomical Observatory of the University of Valencia.

---

[282] *OAUV*, 13/8—36.

[283] See: R. de Montessus de Ballore (dir.), *Index generalis: annuaire général des universités, grandes écoles, académies, archives, bibliothèques ...* (Paris: Gauthier-Villars, 1920).

[284] *OAUV*, 13/8—15.

[285] See for instance: Tomás Almer, 'Estadística de manchas solares del año 1920', *Publicaciones del Observatorio Astronómico de Valencia*, 2 (1927), 17–52.

On 25 July 1919 the University of Valencia asked the Ministry to declare the Astronomical Observatory an Institution of Public Utility.[286] The formal request was raised by the Chancellor of the University on 11 August 1919[287] and, after a favourable report by the Instituto Geográfico y Estadístico,[288] the observatory was declared of Public Utility on 9 January 1920 (and announced in the *Gaceta* on 23 January 1920).[289] For the observatory this declaration meant regular financial support from the Spanish Government.[290]

6.5. University, politics, and research

This chapter has so far has concentrated on Tarazona's personal efforts in founding university observatories. But the foundation of the observatories of the universities of Barcelona and Valencia could also be understood in terms of the important role played by nationalism in science, as it was part of an attempt to project an image of a modern Spain abroad, pursued by successive governments. The attempts by Ignacio Tarazona to modernise the educational system were warmly welcomed by the newly created Ministry of Public Instruction. In fact, the state of the Spanish higher education system, particularly as regards to science, had been used in the past to attack Spain. Good examples are found during the last years of the nineteenth century, around the time of the Cuban War of 1898. For instance, the *New York Times* in 1897 printed the following lines:

> Again the foreign dispatches are filled with warnings in regard to things which a "proud nation" like Spain cannot endure, and again, in consequence, arises the impulse to ask what right or excuse Spain has to be "proud." Surely there is strangely little in her history to explain her haughtiness. Since the expulsion of the Moors—and of learning and civilization with them—her record has been one of great opportunities wasted, of cruel persecutions as the only evidence of power, and of a steady drop in the scale of European influence. Is she proud of Alva on land or of the Armada at sea? Does she point with exultation to the Inquisition as characteristic of her spiritual development, or to the fact that Ptolemaic astronomy was taught in her universities until only a few years

[286] *OAUV*, 13/6—14, 41.
[287] *OAUV*, 13/6—1.
[288] *OAUV*, 13/6—4.
[289] *OAUV*, 13/6—6.
[290] For the exact amounts, see: *OAUV*, 13/8—3, 4, 5, 7, 9, 11, 12.

> ago as proof of her intelligence? Does she claim credit because to-day only three in ten of her population can read or write?[291]

With the end-of-century *regenerationist* movement, scientific research at Spanish universities was promoted. Slowly but surely, economic resources were invested in higher education laboratories, which at the beginning of the twentieth century were unable to compete "even with *second order* European universities".[292] The shortcomings of the practical instruction of science at Spanish universities were gradually addressed and, in the case of astronomy, new university observatories were founded during the first decades of the twentieth century. The main motivation of these new institutions was to provide practical education to students, although they also carried out some research and offered several public services. However, in general, teaching obligations at universities did not allow much time to undertake original research, and even less to publish any results.

Tarazona's interest in research led him to make some relevant contributions to astronomy, but, as with others in his position, his teaching obligations prevented him from making more of a mark as a research astronomer. The most significant scientific contributions from Spanish astronomers during these years were made by individuals and institutions not attached to any university. The foundation of astronomical observatories during these years has to be seen as an attempt, promoted by some individuals and supported by the State in order to modernise the universities in a similar way to what had been done in other, more advanced countries such as Britain, France, and the United States of America. These were the first stages of the transformation of the Spanish university into a centre of scientific research.

---

[291] 'Topics of the Times', *New York Times*, December 28, 1897, 6.

[292] Letter from the Dean of the University of Barcelona's Faculty of Sciences to the Spanish Minister of Public Instruction. *OAUV*, 11/5—27.

Part II

Astronomy in the field: Spain and the total solar eclipses

CHAPTER SEVEN

ASTRONOMY AND ASTROPHYSICS IN THE NINETEENTH AND EARLY TWENTIETH CENTURIES

7.1. Popular and amateur astronomy

During the nineteenth century, astronomy became one of the most popular scientific subjects in western societies. The studies of the heavens by William Herschel (1738–1822) and his son John Frederick William Herschel (1792–1871) with large reflecting telescopes changed the character of astronomy during that period, and a new breed of astronomy popularisers translated the wonders of the Herschels' discoveries for large audiences. Their success opened up "new physical and intellectual horizons that captured the public imagination".[293] Astronomy lectures attracted large audiences, from young to old, some in pursuit of instruction and others looking for the sublime. In fact, astronomy proved to be a remarkably versatile cultural resource, and people used it in many ways to support different interests and commitments.

The second half of the nineteenth century was the golden age of popular science. Public science became more and more important in Western societies during these years. Indeed, the second half of the nineteenth century and the early years of the twentieth was largely a period of great development for popular science and scientific education,[294]

---

[293] H.C. King, *Geared to the Stars* (Bristol: Hilger, 1978), p. xiv.

[294] See for instance: B. Bensaude-Vincent and A. Rasmussen, *La Science populaire dans la presse et l'édition: XIXe et XXe siècles* (Paris: CNRS, 1997). For a useful historiographical survey, see: J. R. Topham, 'Scientific publishing and the reading of science in nineteenth-century Britain: a historiographical survey and guide to sources', *Studies in History and Philosophy of Science*, 31A (2000), 559–612. A

which was linked with social and economic factors such as the growth of an educated middle class and the invention of cheaper, steam-powered printing technology. New scientific educational projects were carried out, and changes in the popularisation of science effected through the direct intervention of governments: for example, the foundation of science museums in many countries after international exhibitions such as London's Great Exhibition of 1851.[295]

The popular interest in science during the second half of the nineteenth century, particularly in astronomy, had important consequences for both professional and amateur scholars. Particularly interesting is the promotion of science by and amongst a new community of amateur scientists, who not only helped to spread scientific knowledge to a broad public, but emphasised how important science was for any society. In Great Britain, new scientific societies such as the British Association for the Advancement of Science opened new doors to amateur scientists. Taking as its model the Gesellschaft Deutscher Naturforscher und Ärzte, the British Association was founded in 1831 by William Vernon Harcourt (1789–1871) to promote and develop science in Great Britain. This institution expected to make available an infrastructure to promote science which was not being provided in the country, neither by the government, nor by the education system or industry. Institutions such as the conservative Royal Society of London regarded science as a preserve of the elite. The British Association, however, provided a forum for discussion into which anyone interested in science was welcomed. Their audiences were in general slightly more specialised and socially elevated than those of the more vocationally driven organisations such as the Mechanics' Institutes.

New societies, which opened scientific activity to a broad and general public, were created all over Europe. That was, for instance, the case of amateur naturalist and astronomical societies such as the Société

---

brief introduction to the development of public science in Britain is: F. M. Turner, 'Public science in Britain, 1880–1919', *Isis*, 71 (1980), 589–608.

[295] There are many books dealing with the foundation of science museums. A good introductory one is T. Bennett, *The Birth of the Museum: History, Theory, Oolitics* (London, New York: Routledge, 1995). See also: S. V. F. Butler, *Science and Technology Museums* (Leicester, London & New York: Leicester University Press, 1992); and R. Brain, *Going to the Fair. Readings in the Culture of Nineteenth-Century Exhibitions* (Cambridge: Whipple Museum of the History of Science, 1993).

Astronomique de France, which was officially constituted in June 1887. The character of this association, promoted by the French astronomer and populariser of astronomy Camille Flammarion, is made clear in the following lines:

> Vous voyez, Messieurs, que la Société Astronomique de France n'est pas une association scientifique ordinaire.
>
> L'intérêt personnel n'y apporte pas son égoïsme glacial, la science n'y dessèche pas le cœur, et le sentiment continue d'y vibrer avec ses énergiques vitalités. Si tout le monde faisait un peu de science—oh! si peu que ce soit—les esprits seraient à la fois meilleurs et mieux équilibrés, et le monde physique et moral ne s'en porterait que mieux. La bonté qui résulte de la vie intellectuelle est sans doute aussi une condition d'heureuse santé, car lorsqu'on est bon et indulgent, on ne se fait pas de bile, et la bile, disait Hippocrate, jaunit le teint.
>
> Développer notre Société, c'est soutenir la science, c'est répandre la lumière, et c'est faire le bien, d'autant plus que les services rendus à l'humanité par la science sont encore loin d'être reconnus comme ils devraient l'être.[296]

In fact, the success of popular and amateur astronomy in France is perfectly illustrated by the figure of Flammarion. In Great Britain the outstanding figures included John Herschel, Robert Goodacre (1777–1835), Thomas Henry Huxley (1825–1895), John George Wood (1827–1889), Richard Anthony Proctor (1837–1888), Agnes Mary Clerke (1842–1907), and Robert Stawell Ball (1840–1913). The large appeal that astronomy held during the nineteenth century remained in the twentieth century.[297] Moreover, the development of astrophysics during the second half of the nineteenth century owed much to amateur astronomers. Well known, for example, are the first studies of sunspots made by Samuel Heinrich Schwabe (1789–1875) during his more than forty years research on intra-Mercurial planets.[298] The systematic observation of sunspots was

---

[296] Flammarion, 'La Société Astronomique de France', *Bulletin de la Société Astronomique de France*, 15 (1901), 205–217, here p. 214.

[297] This is illustrated by many of the answers given by students at the final examination of the General Astronomy course at Berkeley University of 1917, when they were asked to discuss the educational advantages of the study of astronomy: several answers described the great cultural values transmitted by astronomy. See: R. T. Crawford, 'Students' opinions of the educational advantages of the study of astronomy', *Publications of the Astronomical Society of the Pacific*, 29 (1917), 174–176.

[298] The results were presented in 1843, although the merit and importance of these observations was not recognised until the 1850s. See: Agnes M. Clerke, *A Popular*

first undertaken by the wealthy British amateur Richard Christopher Carrington (1826–1875) at his private observatory.[299]

In general terms, new astrophysical techniques were first developed outside the universities. In Britain and the United States of America, amateur astronomers in private observatories innovated and pioneered the application of new techniques such as photography and spectroscopy.[300] They had the resources and freedom to take risks in speculative and experimental research. One of the greatest strengths of the amateur tradition, as opposed to institutional astronomy, was its ability to respond promptly and decisively to new astronomical problems and challenges. Wealthy amateurs, such as Warren de la Rue, Isaac Roberts (1829–1904), Henry Draper (1837–1882), and Andrew Ainslie Common (1841–1903), developed from the 1860s the research technology for the study of the physical and chemical nature of stars and planets.

Organisations such as the Société Astronomique de France not only promoted cooperation between professional and amateur astronomers, but also encouraged original contributions from amateurs to astronomical research. This initiative quickly spread all over Europe. For example, three years after the creation of the Société Astronomique de France, amateur astronomy in Great Britain received an important boost with the formation in 1890 of the British Astronomical Association. Originally, the BAA was set up as an alternative to the Royal Astronomical Society to cater for those who found the subscription of the Royal Astronomical Society too high, or the papers excessively technical, or who, like women, were excluded.[301] The BAA immediately became not only "the voice of amateur astronomy in the United Kingdom",[302] but a prestigious institution for

---

History of Astronomy During the Nineteenth Century (Edinburgh: Alan & Charles Black, 1885) [4th edn., London: A. & C. Black, 1902]. Particularly interesting is the first chapter: 'Foundation of Astronomical Physics', pp. 125–142.

[299] A. J. Meadows, *Early Solar Physics* (Oxford: Pergamon, 1970).

[300] See for instance: A. Chapman, *The Victorian Amateur Astronomer: Independent Astronomical Research in Britain 1820–1920* (Chichester: Praxis, 1998); and John Lankford, 'Amateurs and astrophysics: a neglected aspect in the development of a scientific speciality', *Social Studies of Science*, 11 (1981), 275–303.

[301] British Astronomical Association, 'The history of the British Astronomical Association. The first fifty years', *British Astronomical Association Memoirs*, 42 (1989). See Chapter 1: 'Formation and first session', pp. 7–10.

[302] Words still found today on the website of this institution: http://www.britastro.org

astronomical research. After one year of its existence the BAA had about five hundred members, and in 1895 about twice this number. The membership reached a maximum in 1900, after the success of the eclipse expeditions of 1896 and 1898, and with the prospect of organising a new one that same year. Indeed, the success of eclipse expeditions organised by astronomical amateur societies and associations played a crucial role in the spread of scientific and astronomical knowledge in all Western countries. But now it is time to study more in depth the general context in which solar eclipses became crucial to the development of astronomy.

7.2. Solar physics and the rise of astrophysics

Halfway through the nineteenth century, astronomy was revolutionised by the development of new techniques, and new branches of knowledge and scientific specialities were developed during these years. One of the most interesting of them was solar physics. The origins of solar physics, a discipline devoted to the study of the nature and constitution of the Sun, are found in the seventeenth century. It was not until the nineteenth century, however, that it became recognised as a specialised subject.[303] During the 1860s the results of experimental work by the Germans Gustav Robert Kirchhoff (1824–1887) and Robert Wilhelm Bunsen (1811–1899) established the basis of spectral analysis and the application of spectroscopy to astronomy. In the mid-century, photographic methods also began to be applied to astronomy.[304] The innovative application of

[303] The nineteenth century has been identified by Hufbauer as solar science's 'second epoch'. See: Karl Hufbauer, 'Solar physics', in J. Lankford (ed.), *History of Astronomy: An Encyclopedia* (New York, London: Garland, 1997), pp. 464–471. This work is a very useful introduction to the topic. Particularly interesting is its brief historiographical survey: the text deals with Clerke's first historical approach to nineteenth-century early solar research (*A Popular History of Astronomy during the Nineteenth Century*), Meadows' studies during the 1970s and 1980s (*Early Solar Physics, Science and Controversy: A Biography of Sir Norman Lockyer*), and Hufbauer's works during the last decade of the twentieth century (*Exploring the Sun: Solar Science Since Galileo*). Later contributions by authors such as Pang on British scientific eclipse expeditions (*Empire and the Sun*) should also be taken into account. For a brief and very useful introduction to the development of astronomy and astrophysics during the nineteenth century, see: John North, *The Fontana History of Astronomy and Cosmology* (London: Fontana Press, 1994).

[304] The years 1840 to 1880 saw gradual changes in the nature of the photographic process and rapid increases in the sensitivity of photographic plates. Photographs of the solar corona were taken systematically from 1882. The last two decades of

spectroscopy and photography to astronomy opened the doors to a new understanding of the heavens. "Astrophysics" was the name chosen by the American George Ellery Hale (1868–1938) for these novel astronomical activities.[305] Early astronomical spectroscopy concentrated on the Sun, the brightest object available. This study was thought to provide evidence of solar–terrestrial relations. The series of total eclipses at the end of the 1860s and in the early 1870s were studied using the newly developed astrophysical techniques, and spectroscopic observations made in the eclipse of 1868 in India and Malaya demonstrated the gaseous nature of solar prominences.

This same year, the Englishman Joseph Norman Lockyer (1836–1920) and the Frenchman Pierre Jules César Janssen (1824–1907) obtained spectra of prominences outside an eclipse. Both Lockyer and Janssen were astronomers who started their careers outside established astronomy. Janssen studied mathematics and physics, and graduated with degrees in these subjects from the Sorbonne in 1852.[306] Twenty-four years later, he became director of the new astrophysical observatory at Meudon. Lockyer, on the other hand, worked from 1857 as a civil servant at the British War Office, and pursued his interest in astronomy on an amateur basis.[307] His

---

the nineteenth century and the first half of the twentieth century experienced a rapid progress of astronomical photography with the development of gelatinous silver bromide dry emulsions. See: Gérard de Vaucouleurs, *Astronomical photography: from the daguerreotype to the electron camera* (London: Faber & Faber, 1961).

[305] The term 'cosmochemistry', however, would probably better describe the nature of these early studies. See: Owen Gingerich, 'The development of astronomical theory and practice from the 17th to the 20th century', *Vistas in Astronomy*, ed. Arthur Beer, 20 (1976), 1–9.

[306] Janssen obtained his doctorate in 1860 and remained at the Sorbonne, working at the Faculty of Medicine on the design of medical instruments. He had constructed an observatory on the flat roof of his house in Paris, and in 1862 began his studies of the solar spectrum. In 1865 he was appointed Professor of Physics at the Ecole Spéciale d'Architecture, and three years later was elected to the Académie des Sciences. On Janssen, see for instance: D. Abbott (ed.), *The Biographical Dictionary of Scientists: Astronomers* (London: Blond Educational, 1984).

[307] Lockyer bought his first telescope in 1861, which he installed in the observatory erected at his home in Hampstead. In 1870, eight years after his election as a Fellow of the Royal Society, he was appointed Secretary of the Royal Commission on Scientific Instruction and the Advancement of Science. In 1875 he was given an appointment at the Royal College of Science. See: A. J. Meadows, *Science and*

rising reputation as a solar physicist led him to be appointed Director of the Solar Physics Observatory in South Kensington in 1885. In Spain, it is the figure of the astronomer Comas i Solà who reminds one of Lockyer and Janssen, as they all began their professional careers outside established astronomy institutions. Indeed, all three of them saw how the study of total solar eclipses and the application of new astrophysical techniques to their work provided them with new professional opportunities. As was said about Comas i Solà:

> Amongst his valuable work there stands out, unquestionably, that dedicated to the study of the solar eclipse of 1900, which attracted favourable and universal tributes and congratulations from the most distinguished astronomers.[308]

Many wealthy amateur astronomers from all over the world, but particularly from Britain and the United States of America, with time and the freedom to innovate, promoted the use of new astrophysical techniques during the second half of the nineteenth century and in the early twentieth century.[309] They directed spectroscopes, spectrographs, and photographic telescopes to the heavens. Astronomy was for these elites synonymous with leisure. They regarded the work of institutional astronomers, which was mainly confined to positional astronomy, as too tedious to undertake. William Huggins (1824–1910), in writing about the construction of his private observatory in 1856 at Tulse Hill, South London, had this to say:

> I soon became a little dissatisfied with the routine character of ordinary astronomical work, and in a vague way sought about in my mind for the possibility of research upon the heavens in a new direction and by new methods. It was just at this time … that the news reached me of Kirchhoff's great discovery of the nature and chemical constitution of the Sun from his interpretation of the Fraunhofer lines … Here at last presented itself the very order of work for which in an indefinite way I was looking—namely, to extend his novel methods of research upon the Sun to the other heavenly bodies. A feeling of inspiration seized me: I felt as if I had it now in my power to lift a veil which had never before been lifted; as if a key had been put into my hands which would unlock a door which had

---

Controversy: A Biography of Sir Norman Lockyer (Cambridge, Mass.: MIT Press, 1972).

[308] Alcalde, 'Homenaje a un astrónomo español', *El Mundo Científico*, 7 (1905), 241–245.

[309] Lankford, 'Amateurs and astrophysics …', *Social Studies of Science*, 11 (1981), 275–303.

> been regarded as for ever closed to man—the veil and door behind which lay the unknown mystery of the true nature of the heavenly bodies.[310]

The second half of the nineteenth century saw wealthy and motivated amateurs taking major initiatives in double-star astronomy, the nature of nebulae, and the study of the lunar, solar, and planetary surfaces.[311] The birth of astrophysics was not initially associated with theoretical investigations, but with the design, construction, and use of new specialist equipment. The typical "gentleman amateur" astronomer had a comfortable financial status and was interested in developing research technology for the study of the physical and chemical nature of stars and planets. As in previous centuries, a small part of the population, interested in the development of knowledge, applied the technology available to the practice of science as a leisure activity.

This pattern only began to change by the end of the nineteenth century. In the world of institutional astronomy, it was not until 1873 that William Henry Mahoney Christie (1845–1922), together with Edward Walter Maunder (1851–1928), developed a programme of systematic photography of the solar surface and of spectroscopic observation of the Sun and the stars at the Royal Observatory at Greenwich. They undertook daily sunspot observations and made several attempts to measure the radial velocity of stars visually. However, professional astronomers did not work systematically in solar physics until the 1890s. In fact, the 1880s and 1890s were years of consolidation of this discipline. During these years, many problems were resolved, including controversies such as those related to solar spectral lines. The discussion about the identification of lines belonging to new elements such as helium and "coronium" was particularly interesting. The new astronomical techniques were improved, and a standard approach to solar research was established. During the 1890s the wish to identify the elements present in the Sun led to a new and more accurate photographic map of the solar spectrum by the American Henry Augustus Rowland (1848–1901). These were also years of intensive study and discussion of the nature of sunspots and the solar rotation. Measurements of the heat radiation from the Sun, partly inspired by the wish to determine the temperature of the solar surface, were also made during the second half of the nineteenth century. The 1890s saw the introduction of new methods and techniques, including methods of

---

310 William Huggins (ed.), *The Scientific Papers of Sir William Huggins* (London: William Wesley & Son, 1909), pp. 5–6.
311 Chapman, *The Victorian Amateur Astronomer* (Chichester, 1998).

estimating the Sun's temperature from its spectrum[312] or from the Stefan–Boltzmann law.[313] Finally, the end of the century saw the correction of several misconceptions concerning the properties of the corona. Progress was made in classifying and cataloguing the chromospheric spectrum[314], and the photographic and spectroscopic study of prominences was especially enlightening.

Astrophysics gradually became the domain of professionals, and by 1900 several astrophysical observatories had been founded in Europe, many of them with governmental support. During the 1890s and 1900s Hale revitalised the field by establishing the *Astrophysical Journal*, founding Yerkes Observatory (1897) and Mount Wilson Solar Observatory (1904), organising the International Union for Cooperation in Solar Research (1904), and showing how sunspots are seats of intense magnetic field (1908). As Meadows pointed out, previous decades had been dominated by the classification and identification of the physical factors that caused the spectral differences.[315] The twentieth century saw a major expansion of the astrophysical literature.

---

[312] This is what the Frenchman Henri Louis Le Châtelier did in 1892 when he compared the solar spectrum in the red with radiation of the same wavelength emitted by heated bodies in the laboratory. New results were also obtained after the discovery in the 1890s of how the temperature of a body affected the wavelength at which maximum radiation was emitted.

[313] Josef Stefan and Ludwig Boltzmann deduced this law, which stated that the radiation from a source varies as the fourth power of the source's absolute temperature. At the time it was regarded with some suspicion because of its simplicity. However, in 1894 it was employed by the Irishmen William Edward Wilson and Peter L. Gray to obtain an effective temperature of the Sun of about 8700°C.

[314] Note that the solar atmosphere consists of three regions: the photosphere, the chromosphere, and the corona. The photosphere is the densest and lowest part of the solar atmosphere. Most of the visible light comes from this layer, about 500 km thick. Above the photosphere is the chromosphere. This region is about 2500 km thick and is almost transparent. It is only easily seen in a total solar eclipse, when it appears, immediately prior and after totality, as a thin red ring around the Sun. The outermost layer of the solar atmosphere is the corona, characterized by low densities and extraordinarily high temperatures.

[315] A. J. Meadows, 'The origin of astrophysics', in O. Gingerich (ed.), *Astrophysics and Twentieth-Century Astronomy to 1950* (Cambridge: Cambridge University Press, 1984), pp. 3–15.

As the second half of the nineteenth century progressed, astronomy became increasingly specialised. Sometimes tensions between amateur and professional astronomers became evident. Professional scientists such as the American professor at Harvard Observatory, William Henry Pickering (1858–1938), were openly critical about the contributions of amateur astronomers. For example, while amateur astronomers strongly defended the scientific importance of making drawings of, for example, the eclipsed Sun,[316] Pickering and other professional astronomers argued that these drawings and direct photographs of the corona had little scientific value. In his opinion, "the most useful observation that could be made by the average person would be that of the duration of totality".[317] These attitudes, common amongst some "establishment" scientists, were a consequence of the gradual process of professionalisation of astronomy and astrophysics during the late nineteenth and early twentieth centuries. Other astronomers, however, considered it unfortunate that the work accomplished by amateurs, equipped for example with only small cameras, had for some of their colleagues no scientific value.[318]

7.3. Solar eclipses

As has been shown, solar physics can trace its origins to solar eclipse studies. The stimulus for eclipse studies during the second third of the nineteenth century was threefold: the observations of bright "beads" of light round the limb of the Moon made at the eclipse of 1836 by the Englishman Francis Baily (1774–1844); the agreement, after the eclipse of 1842, that prominences and corona were real phenomena; and the debate about the physical nature of the Sun itself and the chemical elements present there. By the 1860s the intrinsic appeal of solar eclipses, combined with better transport for observers to travel to eclipse sites, had encouraged the methodical and systematic observation of these events. The birth of systematic studies in solar physics was in fact associated with the series of eclipses of 1868 (India and Malaya), 1869 (America), 1870 (Europe), and 1871 (India).[319]

---

[316] See for instance: A. B. Cobham, 'Astronomical drawing', *Journal of the British Astronomical Association*, 15 (1904–1905), 102–103.

[317] W. H. Pickering, 'The total eclipse of 1900, May 28', *Journal of the British Astronomical Association*, 10 (1899–1900), 333.

[318] See for example: William J. S. Lockyer, 'Thorp gratings and small cameras in eclipse work', *Knowledge & Scientific News*, 2 (1905), 117–119.

[319] See for instance: Gingerich, 'The development of astronomical theory and practice …', *Vistas in Astronomy*, 20 (1976), 1–9.

Pang has presented the eclipse of 1870 as a turning-point in the history of eclipse expeditions.[320] For this eclipse there was unprecedented collaboration, particularly in Britain, between societies and the State in the name of national prestige and utility. This cooperation between scientific societies, government agencies, and leading observatories would characterise the organisation of European eclipse expeditions during the second half of the nineteenth century. American eclipse expeditions, in contrast, were planned, funded, and equipped locally by individual astronomers, observatories, and colleges. Governments in Europe could be moved by appeals to national pride. Eclipses drew the attention of the international press, and an expedition became a demonstration, both at home and abroad, of a country's scientific maturity. The eclipse of 1870 is also identified as a turning-point because from then on, amongst the observers on eclipse expeditions were not only gentleman amateurs, but also government-employed professional scientists and academics.

It was in the early twentieth century that international cooperation became crucial for astronomers. This became obvious at the solar eclipses of 1900 and 1905. As Perrine pointed out at the International Congress of Arts and Sciences held at St Louis in September 1904, the progress of astronomy required a number of well-equipped and well-organised expeditions which coordinated plans and ideas before eclipses took place.[321] Good examples are the instructions of the Department of Terrestrial Magnetism of the Carnegie Institution in Washington to make simultaneous magnetic and allied observations during the eclipse,[322] and the proposed cooperation of the International Meteorological Committee and the Union on Solar Research on visual and photographic observations

---

[320] Alex Soojung-Kim Pang, *Empire and the Sun* (Stanford: Stanford University Press, 2002). See in particular his third chapter: 'The experience of fieldwork', pp. 49–82.

[321] 'Our astronomical column: Solar Eclipse Problems', *Nature*, 71 (1904–1905), 329.

[322] Variations of the Earth's magnetism and electricity were thought to be caused by external forces, such as magnetic and electrical phenomena in connection with total solar eclipses. Bauer pointed out that the eclipse of 1905 was a great opportunity to further observe these phenomena and test the theory. See: L. A. Bauer, 'Appeal for Cooperation in Magnetic and Allied Observations during the Total Solar Eclipse of August 29–30, 1905', *Nature*, 70 (1904), 577; and L. A. Bauer, 'Proposed Magnetic and Allied Observations during the Total Solar Eclipse on August 30', *Nature*, 72 (1905), 342.

of the solar surface, the visual observations of prominences, and observations of the solar atmosphere with spectroheliographs.[323]

7.4. Solar eclipses in Spain

Solar eclipses were, and still are, some of the most fascinating astronomical events. But for the historian of science, the study of solar eclipse expeditions is also very valuable in helping to understand how scientific activity interacts with other social and cultural elements and evolves in a particular context. Indeed, the study of total solar eclipse expeditions reveals not only the scientific progress of astronomy and astrophysics, but also the development of, for example, cultural conventions (for instance, literary and visual standards to interpret astronomical observations and to render them persuasive) and innovative social behaviour (such as the development of the careers of women in astronomy).[324] As Pang has explained, focusing on the particular case of Elizabeth Campbell and the Lick Observatory eclipse expeditions, for most of these expeditions to succeed it was necessary to transport all the necessary equipment and build a temporary observatory at the observing site. Study of the technical work involved, as well as the management and administration of the expeditions, also helps to understand how scientific workplaces operate. The links between social life, culture, and work on these expeditions build a more complete picture of scientific practice.[325]

The paths of totality of three solar eclipses crossed Spain in 1900, 1905, and 1912. It is my intention to study such events in order to use them as a window onto the development of astronomy and astrophysics in Spain during the second half of the nineteenth and the first decades of the twentieth century. Each eclipse was an occasion to learn about the nature of the Sun by spectroscopic, polariscopic, and photographic observation. Indeed, total eclipses provided an excellent opportunity to develop and consolidate new astrophysical techniques.[326] There are several reasons why these events excited both scientists and the society of which they

[323] 'International Union for Cooperation in Solar Research', *Nature*, 72 (1905), 563.

[324] Henrika Kuklick and Robert E. Kohler, 'Introduction', in H. Kuklick and R. E. Kohler (eds.). ‚Science in the field', *Osiris*, 11 (1996), 1–14.

[325] Alex Soojung-Kim Pang, 'Gender, culture, and astrophysical fieldwork: Elizabeth Campbell and the Lick Observatory–Crocker eclipse expeditions', in Kuklick and Kohler (eds.), 'Science in the field',*Osiris*, 11 (1996), 17–43.

[326] Meadows, *Early Solar Physics* (Oxford, 1970).

were part: different social and cultural aspects will have to be considered to elucidate the meaning of these expeditions had for those who organised and interacted with them.

As it will be pointed out in a later chapter, the preparations for the eclipse of 1900 led to an improvement in the conditions in which Spanish astronomers worked. In 1900, for example, a telegraphic station was set up at the Observatory of Madrid to help with the time transmission to all the astronomers observing in Spain.[327] The official telegraphic transmission of the time, something characteristic of "civilised countries", was requested by the Observatory of Madrid to the Minister of Public Instruction, García Alix.[328] The Spanish political and economic situation during the second half of the nineteenth century meant that no official eclipse expedition was sent abroad. By the turn of the century calls were being made for the government to support science, particularly to save the honour of the country. This explains how the eclipses of 1900 and 1905 boosted astronomy in Spain. Spanish astronomers did not have to travel abroad, transporting expensive instruments to unfamiliar territories. The total eclipse was visible in their homeland, and the visit of foreign astronomers not only gave Spanish astronomers the chance to learn from them, but also provided them, both professionals and amateurs, with information about the organisation of eclipse expeditions. Spanish astronomers thus steadily became respected members of the international astronomical community. Feelings about Spanish astronomers were gradually changing during these years, as the following words, reproduced in the early 1900s in the *Journal of the British Astronomical Association*, prove:

> That Spain is more ahead of things than many of us have been in the habit of supposing is further shown by the fact that Señor Iñigues [sic] tells me that the circumstances of the eclipse of April 17, 1912, have already been completely calculated at Madrid.[329]

---

[327] *Archivo del Observatorio Astronómico de Madrid* (OAM), 'Eclipse total de Sol, de 28 Mayo 1900', Folder 15.

[328] See: 'Eclipse total de Sol', *El Heraldo de Madrid*, 12 May 1900 [Año 11, nº 3469], 3.

[329] G. F. Chambers, 'The total eclipse of the Sun of 1905', *Journal of the British Astronomical Association*, 12 (1901–1902), 280–281. The characteristics of the eclipse of 1912 had already been calculated by Antonio Tarazona as early as in 1896.

From the scientific point of view, the visit of foreign astronomers and this aim to compete with them in research invigorated scientific research in Spain, in astronomy and also in related subjects such as meteorology and aeronautics. The quality of Spanish astronomy improved rapidly after 1900, mainly as a result of the great effort made to observe the solar eclipses visible from Spain in 1900, 1905, and 1912.[330] Spanish astronomers used this series of eclipses not only to lobby for more official support, but also to promote their own activities amongst the international scientific community. Several descriptive works, published long before the eclipse, were sent to institutions around the world interested in the astronomical event. They were written by authors such as Antonio Tarazona[331] or institutions like the Instituto Geográfico y Estadístico,[332] and were included by foreign astronomers amongst their basic bibliography.[333] As recorded in *The Observatory*, the text published by the Observatory of Madrid for the eclipse of May 1900 contained an "excellent" map of Spain and Portugal with the circumstances of the eclipse, which was of the greatest value to the organisers of foreign expeditions.[334] One of the most relevant examples was probably that of the prestigious Spanish astronomer Lànderer, who in 1902 was asked by Flammarion for a study of the total solar eclipse of 1905 which was published in the *Bulletin de la Société Astronomique de France*.[335] Some

---

[330] Glick, *Einstein in Spain* (Princeton, 1988), p. 49.

[331] See: Antonio Tarazona y Blanch, *Memoria sobre el eclipse total de sol del día 28 de Mayo de 1900* (Madrid: Observatorio de Madrid, 1899); and Antonio Tarazona y Blanch, *Memoria sobre el eclipse total de sol del día 30 de agosto de 1905* (Madrid: Bailly-Bailliere e hijos, 1904). See also the review in *Nature*, 72 (1905), 77.

[332] Dirección del Instituto Geográfico y Estadístico, *Coordenadas geográficas de puntos comprendidos en la zona de la totalidad del eclipse de sol de 30 de agosto de 1905* (Madrid: Instituto geográfico y estadístico, 1905).

[333] See for instance: C. T. Whitmell, 'Contact times of the eclipse on 30 August', *Journal of the British Astronomical Association*, 16 (1905–1906), 220–226. Reference to the work by Lànderer published in the *Bulletin Astronomique* is also made in: A. De la Baume Pluvinel, 'L'Éclipse totale de Soleil du 28 mai 1900', *Bulletin de la Société Astronomique de France*, 14 (1900), 23–33.

[334] '[Notes:] We …', *The Observatory*, 23 (1900), 261–262.

[335] E. Touchet, 'Société Astronomique de France', *Bulletin de la Société Astronomique de France*, 16 (1902), 518–528, here p. 519. Two previous works were printed in the *Astronomische Nachrichten* before the final version was published: Lànderer, 'L'Éclipse totale de soleil des 29–30 août 1905', *Astronomische Nachrichten*, 159 (1902), 125–128; and Lànderer, 'L'Éclipse totale de Soleil des 29–30 août 1905 (Deuxième note)', *Astronomische Nachrichten*, 160

Spanish astronomers also made public previous scientific studies, like those of Ventosa on currents in the upper regions of the atmosphere, which led him to agree with those who attributed the *shadow bands* to their action.[336] Ventosa urged the employment of a method he devised to record the direction and velocity of the wind during the total solar eclipse of 1900 in order to give more definitive data for discussion. Similarly, before the eclipse of 1905, amateur astronomers such as Horacio Bentabol y Ureta (1854–1928) sent their particular ideas and hypotheses to various local and foreign scholars.[337]

(1903), 309–312. The final version was published in the *Bulletin* in 1903: Lànderer, 'L'Éclipse totale de Soleil du 30 août 1905', *Bulletin de la Société Astronomique de France*, 17 (1903), 405–408. It was later complemented by a second article: Lànderer, 'L'Éclipse totale de Soleil du 30 août 1905 au nord de l'Afrique', *Bulletin de la Société Astronomique de France*, 17 (1903), 374–376.

[336] See: V. Ventosa, 'The Dark Fringes observed during Total Solar Eclipse', *Journal of the British Astronomical Association*, 10 (1899–1900), 333; originally published in *Nature* on 24 May 1900. A few minutes before totality, when the image of the Sun is only a thin crescent of a few arc seconds, wide faint ripples of light appear as a result of the sunlight being refracted by the density variation of the Earth's atmosphere, producing a pattern of light and shade which is known as *shadow bands*.

[337] Horacio Bentabol y Ureta, *Preparación é instrucciones para observar con aprovechamiento el eclipse total de sol de 30 de agosto de 1905 estableciendo la existencia y condiciones de la atmósfera lunar y explicando el origen de todos los fenómenos que suelen presentar los eclipses de sol* (Madrid: Est. Tip. De la viuda é hijos de M. Tello, 1905). This book provides some instructions about observations to carry out during the total eclipse of 1905 in order to prove his hypothesis about the existence of a large but very rarefied lunar atmosphere. As William Lockyer pointed out in *Nature*, 'the author makes a number of suggestions to eclipse observers by which the existence of a lunar atmosphere might be detected. Among other matters he suggests that delicate thermometric observations made from some hours before and after the eclipse might exhibit a heat-absorption effect due to the interposition of the lunar atmosphere, between the Sun and the observer, before the actual body of the Moon was interposed. Solar radiation observations might also exhibit the same effect. Photometric observations of the illumination of the sky might show an analogous absorption of light, due to the lunar atmosphere. Exact determinations of the solar diameter would, if the Moon possesses an atmosphere, probably show the results of the refraction due to that atmosphere. The apparent hourly movement of the Sun should become modified, as the lunar atmosphere in interposed, for the same reason. Many other points whereby the existence of such an atmosphere might be tested are given by Señor Bentabol.' See: William J. S. Lockyer, 'The forthcoming total solar eclipse', *Nature*, 72 (1905), 399.

During these solar eclipses Spanish astronomers benefited from their interaction with foreign scientists who presented their research results to different societies and academies on their behalf.[338] The personal contacts they made helped to bring Spanish scholars into the international scientific community and opened new doors to the development of astronomy in Spain. The success of the observations of the eclipses of 1900, 1905, and 1912 led Spanish astronomers to send an official commission abroad, to observe the eclipse of 21 August 1914 in Feodesia, on the Black Sea, for the first time since the emergence of the new astrophysical techniques.[339] The results of this expedition, including the observation of new lines in the solar spectrum, were communicated to the Académie des Sciences in Paris in November 1914. This example illustrates perfectly the slow but continuous process of development of astronomy in Spain during the second half of the nineteenth and the first decades of the twentieth century. The following chapters will deal more in depth with all these crucial events and with the role they played in the development of astronomy in Spain during the early years of the twentieth century.

---

[338] Perhaps the *Comptes Rendus des Séances de l'Académie des Sciences de Paris* is the best example. On the eclipse of 1905, see for instance: José Comas i Solà, 'Observations sur l'eclipse totale du Soleil du 30 aôut 1905', *Comptes Rendus*, 141 (1905), 616–618 [présentée par M. Loewy]; and R. Cirera, 'Observations magnétiques de l'Observatoire de l'Ebre à l'occasion de l'éclipse de Soleil du 30 aôut 1905', *Comptes Rendus*, 141 (1905), 1270–1271 [presentée par M. Mascart].

[339] Sixto Ríos, 'Matemáticas y Astronomía', in José María Jover Zamora (dir.), *Historia de España Ramón Menéndez Pidal, t. XXXIX, La edad de plata de la cultura española (1898–1936). Letras. Ciencia. Arte. Sociedad y culturas* (Madrid: Espasa Calpe, 1994), pp. 475–494.

CHAPTER EIGHT

THE TOTAL SOLAR ECLIPSE OF 1900

8.1. Foreign eclipse expeditions

A good idea of the most important phenomena studied during the eclipse of 1900 can be obtained from the plans for eclipse parties made by the American Eclipse Committee. This Committee, formed by Simon Newcomb, George Ellery Hale (1868–1938), Edward Emerson Barnard (1857–1923), and William Wallace Campbell (1862–1938), included amongst the more relevant topics to be studied during the eclipse those related to the relative position of the Sun and the Moon, the colour of prominences, the distribution of "coronium", the position of the green coronal line, the *flash spectrum*[340], and the heat radiation of the corona; as well as drawing and photographing the corona, a photographic search for an intra-mercurial planet, and photometric observations of the corona and of Mercury.[341] The comparison of large corona photographs with the photographs to be obtained by American astronomers was also considered very interesting.[342] Similarly, the *Journal of the British Astronomical Association* pointed out before the eclipse that the most important observations to make were, according to the opinions of many leading astronomers, studies of the fine structure of the corona, both visually and by means of large-scale photography, photography of the flash spectrum and determination of the wavelength of the green coronal line, the

---

[340] The so-called *flash spectrum* of the chromosphere, also known as the chromospheric spectrum, is seen during the few seconds just before totality, when the chromosphere's bright emission lines suddenly replace the photosphere's dark absorption lines.

[341] Simon Newcomb *et al.*, 'Plans of American eclipse parties', *Astrophysical Journal*, 11 (1900), 314–321.

[342] 'E. H. H.', 'Council notes on some points connected with the progress of Astronomy', *Monthly Notices of the Royal Astronomical Society*, 60 (1900), 376.

measurement of the heat radiation of the corona, and the photographic search for an intra-mercurial planet.[343]

The importance of recording meteorological data during the eclipse was also pointed out by several institutions. For example, before the eclipse of 1900 C. F. Rossovich and C. Honoré, Secretary and Director of the Uruguayan Instituto Solar Internacional, respectively, sent a communication to various astronomers and observatories in order to stress this point: they hoped that the use of thermometers and actinometers, properly coordinated, would confirm their ideas on the nature of the Sun, in particular about the existence of an invisible "interior Sun", an understanding of which would help to explain the characteristics of solar thermal radiation.[344]

The best way to understand the relevance of the work carried out by Spanish observers at the eclipse of 1900 is, however, to compare it with that carried out by foreign astronomers who visited Spain. Table 8-1 provides a partial list of the observations and research undertaken (that will not be further studied) by both foreign amateur and professional astronomers in Spain in 1900.

---

[343] Edwin B. Frost, 'Astronomical publications: The total solar eclipse of 1900, May 28', *Journal of the British Astronomical Association*, 10 (1899–1900), 224. On the use of pictures and images to study this eclipse, see: John Loomis Kennedy, *Representations of the Spectra in Early British Astrophysics* (D.Phil. dissertation, University of Oxford, 2000).

[344] Instituto Solar Internacional, *Eclipses de sol interior: comunicación científica* (Montevideo: Instituto Solar Internacional, 1900).

Table 8-1: Foreign observers in Spain for the total solar eclipse of 1900

Expedition members and commission body	Programme and observations	Location
C. Flammarion and A. De la Baume Pluvinel. Assisted by Mme Flammarion, Mme De la Baume Pluvinel, Moreux, Noël Albanel and the Spaniard Manuel Pritz (from Alicante)[345]	Study of the form and extension of the corona, tone and intensity of light, differentiation between the coronal atmosphere and the circumsolar aureole. De la Baume Pluvinel was in charge of obtaining the photographs. Albanel and Pritz were in charge of meteorological observations. Main result: aureole extension up to 3,500,000 km from the centre of the Sun (485,000 km from the solar surface).	Elche
Expedition organised by the French Bureau des Longitudes: Hamy (Paris Observatory) and Lagarde (Paris Observatory)[346]	Contact times, photographs of the corona, study of the green line at 530 nm. Instruments: three photographic spectroscopes and three photographic cameras. Lagarde assisted Hamy during totality and was in charge of meridian and contact time observations. Seven photographs of the corona showing the characteristics of a period of minimum activity were obtained.	
Joint expedition organised by the universities of Montpellier and Toulouse. Members: Bourget (Observatory of Toulouse) and his wife, assisted by Carrère (Observatory of Toulouse), Joubin (Faculty of Sciences of Besançon), Meslin (University of Montpellier), Moye (University of Montpellier), and Tramblay (University of Montpellier)[347]	Contact times, photographs of the corona, study of the spectrum of the corona, shadow bands, landscape and sky colour, effect of the eclipse on animals, meteorological observations. Instruments: a Secrétan equatorial telescope of 108 mm aperture to observe contact times, and a 160 mm aperture photographic equatorial; two chronometers were employed. Moye was in charge of the general observations, while Meslin was in charge of the photographs of the coronal spectral image. Several photographs of the corona and its spectrum were obtained.	

[345] Camille Flammarion, 'L'Éclipse totale de soleil', *Bulletin de la Société Astronomique de France*, 14 (1900), 289–297.

[346] 'L'Éclipse de soleil du 28 Mai 1900: Observations de M. Hamy, à Elche', *Bulletin de la Société Astronomique de France*, 14 (1900), 326.

[347] 'L'Éclipse de soleil du 28 Mai 1900: Observations de MM. Meslin, Bourget et Lebeuf, à Elche', *Bulletin de la Société Astronomique de France*, 14 (1900), 329–

Members of the BAA: E. W. Johnson, Lady McClure, and E. C. Willis[348]	Several photographs of the corona were secured.	
Donitch[349]	Photographs of the corona and the coronal spectrum.	
Joint expedition organised by the Royal Dublin Society and the Royal Irish Academy. Members: Joly (Royal Astronomer of Ireland), Howard Grubb and his son Rudolph, A. A. Rambaut (Director of the Radcliffe Observatory), W. E. Wilson (Director of the Deramona Observatory), Geoghegan (Royal Society of Dublin), Bogen (Royal College of Cork), and Downing and his wife[350]	Howard Grubb and Rambaut used a prismatic camera to obtain a series of photographs of the spectrum of the corona, including the flash spectrum. Joly had a coelostat arrangement to take direct photographs of the corona. Wilson had a similar arrangement to take photographs of the corona in monochromatic light. Downing was in charge of giving time signals to the photographers as well as of an opera-glass with spectroscope fitted by Thorp. He also had to examine the distribution of coronium with the spectroscopic arrangement, and compare it with the extensions of the corona. Geoghegan observed the shadow bands. Downing studied the possible diffusion of coronium, as the shape of the coronal light seemed to suggest.[351] Mrs Downing was looking for stars.	Plasencia
H. Keatley Moore, Francis Gare, and Captain Alfred Carpenter[352]	Sketches of the corona, and study of the intensity of the coronal light, shadow bands, and the colour of the corona. Moore and Carpenter were in charge of drawing half of the corona each. Gare was in charge of measuring photographically the intensity of the light of the corona. A combined drawing made	Manzanares

331; Marcel Moye, '[Solar eclipse of May 28:] Naked-eye observations', *The Observatory*, 23 (1900), 279–281.

[348] 'Report of the meeting of the Association, held on June 27, 1900, at Sion College, Victoria Embankment', *Journal of the British Astronomical Association*, 10 (1899–1900), 341–351, here p. 350.

[349] A. Kononowitsch, 'Beobachtung der Sonnenfinsterniss 1900 Mai 28', 154 (1901), 15–16.

[350] See: 'Report of the meeting … held on June 27, 1900 …', *Journal of the British Astronomical Association*, 10 (1899–1900), 341–351, here pp. 344–345; and 'The total solar eclipse of May, 1900', *Nature*, 69 (1903–1904), 160.

[351] E. Walter Maunder, 'The total solar eclipse of 1900, May 28 (Second paper)', *Knowledge*, 23 (1900), 175–179.

[352] 'Report of the meeting … held on June 27, 1900 …', *Journal of the British Astronomical Association*, 10 (1899–1900), 341–351, here p. 350.

	from the sketches of Moore and Carpenter was drawn after totality.	
S. J. Johnson[353]	General observations, meteorological observations. Instruments: 2-inch (51 mm) aperture telescope.	Navalmoral
C. T. Whitmell[354]	Observation of stars, sketch of the corona.	

The eclipse expeditions of the late nineteenth century were opportunities to innovate and apply the latest technological advances to the study of astronomical phenomena. For example, the Irish official expedition employed a kinematograph to take twelve spectra at second contact and twelve more at third contact, with no interval between successive exposures.[355] This device was similar to the "animatograph" made by Nevil Maskelyne,[356] later referred to as a "kinematograph telescope", an instrument that was taken to India for the eclipse of 22 January 1898 and to the United States of America on 28 May 1900. Successful results were obtained, however, only from the expedition to Wadesborough (North Carolina).[357]

One of the best represented countries in Spain in 1900 was Great Britain. In fact, the Joint Permanent Eclipse Committee of the Royal Society and the Royal Astronomical Society had been preparing expeditions to observe the solar eclipse since February 1899.[358] Under the auspices of this Joint Committee, several observers were commissioned to

---

[353] S. J. Johnson, 'Remarks on the total eclipse of the Sun, 1900 May 28, observed at Navalmoral (Spain)', *Monthly Notices of the Royal Astronomical Society*, 60 (1900), 590–591.

[354] 'Report of the meeting … held on June 27, 1900 …', *Journal of the British Astronomical Association*, 10 (1899–1900), 341–351, here p. 351.

[355] 'Astronomical publications: The total solar eclipse of May 1900', *Journal of the British Astronomical Association*, 14 (1903–1904), 140.

[356] Nevil, son of the British magician and illusionist John Nevil Maskelyne (1839–1917) and descendant of the Royal Astronomer Nevil Maskelyne (1732–1811), was to be a frequent scientific collaborator of the British scientific lecturer, aeronaut and astronomer John Mackenzie Bacon (1846–1904).

[357] British Astronomical Association, 'The history of the British Astronomical Association …', *British Astronomical Association Memoirs*, 42 (1989). See Chapter 10: 'Four eclipse expeditions 1896, 1898, 1900, 1905', pp. 50–54. See also: E. Walter Maunder, 'The coming eclipse of the Sun', *Knowledge*, 23 (1900), 49–51, here p. 50.

[358] 'E. H. H.', 'Council notes on some points connected with the progress of astronomy', *Monthly Notices of the Royal Astronomical Society*, 59 (1899), 288.

expeditions to observe the solar eclipse since February 1899.[358] Under the auspices of this Joint Committee, several observers were commissioned to observe the total solar eclipse of 1900 from Ovar (Portugal), Santa Pola (Spain), Algiers (Algeria), and Mazapan (Algeria).

From Portugal, W. H. M. Christie took photographs of the corona with a Thompson photoheliograph, Davidson obtained small photographs showing the extension of the corona, and Dyson secured a series of slit spectroscope photographs showing the spectrum of the Sun's limb and of the corona, while Norman Lockyer and his party in Santa Pola obtained a complete record of the spectrum of the chromosphere at various depths using a prismatic camera. Similarly, Ralph Copeland (1837–1905), based at Santa Pola, took photographs with a 4-inch (102 mm) telescope fitted with a small Iceland spar and quartz prismatic camera.[359] These photographs of the corona made by British official observers were all standardised by the imprinting upon the plates of a series of squares representing known light values.[360] To better understand what these official parties organised for the eclipse of 1900, I shall explore more in depth the programme of observations and the results obtained by these two expeditions sent to Santa Pola and the French official commission to Argamansilla de Alba.

8.2. The Solar Physics Observatory eclipse expedition

Most of the work carried out by both amateur and professional astronomers could be considered regular routine observations. However, as solar eclipse expeditions show, strong efforts were also made to push forward enquiries in various directions as regards to the nature of the Sun and its surroundings. A good example is Norman Lockyer's attempt in 1900 to gain a more intimate and detailed knowledge of the flash spectrum. The object of the expedition that he led to Spain was to obtain a complete record of the chromosphere and coronal spectra and to acquire data on the distribution of different elements.[361] For this purpose, two

---

[358] 'E. H. H.', 'Council notes on some points connected with the progress of astronomy', *Monthly Notices of the Royal Astronomical Society*, 59 (1899), 288.

[359] 'E. H. H.', 'Council notes on some points connected with the progress of Astronomy', *Monthly Notices of the Royal Astronomical Society*, 61 (1901), 251.

[360] Maunder, 'The total solar eclipse of 1900 …', *Knowledge*, 23 (1900), 175–179.

[361] See for instance: Vicente J. Molina García and Enrique P. Oriola Lafuente, *Historia del eclipse total de sol del 28 de mayo de 1900 observado desde Santa Pola* (Santa Pola: Ajuntament de Santa Pola, 2000).

main instruments were taken to Spain: a 6-inch (152 mm) camera with two prisms to obtain data comparable with those from previous eclipse photographs, and a Taylor triple lens of 6 inches aperture and 20 ft (6.09 m) focal length purchased by the Royal Society to find comparative lengths of the various spectral arcs. Lockyer employed such a long focal length to give a large image of the Sun in order to find more details about the level or levels above the Sun's surface of the gases which give the flash spectrum.[362] Other points of interest for the members of this expedition were described by Lockyer as follows:

> For comparison with the spectroscopic pictures of the corona given by the prismatic cameras, it was considered desirable to attempt to secure direct photographs of the corona with instruments having lenses of focal lengths nearly the same as those of the prismatic cameras, and arrangements were accordingly made to use coronographs of 16 feet and 8 feet focal lengths for this purpose. Other coronographs, of shorter focal lengths, were also provided … Some importance was also attached to visual telescopic observations of the inner corona, in order to determine whether the filamentary structure observed in the eclipse of 1871, at a time of maximum sun-spots, was also a feature of the inner corona at a time of sun-spot minimum.[363]

This expedition, mounted by the Solar Physics Observatory at South Kensington, was one of the largest ones. More than a hundred people were trained to carry out the programme of observations. In fact, Lockyer had the 120 officers and men of H.M.S. *Theseus* at his disposal. They assisted the astronomers in time-keeping, as well as working with coronographs and occulting discs, making sketches of the corona, and observing both visible stars and the corona during the eclipse itself, with 3¾-inch (95 mm) telescopes. They also observed landscape colours, shadow phenomena, and shadow bands, and made contact and meteorological observations. Table 8-2 lists the expedition's instruments and observers.

---

[362] E. Walter Maunder, 'The total solar eclipse of May 28, 1900', *Knowledge*, 23 (1900), 145–147.

[363] Norman Lockyer, 'Total eclipse of the Sun, May 28, 1900. Account of the observations made by the Solar Physics Observatory eclipse expedition and the officers and men of H.M.S. "Theseus", at Santa Pola, Spain', *Philosophical Transactions of the Royal Society of London. Series A*, 198 (1902), 375–415, here p. 377.

Table 8-2: Instruments and observers on the H.M.S. *Theseus*

Instruments and observations	Observers
Time-keepers	Officers and men of the *Theseus*: Lieut. F. A. Andrews, Boughey, Lambert, J. Wale, W. Webb, B. Sneller
6-inch prismatic camera	W. Lockyer Officers and men of the *Theseus*: S. Birley, J. Green, C. Fishenden, C. Willmott, A. Humphries, G. Hyatt
20-foot prismatic camera	Fowler Officers and men of the *Theseus*: W. F. Cox, A. Whitbourne, F. Burt, A. Maskell, E. Davies, H. Christopher, W. Harrison
4-inch equatorial	N. Lockyer Officers and men of the *Theseus*: C. C. Lambert
3¾-inch equatorial	Officers and men of the *Theseus*: Lieut. H. M. Doughty, A. G. N. Lane
Long-focus coronograph	Payn Officers and men of the *Theseus*: T. McGowan, E. Woodland, H. Eary, W. Mann, H. Brooks
Graham coronograph	Officers and men of the *Theseus*: Asst Engr. W. J. S. Perkins, W. Walker, J. Knowles
De la Rue coronograph	Officers and men of the *Theseus*: Asst Eng. H. W. Portch, W. Waterfield, H. Frost
Dallmeyer coronograph	Officers and men of the *Theseus*: Surgeon J. Martin, E. Buckingham, R. Quint
Occulting discs	Officers and men of the *Theseus*: J. B. Bateman, W. Fraser, R.S. Bradbrooke, J. A. Daniels, G. Fair, E. Gordon; H. W. Richardson, E. Voyle, T. Orange; W. Tucker, W. Brewer, B. Salmon, A. Mason, A. Steven, C. Paul, A. May, H. Bailey, J. Entwistle
Sketches of the corona without discs (on shore)	Officers and men of the *Theseus*: W. Butt, G. Guilliame, H. Meacher, H. Schmidtael
Sketches of the corona without discs (on board)	Officers and men of the *Theseus*: W. Baxter, W. Butts, C. Jacob, J. Wheeler, E. Willis
Observations of stars (on shore)	Officers and men of the *Theseus*: Clerk Bennett, W. Riches, A. Pontifex, W. Boaworth, H. Angus, W. Kinvett, W. Oliver
Observations of stars (on board)	Officers and men of the *Theseus*: Rev. G. B. Robinson, H. Croxon, A. Phillips, R. Vigus, E. Price, E. Hammond, G. Andrews, G. Nightingale, S. Wilson, E. Savage

Observations of shadow bands (on shore)	Officers and men of the *Theseus*: Commander Hon. R. F. Boyle, T. Slator, J. G. Walsh, F. C. Skinner
Meteorological observations (on shore)	Officers and men of the *Theseus*: Lieut. Pattrick, G.S. Hallowes
Meteorological observations (on board)	Officers and men of the *Theseus*: G. Donnelly, E. Gant, A. Enstidge, W. Hearne, J. Beach
Landscape colours (on shore)	Officers and men of the *Theseus*: Capt. F. V. Whitmarsh, D. Green, Lance-Corpl. Wade, W. Birkett
Landscape colours (on board)	Officers and men of the *Theseus*: Fleet Paymaster A. W. Askham, Lieut. W. J. Frazer
Shadow phenomena (on shore)	Officers and men of the *Theseus*: C. Prynn
Shadow phenomena (on board)	Officers and men of the *Theseus*: Lieut. H. R. Shipster
Photography	Officers and men of the *Theseus*: J. Knight, B. Bulbrook

The training of the officers and men of the *Theseus* was initially left in the hands of the chaplain of the steamer, who delivered on board three general lectures of about an hour and a half, on 6, 7, and 13 May 1900. When Norman Lockyer arrived on board the *Theseus*, he ordered more lectures to be delivered, in three sessions, organised around the three main tasks that they had to help with during the eclipse: work with occulting discs (for cutting the bright light of the inner corona) to detect long extensions of the corona, sketching the corona, and star charting.[364] The party arrived at Santa Pola on 17 May 1900. Once the site previously selected by Payn was approved and the meridian line laid down by him was confirmed with a large azimuth compass, the foundation for one of the siderostats was built. The following day the whole party landed and set up brick piers and some of the instruments. From 19 to 21 May instruments were set up and adjusted, and on 22 May the training commenced. Training in using specific instruments took place over the next two days, and general training also took place.[365]

The day of the eclipse, Howard Payn took several photographs of the corona using a 4-inch (102 mm) Cooke photovisual objective of 16 ft (4.88 m) focal length. The telescope was fed by a coelostat loaned by the Royal Astronomical Society which had previously been used in India. Five

[364] *Ibid.*, pp. 383–384.
[365] *Ibid.*, pp. 385–386.

bluejackets assisted Payn at the instrument: four to hand and receive the plates and to carry and return them to their covers, and the fifth to make the exposures by uncovering and covering the mirror with a piece of millboard.[366] William James Stuart Lockyer (1868–1936), Norman Lockyer's fifth son, and Alfred Fowler (1868–1940) operated two prismatic cameras. Ten plates were obtained from the exposures in the 20 ft prismatic camera, and another ten from the exposures made with the 7 ft and 6 inches (152 mm) aperture camera.[367]

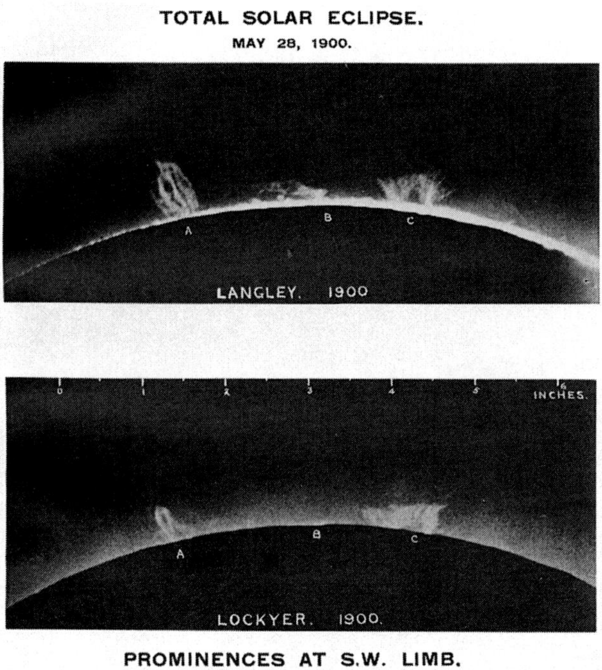

Fig. 8-1: Photographs of the prominences at the solar south-west limb obtained by Langley and Lockyer on 28 May 1900. From the *Philosophical Transactions of the Royal Society of London. Series A*, 198 (1902).

[366] *Ibid.* See: 'Part III. Photographs of the corona taken with a 4-inch Cooke lens of 16 feet focal length. By Howard Payn', pp. 401–404.
[367] *Ibid.* See: 'Part IV. The prismatic cameras. By W. S. J. Lockyer, M.A., Ph.D., and A. Fowler', pp. 405–408.

As regards the spectra of the chromosphere obtained with the prismatic cameras, Norman Lockyer pointed out that the series of photographs greatly resembled those taken in India in 1898. Unfortunately, the observations did not include measurements of the wavelengths of the chromospheric spectral arcs, and so the origins of the arcs remained a mystery. An analysis of line intensities in different photographs showed that while strontium extended down to the Sun's limb, helium existed only in an elevated shell concentric with the photosphere. A similar pattern was noted in the lines of "asterium" and in the unknown line at 4687 Å.[368]

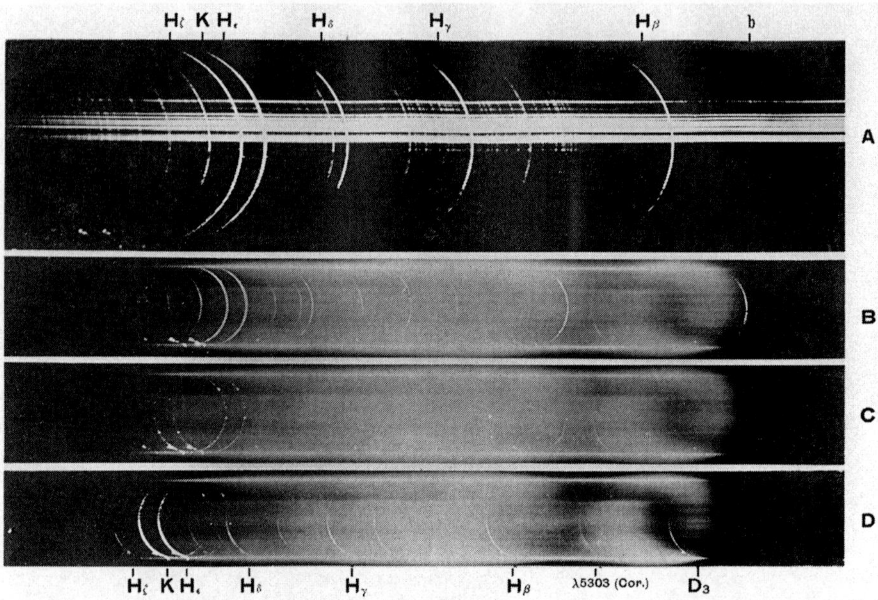

Fig. 8-2: Spectra obtained with a 20 ft prismatic camera near beginning of totality (A) and with a 7 ft prismatic camera near the beginning (B), middle (C), and end (D) of totality. From the *Philosophical Transactions of the Royal Society of London. Series A*, 198 (1902).

---

[368] *Ibid*., p. 409. Asterium, like coronium, was thought to be a new gaseous element. After helium was isolated in terrestrial laboratories and proved to be chemically inert, Carl Runge and Louis Paschen suggested that helium consists of a mixture of two gases, one of them with a yellow spectral line and the other with a green line. They proposed to name the second gas asterium. (Ramsay and Travers proved that this was erroneous, since the colour of the spectral line of helium depends on the gass pressure.)

Spectra of several prominences were also obtained, as well as measurements of the lengths of the arcs photographed at the commencement of totality in order to determine the heights above the photosphere to which corresponding elements were detectable. Finally, the following lines of the spectrum of the corona were observed and studied: 3800 Å, 3987 Å, 4231.3 Å, 4359.5 Å, 4568.5 Å, and 5303.7 Å. These wavelengths corresponded to the main spectral lines or rings recorded in 1898. Other lines were suspected (one of them, near 5537 Å, not previously recorded), but no satisfactory measurement was made.[369]

8.3. The Scottish expedition from the Royal Society of Edinburgh

The Royal Society of Edinburgh also sent to Santa Pola an expedition led by Ralph Copeland (Astronomer Royal for Scotland) and Thomas Heath (of Edinburgh Royal Observatory).[370] On the afternoon of 17 May 1900 they arrived at the camp already set up for Norman Lockyer's party. Although space had been left for them and their equipment, on close examination of the ground they decided to select another site, La Casa del Pleito. As Copeland himself wrote, "it served in the threefold capacity of a store-place for our empty boxes, a photographic laboratory, and a most welcome retreat from the burning rays of the noonday sun".[371] The main instruments brought by this commission were a 40 ft (12.2 m) horizontal telescope of 4 inches (102 mm) aperture, which had already been used on eclipse expeditions to Norway and India. The light was reflected into it by a fixed mirror, and the plate was made to travel instead of the telescope.[372] The Scottish party also brought a small Iceland spar and quartz prismatic camera of about 2 inches (50 mm) effective aperture.

Heath's main instrument was a 6-inch (152 mm) photovisual telescope, made by T. Cooke & Sons, arranged to photograph the corona in the primary focal plane. John Franklin-Adams (1843–1912), a member of the expedition, was left in charge of several very accurate thermometers. Franklin-Adams also had a number of cameras, several of them of large

[369] *Ibid.*, pp. 409–413.
[370] Ralph Copeland, 'Preliminary note on observations of the total solar eclipse of 1900 May 28, made at Santa Pola (Casa del Pleito), Spain', *Proceedings of the Royal Society of London*, 67 (1900), 385–391.
[371] *Ibid.*
[372] Maunder, 'The total solar eclipse of 1900 …', *Knowledge*, 23 (1900), 175–179.

aperture, to obtain pictures of coronal rays and the Sun's surroundings. Three successful negatives from photographic techniques were obtained of the prominences and the corona with the 40-foot telescope, with exposures of about 5, 19, and 3 seconds, respectively, as well as 24 spectrograms taken with a direct-vision prism positioned in front of the object glass of the 40-foot telescope. Of these spectrograms, two groups of ten each were secured, one set as totality was coming on and the other immediately after it had ended. The remaining four spectrograms were taken after totality with the 40-foot acting as a prismatic camera. The two early sets showed the H and K lines and other lines extending beyond the continuous spectrum. Assistance was provided by members of the *Theseus*, such as William Slaughter, who with the 2-inch prismatic camera took six spectrograms, several of which contained ultraviolet lines from the chromosphere or the lower layers of the corona. Copeland also used the telescope of 40 feet focal length and a telescope with an object-glass of Iceland spar to obtain several photographs of the corona. The extension, the structure of the lower corona, the polar plumes, and the amount of coronal light could be easily compared with photographs from previous eclipses.

8.4. The French expedition from the Observatory of Meudon

Like their British colleagues, many French astronomers visited Spain to study the total solar eclipse of 1900. The French expedition from the Observatory of Meudon, for example, consisted of Deslandres and Lucien d'Azambuja (1884–1970), as well as Gaston Millochau, Albert Charbonneaux, Vital Bouson, and the Spaniard Fernando Aguilar, a doctoral student from the Faculty of Sciences and the Observatory of Madrid.[373] Their destination was the village of Argamansilla de Alba, chosen by Deslandres when he knew that no foreign expeditions were going there, as he wanted to keep secret the work undertaken by the expedition that he led. Deslandres was the first, in 1893, to try to determine the velocity of rotation of the corona by spectroscopic methods, and he repeated again this study in 1900 from Argamansilla. To confirm

[373] Ángel Pereira Cabrera, 'Eclipse total de Sol', *El Heraldo de Madrid*, 17 May 1900 [Año 11, nº 3474], 3.

this spin of the solar corona and to obtain its heat intensity, he brought to Spain fifteen spectroscopes.[374]

He also planned to observe the ultraviolet spectrum of the corona in the region from 3500 Å to 3000 Å, which is absorbed by ordinary glass. He had previously obtained such a spectrum for the first time in 1893, though in an incomplete form. Deslandres also attempted to study the ultraviolet spectrum of the reversing layer in that same region and to photograph the corona directly with slow plates of very fine grain.[375] Since 1893, some photographs of the spectrum of the reversing layer had been secured by, among others, Lockyer, Fowler, and Baxandall. These observations were later complemented by those by astronomers such as the Americans Edwin Brant Frost (1866–1935) and Samuel Alfred Mitchell (1874–1960), with spectroscopic records from 6000 Å to 3650 Å. However, only Evershed recorded the spectrum as far as 3350 Å. One of the main objectives of Deslandres's expedition was to extend the ultraviolet spectrum to the extreme limit (3000 Å), and to obtain even more complete and useful photographs of the flash spectrum than those already existent. In the former task he was assisted by the astronomer Millochau, and in the latter by the volunteer astronomer Fallot. He obtained a spectrum from 5000 Å to 3000 Å. In the newly recorded region from 3400 Å to 3000 Å there were 157 lines, the strongest of which were of titanium, vanadium, and iron.[376] An early type of movie camera named chronophotographe was also employed to study the flash spectrum. It allowed the second and third contacts to be photographed in the region from 3770 Å to 3570 Å (six to ten photographs per second).

8.5. The eclipse expedition of the Observatory of Madrid

The words used by astronomers when seeking funding suggests that the involvement and support of governments in eclipse expeditions was in part moved by appeals to national pride and obligation. Eclipses attracted the attention of both the national and the international press, and official

---

[374] Ángel Pereira Cabrera, 'Acerca del Eclipse. En Argamansilla de Alba', *El Heraldo de Madrid*, 22 May 1900, 3.

[375] H. Deslandres, 'Observations of the total solar eclipse of May 28, 1900, at Argamasilla, Spain', *Astrophysical Journal*, 12 (1900), 287–290.

[376] H. Deslandres, 'Spectres ultra-violets de la couche renversante pendant l'éclipse totale du 28 mai 1900', *Comptes Rendus*, 141 (1905), 409–414.

expeditions were a manifestation of a country's scientific maturity.[377] In 1895 the Spanish Government approved an extra budget of 300,000 pesetas for the Observatory of Madrid to acquire new astronomical instruments for the progress of astronomy in the country. However, because of the political instabilities of the time, reflected in the last Spanish colonial conflicts, that money was never received. On 21 February 1899, Ventosa, as interim director of the observatory, asked for half of the money in order to buy some of the equipment needed to observe the total solar eclipse of 1900:

> In my communication to the Minister of Promotion, after clarifying the current state of the issue and that Spain's honour could be damaged if the fateful event [the eclipse of 1900] came without the Observatory being suitably prepared for the study of the phenomenon that, with so much interest on the part of the general public, was expected, I made him see, not only the urgency of obtaining the necessary funds for the acquisition of the most indispensable instruments …, but also the need to send abroad a Commission consisting of two astronomers of the Observatory to visit the main establishments of this kind and the most accredited workshops, in order to be suitably oriented in such delicate matters before making any commitment; and besides, the indispensable thing, to my judgement, that one of the astronomers should also visit other countries (where eminent scholars would provide us with all sorts of facilities) to study during a period the management of instruments and procedures of celestial photography and spectroscopy, new to us, so that he could instruct his colleagues with sufficient time in those practices both useful and necessary.
>
> The main reasons that I adduced were, apparently, properly received, and in all the official institutions the urgency of the case was recognised; but because of the political oscillations in our country, personal and short-term expediency is often preferred to long-term interests, and I came up against unexpected and insuperable difficulties; and … it was not possible for me to have any amount of money during the interim of my control.[378]

The Observatory of Madrid was the main astronomical institution in Spain and, as a consequence, it was a reference point for both local and foreign astronomers in the organisation of their expeditions to study the solar eclipses of 1900 and 1905. The image of Spain projected abroad had to be closely linked to that projected by the Observatory. Nevertheless, there was a lack not only of astrophysical instruments at the institution, but

[377] Pang, *Empire and the Sun* (Stanford, 2002), pp. 49–82.
[378] *Memoria del primer astrónomo del Observatorio de Madrid ... 1899–1900* (Madrid, 1902), pp. 67–68.

also of qualified staff to operate them. The problem had been reported to the Spanish authorities, but no immediate solution was proposed. Not even by September 1899 had the new director, Francisco Iñiguez, received an official answer to the request raised by Ventosa seven months before.

Even during this period of uncertainty, the Observatory of Madrid still developed an observational programme. They wanted to avoid bureaucratic problems delaying the organisation of the expedition to observe the eclipse of 1900. In fact, the observational programme was finalised by September 1899, and various instrument-makers were asked to make instruments that could be employed in the observation of the eclipse and later set up at the Observatory of Madrid. Finally, a concession of 150,000 pesetas was obtained by Royal Order, published on 10 October 1899,[379] by which time several instrument-makers were already working in the construction of the instruments, so that they would be finished in time.[380] The money was spent both in paying for the apparatus purchased and in sending the official commission of the Observatory of Madrid to observe the total solar eclipse of 1900.

The programme of observations was divided into three parts: determining the contact times of the eclipse; obtaining photographs of the several phases of the eclipse (and of the solar corona and the surroundings of the Sun during totality); and observing the spectrum of the solar corona, together with spectrometric measurements to obtain the wavelength of the famous coronal green line—to be discussed later more in depth.[381] A fourth part of the programme was removed from the official record written after the eclipse by the Director of the Observatory of Madrid. The assistant astronomer, Gonzalo Reig y Soler, was in charge of the study of the rotation of the solar corona during the eclipse, with the spectroscope of the visual equatorial made by Grubb. However, as Ventosa wrote, it seems

---

[379] See the Royal Order signed by the Minister of the Treasury in: 'Ministerio de Hacienda. Reales Decretos', *Gaceta de Madrid*, 14 October 1899 [Año 238; num. 287; Tomo IV], 149; and that signed by the Minister of Promotion in: 'Ministerio de Fomento. Real Decreto', *Gaceta de Madrid*, 8 November 1899 [Año 238, num. 312, Tomo IV], 477.

[380] Observatorio Astronómico y Meteorológico de Madrid [Francisco Iñiguez], *Observaciones del eclipse total de sol del 28 de Mayo de 1900 verificadas en Plasencia por la Comisión Oficial* (Madrid: Sucesores de Rivadeneyra, 1900), pp. 9–10. See also: OAM, 'Eclipse total de Sol, de 28 Mayo 1900', Folder 15.

[381] *Ibid.*

that the Observatory of Madrid was not properly equipped to carry out that investigation:

> It is relevant that such an important problem has not been mentioned by Your Excellency [the Director of the Observatory, Francisco Iñiguez], since undoubtedly it was included in the preliminary plan of observations, according to the words of [Señor] Reig: to this forgetfulness may have contributed the circumstance of the little success in his attempt, as with laudable naiveté [Señor] Reig himself confesses, because of the defective conditions in which he had to observe it, outdoors and without a tent that protected him against the currents of light in which, still during the totality of the eclipse, he necessarily had to be involved. And it could not be otherwise, being such a delicate observation, that even observers of M. Deslandres's prestige, with his long experience and skills, with a well defined programme and with instruments prepared expressly and without meanness for that purpose, have not managed yet to observe it with positive results, since the same Deslandres has not hesitated, believing, deservedly, that in explaining what happened there was no dishonour for himself. But without this considered preparation, and without the necessary means, it seems to me a madness to have tried it, however great the good will and the habitual conscientiousness of the observer.[382]

During the months leading up to the eclipse, most of the instruments that had been ordered were delivered to the Observatory and were tested by the astronomers. The first order, a theodolite and meridian circle both made by A. Salmoiraghi, arrived at the Observatory by the end of March. The former, which cost 4000 French francs, was tested by Tarazona and Vela,[383] and the latter, purchased for 5050 French francs, was tested by Tarazona in April 1900 in the Observatory's library.[384] The delivery of these two instruments ensured that the first part of the observational

---

[382] *Memoria del primer astrónomo del Observatorio de Madrid ... 1899–1900* (Madrid, 1902); p. 70.

[383] Antonio Vela y Herranz, *Estudio del teodolito y del anteojo de pasos Salmoiraghi y determinación de la latitud de la hora en las estaciones de Plasencia y Burgos* (Madrid: Imprenta de la Dirección General del Instituto Geográfico y Estadístico, 1906). Other sources, however, say it was during the winter nights that Tarazona and Vela tested it, extending Tarazona his study of the instrument to February and March, although it seems that there is no record of what exactly was done. See: Observatorio Astronómico y Meteorológico de Madrid, *Observaciones del eclipse total de sol del 28 de Mayo de 1900 ...* (Madrid, 1900), pp. 11–14.

[384] Vela y Herranz, *Estudio del teodolito ...* (Madrid, 1906). Note that by 1914 the exchange rate was 1 peseta to 1.3 French francs.

programme and part of the third could be realised. The scientific instruments commissioned from the Swiss firm Peyer, Faverger & Co. of Neufchatel, founded by Matthaeus Hipp (1813–1893) in 1860, and the Irish firm of Grubb were finally received in May. Grubb's instruments were two equatorials and a coelostat (to complete the spectroscope). The instruments for making spectroscopic observations arrived at Madrid with no time to test them, and they had to be tested on site in Plasencia, the location chosen by the Observatory for their expedition. Worse was to come with the photographical equatorial: the crates which contained it were sent straight to Plasencia, where the astronomers of the Observatory of Madrid were by the time the instrument had reached the main train station in Madrid.[385]

On 30 April 1900, the Director of the Observatory of Madrid, Francisco Iñiguez, together with the astronomers Antonio Tarazona and Antonio Vela, the assistant Pedro Jiménez, and the Professor of the University of Barcelona, Ignacio Tarazona, left Madrid for Plasencia. Once they had chosen the final location for the party, Iñiguez returned to Madrid, leaving Tarazona and Vela in charge of determining the latitude of the place and the local time with the Salmoiraghi theodolite and the meridian circle. At the same time, the assistant Jiménez was left in charge of setting up the meteorological station. The rest of the staff of the Observatory had travelled to Plasencia during the previous days at intervals and in small groups.

On 16 May 1900 the auxiliaries Miguel Aguilar and Gonzalo Reig, and the engineer of the Observatory, Gumersindo Cobo, left Madrid together with Rudolph Grubb, son of Howard Grubb, sent by his father to help with the installation of several scientific instruments purchased from their firm.[386] The visual equatorial, 200 mm in aperture and of 3 m focal length, was set up on 20 May 1900. Attached to it was a six-prism spectroscope, located under a special tent. On 23 May 1900, Iñiguez, the assistant Cos, and the Commission sent by the Royal Society of Dublin, left Madrid.

---

385 Observatorio Astronómico y Meteorológico de Madrid, *Observaciones del eclipse total de sol del 28 de Mayo de 1900* ... (Madrid, 1900), pp. 11–14. This equatorial was set up, after the eclipse, at the Observatory of Madrid. The assistant Pedro Jiménez did much work with this telescope during the years that followed, until he finished his doctoral studies. See: Pedro Jiménez Landi, *Fotografía Celeste. Tesis de doctorado* (Madrid: Bailly-Bailliere, 1906).

386 Observatorio Astronómico y Meteorológico de Madrid, *Observaciones del eclipse total de sol del 28 de Mayo de 1900* ... (Madrid, 1900), pp. 15–18.

They arrived at Plasencia to find that the installation of not only the visual but also the photographic equatorial was almost complete.[387] During the following days, while Reig made some spectroscopic observations with the former instrument, Aguilar worked with the latter. The coelostat and the attached spectroscope were set up on 25 May, and on that night and the following they were used to focus the equatorial cameras. The coelostat sent the Sun's rays horizontally to a Steinheil telescope of 120 mm aperture and 1.8 m focal length to project a solar image on the spectroscope slit, under a darkened tent, for observation of the spectral lines.

On 26 May 1900 Vicente Ventosa and Puente, astronomers from the Observatory of Madrid, travelled to Plasencia, together with the amateur astronomer Octaviano Romeo. The regulation of the clock movement was done by Howard Grubb on the morning of 27 May 1900. That same day there arrived in Plasencia the assistant Fernández Ascarza, and the British astronomer Arthur Matthew Weld Downing (1850–1917) and his wife.[388] That day, the Spanish party held a meeting to decide who would make which observations. It is not possible to establish whether this meeting, only one day before the eclipse, was a consequence of the lack of organisation that ruled the official expedition or it had been planned in advance. If the former is true, then the Director of the Observatory would seem to have been acting imprudently on a matter that should have been resolved months before. If the latter, one is tempted to suggest that Iñiguez was applying a strategy to punish those astronomers who criticised him—as is suggested by the distribution of tasks, described below. Whatever the reason, this case shows again the severe problems that some astronomers had with the Director of the Observatory.

At the meeting it was decided that Ventosa and Puente would observe the contacts. Ventosa, who was in open disagreement with the Director, would have to project solar images through a comet browser, while Puente would make direct observations with a Merz telescope of 90 mm aperture and 1.5 m focal length. Ventosa would sketch the solar corona during totality, and Puente and Jiménez were designated to draw any relevant aspects of the corona. Antonio Tarazona was placed in charge of observing

---

[387] The photographical equatorial had one camera of 200 mm aperture and 2 m focal length, a second camera of 160 mm aperture and 1 m focal length, and a lens of 160 mm aperture and 2 m focal length.

[388] Observatorio Astronómico y Meteorológico de Madrid, *Observaciones del eclipse total de sol del 28 de Mayo de 1900 ...* (Madrid, 1900), pp. 19–21.

shadow bands and visible stars. Cos and Ignacio Tarazona were in charge of time-keeping. Ascarza, Reig, and Romeo would make spectroscopic observations, while Vela, Aguilar, and Iñiguez were in charge of photography. For the duration of the eclipse, Cobo had to expose a photographic plate inside a box every ten minutes, for a few minutes at a time. He was in charge of these photometric observations, and of announcing the remaining time of totality. The director of the school at Plasencia, Maximino Martínez Cuesta, who assisted the Spanish official commission, was placed in charge of the meteorological observations. There is also evidence that the theosophist and amateur astronomer Mario Roso de Luna helped this official expedition.[389]

The astronomer Ramón Escandón was the only member of staff of the Observatory of Madrid who stayed in Madrid on the day of the eclipse; with the help of several students from the Faculty of Sciences, he had to complete some minor observations. That day, in Plasencia, thirty Civil Guards prevented curious people from getting closer than two hundred metres to the location chosen by the astronomers from the Observatory of Madrid. At 2.18 p.m., five minutes before totality began, the members of the party took up their positions. When Baily's beads appeared, the astronomers started their work and exposed their first photographic plate. Five photographs were obtained in total: three photographs were taken with a camera of long focal length and two with a camera of short focal length. Particularly interesting were the first and fourth photographs, which showed that the corona can be visible even when the solar disc is still visible, and recorded the extension of the solar corona up to three solar diameters.[390] All the photographs showed a major extension of the solar corona at the equator. These photographic plates were successfully developed the next morning.[391]

Spanish astronomers tried to advertise the results of their observations to the international community in order to project abroad a modern image of Spain, to justify their scientific work, and to promote the astronomical activity in their country. They sent their results to various leading scholars and institutions, and their observations were, if not published, at least

---

[389] *Ibid*., pp. 23–28.

[390] 'Astronomical publications: Spanish observations of the eclipse of May 28', *Journal of the British Astronomical Association*, 11 (1900–1901), 127.

[391] Observatorio Astronómico y Meteorológico de Madrid, *Observaciones del eclipse total de sol del 28 de Mayo de 1900 ...* (Madrid, 1900), pp. 28–32.

reviewed in several international journals.[392] However, it seems that the new Director of the Observatory of Madrid, Francisco Iñiguez, was obsessed with drawing significant conclusions from the astronomical work pursued at the Observatory, which led in some cases to amusing and even problematic outcomes. Particularly interesting is the case, explained by Ventosa, of the photographs of the solar corona obtained by the official party to Plasencia:

> The photographs of the corona turned out to be excellent, though they did not have as many details as Your Excellency [Francisco Iñiguez] wished; Your Excellency tried to retouch them, from which I tried to dissuade Your Excellency, believing that with observations and calculations one has to keep the most scrupulous fidelity if the results are to contribute to the progress of the Science and earn the confidence of those who want to use them.[393]

The problems between the Director of the Observatory of Madrid and his first astronomer are also well presented in these lines written by Ventosa. As well as the disapproval of the work done for the eclipse of 1900, described above by Ventosa, Iñiguez was criticised for his conclusions on how to use the local weather conditions to select the observing site—by, in Ventosa's words a "special procedure" unknown to anyone else.[394] Ventosa also questioned the Director's description of the weather conditions during the day of the eclipse, the criteria he used to judge the quality of an instrument,[395] and his knowledge of astronomy and competence as an astronomer. In fact it was, in Ventosa's opinion, Howard Grubb and his son Rudolph who "in the eyes of Spaniards, saved the national honour, though we do not know whether the foreigners believed that honour had been saved".[396]

---

[392] See for instance the review of the preliminary report of observations made by Iñiguez, 'Observaciones del Eclipse Total de Sol del 28 de Mayo de 1900', *Journal of the British Astronomical Association*, 11 (1900–1901), p. 84.

[393] *Memoria del primer astrónomo del Observatorio de Madrid ... 1899–1900* (Madrid, 1902); pp. 77–78.

[394] *Ibid*, p. 77.

[395] Ventosa pointed out that the beauty of an instrument had been considered, in an official publication of the observatory, a relevant criterion and that the apparent agreement in values obtained by two observers could indicate the quality of their instruments, without considering their personal equations. See: *Ibid.*, pp. 71–72.

[396] *Ibid.*, p. 68.

The disagreements between Iñiguez and Ventosa are a good illustration of how official accounts do not always give a true picture of the progress of science. Ventosa's complaints go beyond the limits of personal differences and illustrate well the problems of astronomical research and the importance of publishing relevant results. Publication of important results gains not only prestige but also attracts official funding. At the same time, the arguments presented by the first astronomer of the Observatory of Madrid show the fragility of these scientific enterprises. The Spanish expedition carried out drills only on the last day before the eclipse—in contrast to the meticulous organisation of British and French commissions. The lack of appropriate organisation on these expeditions could lead to chaotic situations, which often had to be resolved in an improvised way.

8.6. The mystery of "coronium"

The study of the green coronal line was one of the main objectives of the party sent by the Observatory of Madrid to Plasencia in 1900. Some prior history is needed to understand the significance of this investigation. It was in 1869 that the American astronomers Charles Augustus Young (1834–1908) and William Harkness (1837–1903) independently observed for the first time and during a total solar eclipse an emission line of feeble intensity in the green part of the spectrum of the corona. Each spectral line of the photosphere and the chromosphere was associated with chemical elements found on Earth, but the line from the coronal spectrum matched nothing that could be reproduced in the laboratory. Although weather conditions prevented Harkness from observing the line again from Syracuse (Italy) during the solar eclipse of 1870, Young succeeded in his observations and reaffirmed his previous results that identified the green coronal line with the iron line numbered 1474 of Kirchhoff's catalogue (1474 K), at about 5317 Å. However, this classification presented some problems: why was only one of the hundreds of iron lines in the coronal spectrum being observed, and why did it appear at such great heights?

The eclipses which followed that of 1869 gave evidence of other lines which could not be satisfactorily identified and which were thought to have something to do with the green coronal line. The green line was observed again in 1871 by Janssen, Abbay, Lockyer, and Respighi; no accurate measurements were made, but Abbay determined that the line he observed was between lines 1464 and 1494 of Kirchhoff's catalogue. Seven years later, in 1878, Young observed the green line again and confirmed once more his original observation. By the late 1870s some scholars thought that the green line was a consequence of two elements, one next to each other in the spectrum. On this basis, in the early 1880s the British scientists George Dawning Liveing (1827–1924) and James Dewar (1842–1923) made a comparison of the iron spectrum in the laboratory with the solar spectrum, and their results seemed to show that the coronal line was identical to one of the iron doublets. Observations were still made at every opportunity in order to increase the small amount of available data for theoretical and for laboratory work on identification. Finally, on 17 May 1882, Lockyer, Tachini, Trepied, and Thollon sent a telegram from El Cairo (Egypt) confirming that the green line was line number 1475 in Kirchhoff's catalogue.

Even when the green line was observed during the eclipses of 1883, 1886, 1887, and 1889 it was not measured. In fact, no further reliable spectral observations were obtained until 1893. In 1898 both Norman Lockyer and William Wallace Campbell observed the green line photographically from India.[397] They concluded that the green coronal line was about 14 Å towards the violet end of the spectrum from its previously accepted measured position. Their more accurate determination of the wavelength of the line as 5303 Å was possible because they were using instruments of better definition and higher dispersion. The publication of a note by Campbell in the *Astrophysical Journal* led to a brief note from Young in which he acknowledged that in his observations of the coronal line in 1870 and 1878, totality ended before he was able to apply the micrometer to measure the position of the green line.[398]

In 1898 it was proposed that the mysterious green coronal line was due to a substance unknown on Earth, and "much lighter than any known terrestrial substance". It was named "coronium".[399] In fact, it was not until

---

[397] Fowler also obtained, at the solar eclipse of 1898, photographs of the coronal spectrum that showed that the 1474 line had been wrongly measured.

[398] See: W. W. Campbell, 'The wave-length of the green coronal line and other data resulting from an attempt to determine the law of rotation of the solar corona', *Astrophysical Journal*, 10 (1899), 186–192. See also: S. P. Langley, 'The new spectrum', *Philosophical Magazine*, 2 (1901), 119–130; and J. Evershed, 'Wave-length determinations and general results obtained from detailed examination of spectra photographed at the solar eclipse of January 22, 1898', *Proceedings of the Royal Society of London*, 68 (1901), 6–9.

[399] W. W. Bryant, *A History of Astronomy* (London: Methuen & Co., 1907). See in particular the ninth chapter, 'Solar Eclipses: Spectroscopy', pp.169–182.

the late 1930s and early 1940s that the Swedish Bength Edlén and the German Walter Robert Wilhelm Grotrian identified some of the bright lines observed in spectra of solar corona during eclipses as forbidden lines of various elements. The green emission at a wavelength of 530 nm was identified as very highly ionised form of iron, an iron atom that has lost 13 of its 26 electrons (Fe-XIV).

Many observations were made to obtain better determinations of the wavelength of the coronal lines, especially by British and American astronomers, and in particular of the green line, which Lockyer and Campbell had measured to be at 5303 Å. In January 1900 the American Eclipse Committee, formed by Newcomb, Hale, Barnard, and Campbell, reported their plans for the American eclipse parties that were to observe the eclipse in May of that year. The instructions, followed by more than fifteen American parties, included not only studies of the relative position of the Sun and the Moon, drawings of the corona, observations of the colour of prominences, photographs of the corona, and a photographic search for an intra-mercurial planet, but also studies on the distribution of coronium, the position of the green coronal line, the flash spectrum, the heat radiation of the corona, and photometric observations of the corona and Mercury.[400]

Amongst the most important astronomers who intended to study the green line during the eclipse of 28 May 1900 were Gautier, Tachini, and Riccò in Algeria, and Young, Russell, Huff, and Reed in America.[401] But Gautier, Director of the Observatory of Geneva, was unable to measure the coronal line. Tachini and Riccò did not find the green line amongst the 36 lines of the spectrum that they photographed. At the same time, Hugh Frank Newall (1857–1954) did not find the coronal ring perceptible on a plate exposed for 46 seconds. Young and Russell, observing from Wadesboro (North Carolina), did not obtain any record of the green line. Similarly, the photographic observations of the green line made by Dyson from Ovar (Portugal) were unsuccessful. In Spain, several prestigious astronomers such as Lockyer, Copeland, De la Baume Pluvinel, Hamy, and Deslandres planned similar observations.[402] Lockyer observed from Santa Pola a weak green line, while Copeland did not distinguish anything.

---

[400] Newcomb *et al.*, 'Plans of American eclipse parties', *Astrophysical Journal*, 11 (1900), 314–321.

[401] Fernández Ascarza, 'Eclipse total de sol de 1900 …', *Revista General de Marina* (April 1901), 1–15.

[402] *Ibid.*, pp. 12–15.

From Elche, Hamy could obtain no photographic record of the green line, and De la Baume Pluvinel could observe no image in a six-prism spectroscope. Deslandres, in Argamansilla, observed a weak, short, wide, and diffuse green line.

8.7. Spanish observations of the green coronal line

The Committee of the Observatory of Madrid considered the observation of the green line as one of the main studies to be pursued during the total solar eclipse of 28 May 1900. However, the first astronomer, Ventosa, believed that it was a mistake to assign two assistants to these observations.[403] As he pointed out, the two assistants chosen, Fernández Ascarza and Reig, had no previous experience in spectroscopic studies. Their first spectroscopic observations were made three months before the eclipse. This lack of experience might partially explain why Reig did not succeed in his measurements—he only detected an absorption line next to the C line of the spectrum.[404] However, the observations and measurement of the coronal green line made by Fernández Ascarza in Plasencia were advertised as very successful and important to the international community. The results of these observations were published not only by Iñiguez, in his *Observaciones del Eclipse Total de Sol del 28 de Mayo de 1900*, but also by Fernández Ascarza in journals such as *Revista General de Marina* and *Astronomische Nachrichten*. These results also came to the attention of the international community through their

---

[403] *Memoria del primer astrónomo del Observatorio de Madrid ... 1899–1900* (Madrid, 1902), pp. 79–97.

[404] The dark lines of the solar spectrum, known as absorption lines, are caused by specific chemical elements in the Sun's atmosphere. For example, Fraunhofer's C line is produced by hydrogen and is located in the red region of the spectrum, at 656.3 nm; the D line is produced by sodium and is located in the yellow region of the spectrum, at 589.2 nm; and the F line, also produced by hydrogen, is located in the blue region of the spectrum, at 486.1 nm. These two hydrogen lines are also named Hα and Hβ, as they are the first two lines of the four characteristic lines of the hydrogen spectrum originally described by the Swiss Johann Jakob Balmer. The E and G lines are related to Iron and are located in the green and violet regions of the spectrum, respectively, while H and K are identified with calcium. At the same time, D3 is related to helium and b to magnesium. Further information on how various types of spectra, especially the Sun's Fraunhofer lines, have been recorded, displayed, and interpreted is found in: Klaus Hentschel, *Mapping the Spectrum: Techniques of Visual Representation in Research and Teaching* (Oxford: Oxford University Press, 2002).

publication in other prestigious journals such as the *Journal of the British Astronomical Association*.[405]

As has been already mentioned, the instruments employed for this investigation were a coelostat and a six-prism spectroscope made by Duboscq. In order to avoid too much light being absorbed by the spectroscope, the observation was made with only three prisms. The setting up of the device was finished on 27 May 1900. To obtain numerical results from the observations of the green coronal line, three well defined lines of the solar spectrum were studied: Fcα (5327 Å), E (5269 Å), and b1 (5183 Å).[406] Ignacio Tarazona, Octaviano Romeo, and Victoriano Fernández Ascarza spent the morning of 28 May 1900 fixing the relative positions of these lines and the intervals between them in terms of the graduations on the micrometer. These positions gave 62 units for the line at 5327 Å, 208 units for the line at 5269 Å, and 433 units for the line at 5183 Å. Official records explain that during totality Tarazona was in charge of writing down the results of the observations dictated from inside the tent, while Romeo was given only the observation of the spectrum during the last 10 seconds of totality. Fernández Ascarza was chosen to observe the spectrum from the beginning and make micrometric measurements. He observed, however, a diffuse green line and made two measurements of each of its edges (four measurements in total). The average was 136 micrometer units. Applying the interpolation formula of Gibbs gave a wavelength for the green line of 5297.3 Å. This result was obtained making use of wavelength values determined by Anders Jonas Ångström (1814–1874). The results were later calculated for the values determined by Rowland of F (5328.696 Å), E (5270.108 Å), and b1 (5183.792 Å). These new values led to a wavelength for the green line of

---

[405] Reference to a review in *Nature* (20 December 1900) was made in 'Astronomical publications: Spanish observations of the eclipse of May 28', *Journal of the British Astronomical Association*, 11 (1900–1901), 127.

[406] The E line was given by Ångstrom as double: E1 at 5269.5 Å, and E2 at 5268.5 Å. However, with just three prisms in use, the spectroscope did not separate these two lines well. The average was taken, giving 5269 Å. The complete method and results are found in: Fernández Ascarza, 'Eclipse total de sol de 1900 …', *Revista General de Marina* (April 1901), 1–15; and Victoriano Ascarza, 'Nota sobre la longitud de onda de la raya verde (1474 K) del espectro de la corona solar', *Astronomische Nachrichten*, 155 (1901), 23–24.

5298.7 Å, as compared with the previous measurements by Young (5317 Å) and by Lockyer and Campbell in 1898 (5303 Å).[407]

How Fernández Ascarza's observations were interpreted illustrates the lack of experience of many other Spanish astronomers who, like him, attempted spectroscopic observations. But problems arose not only from practical but also from theoretical issues. For example, Fernández Ascarza originally calculated the wavelength of the green coronal line from the map of the solar spectrum produced by Ångström in 1869. These results, however, had to be later recalculated in terms of the new and more accurate photographic map made by Rowland in the 1890s.[408] At the same time, these results were very controversial. First of all, the average values of the positions of the three spectral lines that had to be compared with the observations were taken in round figures. As has been already mentioned, these lines were located, in terms of the micrometer scale, at positions 62, 208, and 433. As Ventosa pointed out, this may have had important repercussions in the final results: a reading error of a single unit on the micrometer scale would have produced a significant variation in the wavelength, half a unit corresponding to one ten-millionth part of a metre.[409] There were also some problems related to the installation of the device and the method of measurement. As mentioned above, Fernández Ascarza explained how he measured the diffuse green line using the tangential method. But how could a diffuse line with no borders (by definition) be measured by that method? Other Spanish scholars such as Fontserè seemed very reluctant to accept the results obtained by the Spanish official commission and described them as mere "spectroscopic essays".[410] There was even more criticism when it became known that Fernández Ascarza's results contradicted the observations made in 1898 by Lockyer and Campbell. Ventosa summarised his opinion of these results in the following words:

> It is enough … to see the conscientiousness with which Mr. Lockyer and Mr. Campbell made their observations, the means they had to carry them

---

[407] That same value of 5298.7 Å was obtained with the interpolation formula of Hartmann.

[408] Meadows, *Early Solar Physics* (Oxford, 1970); particularly relevant is the second chapter, 'The new astronomy (1850–1900)', pp. 20–82.

[409] Consequently, the error for a wavelength stated in angstroms ($1 \text{ Å} = 10^{-10}$ m $= 10^{-7}$ mm) would be $0.5 \times 10^{-7} = 500$ Å (!!)

[410] Eduardo Fontserè, 'Resultado generales del eclipse de Sol de 28 de Mayo de 1900', *El Mundo Científico*, 2 (1900), 419–422, here p. 420.

> out, and the extreme meticulousness with which they made their calculations, in order to not be surprised at the notable conformity that they reached in their results. How, after this and with the indisputable authority that those gentlemen have, has Mr. Ascarza dared to correct them by publishing his first results (which he later had to amend) as definitively exact?[411]

Octaviano Romeo also criticised the results obtained by Fernández Ascarza. Romeo, a lawyer and Doctor of Sciences, had long experience in spectroscopic studies, at the Jesuit Observatory of Manila (Philippines). He stated that, when Fernández Ascarza finished his observations, he had about 25 seconds of totality left, in which he distinguished a green line, badly defined and of uncertain limits. This observation persuaded Romeo of the uselessness of micrometric measurements for determining the wavelength of the green line.[412] He also complained about the lack of suitable equipment for pursuing such an important study: no prismatic cameras, photographical spectroscopes, or quartz prisms were available.[413] Ventosa held the Director of the Observatory responsible for the publication of these dubious results and of the bad reputation that the Observatory might acquire as a consequence. Together with the poor behaviour of Ramón Escandón, who remained at the Observatory on the day of the eclipse and not only made inadequate use of the instruments when he conducted visiting politicians and journalists, but openly expressed his opinions about the lack of facilities and the work at the institution in front of them,[414] this issue seems to have led to some malicious and mocking words in Spanish journals and periodicals. As Ventosa wrote:

> Was the green line such a *perfectly outlined line on the continuous spectrum of the solar atmosphere, capable of being measured*, according to the opinion of Mr. Ascarza and admitted by Your Excellency [Francisco Iñiguez]; or was a *broad band, badly defined and of confused limits, highly difficult to make any micrometric measurement of its position*, as noticed

---

411 *Memoria del primer astrónomo del Observatorio de Madrid ... 1899–1900* (Madrid, 1902), p. 91.

412 Published in: *Madrid Científico*, 16 September 1900, and cited in: *Memoria del primer astrónomo del Observatorio de Madrid ... 1899–1900* (Madrid, 1902), p. 94.

413 'El eclipse total de Sol de este mes', *El Mundo Científico*, 7 (1905), 527–534, here p. 528.

414 *Memoria del primer astrónomo del Observatorio de Madrid ... 1899–1900* (Madrid, 1902); p. 97–107.

> Mr. Romeo, in agreement with what the foreign observers saw? Your Excellency, with his superior criterion and his more exact knowledge of all the precedents of the issue, will have well weighed the reasons that led him to adhere to one of the mentioned opinions.[415]

Similarly to what the Spanish commission attempted, De la Baume Pluvinel tried to observe the coronal line with a six-prism spectroscope in 1900, from Elche, but he was unsuccessful.[416] The six prisms weakened the light so much that no images were obtained. De la Baume Pluvinel obtained at this eclipse nine photographs of the corona. He also made spectral observations with a one-prism spectroscope. Hamy, also at Elche, failed to observe the green line, and the planned studies of the physical nature of the line were not made.[417]

8.8. Other Spanish observations of the eclipse of 1900

Many Spanish amateur scientists observed the event and sent their results to different institutions and prestigious scholars. Amateur astronomy was regarded in Spain during the late nineteenth and early twentieth centuries as a crucial part of the development of astronomy and astrophysics in the country. In fact, several scholars promoted amateur science in Spain, in particular, amateur astronomy on the occasion of the total solar eclipse of 1900.[418] These initiatives were very successful, as shown by the amount of astronomical work carried out in Spain during that eclipse. Table 8-3 lists some of the most interesting observations made by Spanish amateur observers, and illustrates the fascination awoken by the total eclipse of 1900 and how amateur astronomy spread all over Spain in that same year.[419]

---

[415] *Ibid.*, p. 96.

[416] 'L'Éclipse de soleil du 28 Mai 1900: Observations de M. de la Baume Pluvinel, à Elche', *Bulletin de la Société Astronomique de France*, 14 (1900), 331–332.

[417] 'L'Éclipse de … 1900: Observations de M. Hamy …', *Bulletin de la Société Astronomique de France*, 14 (1900), 326.

[418] See for example: Eduardo Fontserè, 'El eclipse total de sol del 28 de mayo de 1900', *El Mundo Científico*, 2 (1900), 259–263.

[419] Data obtained from: Camille Flammarion, '[L'Éclipse totale de soleil] Notes complémentaires', *Bulletin de la Société Astronomique de France*, 14 (1900), 298–307; Observatorio Astronómico y Meteorológico de Madrid, *Observaciones del eclipse total de sol del 28 de Mayo de 1900 …* (Madrid, 1900), pp. 33–43; Eduardo Fontserè, 'Eclipse total de 28 de mayo de 1900', *El Mundo Científico*, 2 (1900), 403–410, here p. 410; 'En vísperas del eclipse', *El Heraldo de Madrid*, 26 May

Table 8-3: The eclipse of 1900. Spanish amateur observers

Observers	Observations	Location
Cabrera Navarro	Shadow bands, meteorological observations	Aguilas (Murcia)
Professors from the Secondary School of Albacete and volunteers: E. Alonso, R. López, J. Torres Reina, S. do Rodrigo, T. Sabras, B. Alaraz, E. Navarro, J. Carrilero, and R. Serrano	Contact times, sunspots, shadow bands, meteorological observations, photographs and sketches of the corona	Albacete
J. Marco and J. González Gómez	Contact times, observation of the corona, effect of the eclipse on plants and animals	Alcaraz (Albacete)
G. Cea	Effects of the eclipse on ants	
Senante and Pericás	Contact times, sketch of the corona, meteorological observations, visual observation of stars, effect of the eclipse on animals	Alicante
Amateur astronomers from Majorca: N. Roca, etc.	Meteorological observations, Baily's beads, sketches of the corona, visual observation of stars	
Amare	Photographs of the eclipse	
Members of the Real Academia de Ciencias y Artes de Barcelona: Mundi, Lozano, Rouviére, Fontserè, Nacente, Estalella, etc.	Study of the relationship between the intensity of light and heat radiation (photometric observations), meteorological observations, 40 photographs of the partial phases of the eclipse, contact times	Barcelona
S. Raurich, F. Sluch, and A. Torres	Shadow bands	

1900 [Año 11, nº 3483], 3; and *Archives of the Observatory of Madrid*, 'Eclipse total de sol del 30 de agosto de 1905', Folder 16.

F. Calderón	General observations with a small telescope	Cartagena (Murcia)
A. de Soto	Shadow bands, study of the intensity of light	Casavieja (Ávila)
M. Prieto	Confirmation of totality where the eclipse was observed	Ciudad Rodrigo (Salamanca)
E. Merino and O. León	Meteorological observations, shadow bands, sketch of the corona	Coria (Cáceres)
A. Fernández Miranda, and F. Sánchez Fernández	Meteorological observations, contact times, effect of the eclipse on plants and animals	
G. Gillmán, Director of the Compañía de los Ferrocarriles de Lorca á Baza	Photographs of the corona	Elche (Alicante)
A. Meseguer	Study of the colour of the sky	Murcia
Secondary School of Toledo: Professors Hoyos, Olavarrieta, Vega, Comendador, Fraile, Aragonés, Tebar, Besteiro, and Ferrer, Román (Director of the School), Lieut.-Col. Villalba, Cmdt. García Moreno, Capt. Maldonado, Capt. García Miranda, and Lieut.-Col. Solás	Contact times, meteorological observations, sketches and photographs of the corona, shadow bands, effect of the eclipse on plants and animals	Navahermosa (Toledo)
M. Gil, lecturer at the Central University of Madrid, together with other lecturers of that faculty	Photographs of the corona	Navalmoral de la Mata (Cáceres)
Students from the Escuela Normal Central de Maestros: Miralles, Sensat, La Pouide, etc.	Meteorological observations, sketches of the corona, totality of duration, study of the intensity of light	
Broussain	Effects of the eclipse on plants and animals	Plasencia (Cáceres)
M. Caparyon, J. Lopez, Aquero, and M. Camaillad	Confirmation of totality where the eclipse was observed	Toledo

Conde of Cañete and his daughter	General observations with a sextant and a small equatorial	Tomelloso (Ciudad Real)
J. Bru Marin	Effects of the eclipse on plants and animals	Torrevieja (Alicante)
R. Troula	Barometric observations, study of the intensity of light	Villena
J. M. Carpena	Shadow bands, meteorological observations, effects of the eclipse on animals, sketch of the corona	Yecla (Murcia)
Lieut.-Col. F. Figueroa (with a Mailhat telescope of 50 mm millimetres aperture)	Lunar observations during the partial phases of the eclipse, meteorological observations, effect of the eclipse on animals	Zafra (Badajoz)

Other outstanding observers, such as Augusto Arcimis, Director of the Central Meteorological Institute at Madrid, observed the eclipse from Argamansilla. Arcimis was assisted by Castellarnau, Breñosa, and Breñosa's daughter, and his reputation as a meteorologist led Camille Flammarion to ask him for the meteorological observations that he undertook during the eclipse.[420] Amongst the astronomical work of the expedition led by Arcimis, his work on the luminosity of light during the eclipse and on shadow bands stands out. The Observatory of San Fernando also sent a party to Elche, led by its Director, Juan Viniegra.[421] This Spanish official party employed several instruments, including a two-prism spectroscope, to obtain two photographs of the spectrum of the corona. The first photograph, that of the normal spectrum, was compared with the second one, a photograph of the coronal spectrum taken about 34 minutes of arc from the solar limb, which showed a continuum spectrum up to the G line and a discrete spectrum beyond that line.[422]

420 Flammarion, '[…] Notes complémentaires', *Bulletin de la Société Astronomique de France*, 14 (1900), 298–307, here pp. 300–303.

421 The telegraphic communication between this party and the Observatory of San Fernando itself was of great assistance in time-keeping for the joint expedition organised by the universities of Montpellier and Toulouse. See: 'L'Éclipse de … 1900: Observations de MM. Meslin …', *Bulletin de la Société Astronomique de France*, 14 (1900), 329–331.

422 Fontserè, 'Resultado generales del eclipse de Sol de 28 de Mayo de 1900', *El Mundo Científico*, 2 (1900), 419–422; p. 420.

All in all, the most outstanding work by Spanish astronomers during the eclipse of 1900 came from Elche, where astronomers such as Comas i Solà and Lànderer set up their instruments. The nature of their observations helps one to understand the slow but continuous progress that astronomy experienced in Spain during the last years of the nineteenth and the early twentieth century. For example, Comas i Solà observed the eclipse of 1900 with a prismatic camera attached to a telescope of 55 mm aperture and 0.8 m focal length, together with a photographic camera of 110 mm aperture and 1.6 m focal length, both mounted equatorially.[423] He not only made meteorological observations, but also obtained two spectra of the chromosphere and two photographs of the corona. Although he did not have an appropriate device to measure the spectra, the first one, obtained at the beginning of totality in a 2-second exposure, recorded more than 120 lines, prominent amongst which were the H and K lines, those of

Fig. 8-3: Photograph of the solar corona and Mercury, obtained from Elche by Comas i Solà in 1900. Published in *El Mundo Científico*, 22 (1900).

[423] José Comas i Solà, 'El eclipse de Sol del 28 mayo 1900', *Astronomische Nachrichten*, 153 (1900), 85–88.

hydrogen (both in the visible and in the ultraviolet), the D3 line, the double b line of magnesium, some titanium and iron lines, and, very weak, the 1474 K coronal line. The second spectrum, a 4-second exposure obtained 25 seconds after the beginning of totality, showed mainly the H and K lines. Comas i Solà concluded that, because it was the only element known to produce these lines, calcium was the most elevated element in the chromosphere, with an average thickness of 22 seconds. In other words, the calcium layer was no lower than 18,000 km above the photosphere.[424] The two photographs of the corona showed what would be expected at a period of minimum solar activity: big equatorial expansions, formed by filaments and beams of light converging to the solar equator, and fan-shaped light at the poles. In his photographs, the corona extended to about three solar radii. Lànderer also observed the eclipse from Elche. He studied the polarisation of the light from the solar corona, employing a Cornu photopolarimeter.[425] Lànderer was assisted in his observations by two local physicians, Llopis and Palomares, who were in charge of the chronometer and the illumination of the division of the circle, respectively.

8.9. Local versus foreign astronomy

The eclipse of 1900, and in particular the role of the Observatory of Madrid, can be seen as a turning point in the development of astronomy and astrophysics in Spain. On the expeditions sent in 1860 by the Observatory to El Desierto de las Palmas (Castellón) and Moncayo (Zaragoza), Spanish official astronomers had worked as assistants of foreign astronomers such as Secchi and LeVerrier—as had almost every Spanish scholar who observed the eclipse.[426] However, things were different in 1900. The Observatory of Madrid not only organised a large and important expedition to Plasencia but also developed its own research project, for which it renewed its astronomical instruments, a first step in catching up with more developed countries. For example, the two 4½-inch (114 mm) Stenheil telescopes acquired for the eclipse of 1860 were

[424] Fontserè, 'Resultado generales del eclipse de Sol de 28 de Mayo de 1900', *El Mundo Científico*, 2 (1900), 419–422, here p. 420.
[425] J. J. Lànderer, 'Sur la proportion de lumière polarisée de la couronne solaire', *Comptes Rendus*, 130 (1900), 1524–1525.
[426] On the role of Spanish astronomers in the eclipses of 1860 and 1870, particularly as regards to the commissions organised by the Observatory of Madrid, see: López Arroyo, *El Real Observatorio Astronómico de Madrid (1785–1975)* (Madrid, 2004), pp. 115–122.

obsolete by 1900, so new visual and photographic Grubb telescopes were acquired.

This awakening of Spanish astronomy by the turn of the century is illustrated by comparing the research programmes and observational techniques employed by both local and foreign astronomers at the eclipse of 1900. As mentioned above, the main objectives of all the official expeditions were to study of the corona and the solar spectrum. The former was done photographically in Spain by the Observatory of Madrid, the Universities of Montpellier and Toulouse, the Solar Physics Observatory, and the Irish and Scottish commissions. Spanish astronomers from the Observatory of Madrid employed two cameras, of 200 and 160 mm aperture, attached to the Grubb photographic equatorial. These instruments are comparable to the 160 mm photographic equatorial and the 150 and 100 mm photovisual telescopes used by the commission of the Universities of Montpellier and Toulouse, the Solar Physics Observatory, and the Scottish commission, respectively, to photograph the corona. The attempt by the Observatory of Madrid to study the rotation of the solar corona with a spectroscope attached to the Grubb visual equatorial, although it produced no results, was very original, as the secrecy with which these studies were carried by the expedition from the Observatory of Meudon shows. This absence of results is a good example of how Spanish astronomers, though they were catching up with their foreign counterparts, still had a gap to close.

This disparity is even clearer in the case of the attempt by the Observatory of Madrid to measure the green line in the solar spectrum. Several institutions tried to photograph the spectrum during the eclipse, including the Irish and Scottish commissions and the expeditions from the Solar Physics Observatory, the Observatory of Meudon, and the Universities of Montpellier and Toulouse, but the Observatory of Madrid attempted to observe it visually. Even the Observatory of San Fernando, with its very modest programme, employed a two-prism spectroscope to photograph the spectrum of the corona. It is not surprising that Octaviano Romeo criticised Madrid's lack of prismatic cameras, quartz prisms, and photographical spectroscopes to study the spectrum. The Solar Physics Observatory and the Scottish commission both employed several prismatic cameras to photograph the spectrum of the corona and the chromosphere. In conclusion, the Spanish eclipse expeditions of 1900 are interesting for the way in which they aspired to emulate the top astronomical research pursued by foreign astronomers, but still fell short.

CHAPTER NINE

THE TOTAL SOLAR ECLIPSE OF 1905

9.1. Introduction

At the eclipse of 30 August 1905 the Sun's shadow crossed Europe again, but as in 1900, Spain was once more the only country on the continent where the eclipse was seen as total. The path of totality ran from Labrador to North Africa, and many official expeditions from various countries were sent abroad to stations all along the shadow path. The best conditions, however, were thought to be in Spain, with more than three and a half minutes of estimated totality at locations closest to the central line. As William Lockyer pointed out four days before the eclipse:

> When it is recalled that since the general introduction of photographic methods into eclipse work the sun has only lasted for about half an hour, that the duration of totality in the coming event exceeds 3¾ minutes, and that no favourable opportunity will occur again until 1912, when totality will only last for about 60 seconds, some idea may be obtained of the anxiety of those observers who are fortunate enough to take an active part in next Wednesday's observations.[427]

At the same time, the low altitude of the Sun during totality at Labrador and Egypt (27° and 24°, respectively) made Spain the most favourable location for the best observations.[428] As a consequence, scholars of almost all nationalities went to Spanish territory to observe the event. As Mitchell pointed out:

> Spain was chosen by the majority of astronomers, both because the duration of totality was longer, and because the promise of good weather conditions was better; and here in a path of one hundred and twenty miles

---

[427] Lockyer, 'The forthcoming total solar eclipse', *Nature*, 72 (1905), 399.

[428] See: William J. S. Lockyer, 'The approaching total solar eclipse of August 30', *Nature*, 71 (1905), 393–395.

width running diagonally across the peninsula, hundreds of astronomers, American and European, were gathered.[429]

Fig. 9-1: The total solar eclipse of 1905 in Spain. From 'L'Éclipse totale de soleil du 30 août 1905', *Bulletin de la Société Astronomique de France*, 18 (1904).

The shadow path stretched from the north coast of the Iberian Peninsula to the Balearic Islands. As the circular of the U.S. Naval Observatory reported:

> The middle of the shadow crosses the coast line about 100 geographic miles west of the city of Santander and sweeps southeastwardly across the Spanish Peninsula, leaving Madrid about 40 miles and Valencia about 2 miles distant from the southern edge and outside of the shadow path. Some of the cities within the limits of the shadow path are Leon and Burgos, not far from the central line, Valladolid, near the southern edge, and Zaragoza, near the northern edge, while the old city of Sagunto lies about 20 miles within the southern edge … After leaving the Spanish coast, the shadow sweeps over the Columbretes … the shadow passes over two islands of the Balearic group—Ivica [sic], just with the southern edge, and Majorca, within the northern edge—but the only satisfactory harbor is the port of Palma … about 40 miles north of the central line.[430]

Official expeditions from countries such as France, Great Britain, the United States of America, Russia, Holland, Germany, and Switzerland were sent to Spain to observe that solar eclipse. Furthermore, astronomers

---

[429] S. A. Mitchell, *Eclipses of the Sun* (New York: Columbia University Press, 1923). See: 'Chapter XI. The Spanish eclipse of 1905', pp. 185–194, here p. 185.

[430] Colby M. Chester, 'The total solar eclipse of August 29–30, 1905', *Science*, 21 (1905), 635–637.

from many other countries went to Spain as independent individuals or were invited by the organisers of these parties. There is evidence of hundreds of observers from all over the world travelling to more than twenty localities to study the eclipse. These numbers increase dramatically if we include local astronomical parties and observers.

9.2. Foreign astronomical research

The eclipse of 1905 was a good opportunity to compare the results obtained by the different expeditions located on opposite sides of the Atlantic. As William Wallace Campbell wrote:

> The interval of two hours and one-half between the instants of totality in Labrador and Egypt offers an unusual advantage for obtaining large-scale photographs of the solar corona, with a view to determining changes in the forms and positions of the delicate details of structure. The opportunity to bring the search for intra-mercurial planets to a satisfactory conclusion is also exceedingly promising.[431]

However, in general terms, observations from the western side of the Atlantic were unsuccessful due to adverse weather conditions.

Astronomers planned to investigate, among other things, the relative position of the Sun and the Moon at the time of the eclipse, the search for intra-mercurial planets, and the corona. In 1905 the existence of intra-mercurial planets was still a problem to be solved, and something that needed "to be eliminated from eclipse programs".[432] In fact, many efforts were made to resolve the question. Little credit was given to the presence of bodies of appreciable size, although the existence of a ring of very small bodies within the orbit of Mercury was considered. If there turned out to be no intra-mercurial planets, mathematical astronomers would need to find an alternative cause of the discrepancies in Mercury's motion. Indeed, the movement of Mercury's perihelion was only later explained with the revolutionary theory of (general) relativity presented by Albert Einstein (1879–1955).

---

[431] W. W. Campbell, 'The Crocker eclipse expeditions in 1905', *Astrophysical Journal*, 20 (1904), 235–236.

[432] C. D. Perrine, 'Some total solar eclipse problems', *Astrophysical Journal*, 20 (1904), 331–337 [Read at the International Congress of Arts and Science, St. Louis, 23 September 1904].

Studies of the corona were considered of the highest importance. They included photographs with long- and short-focus lenses to obtain, respectively, large-scale pictures of the inner corona, and pictures of the corona as a whole embracing its farthest extensions. Also very important were spectroscopic investigations of the reversing layer, the chromosphere, the prominences, and the corona. Attention was also to be paid to photometry, shadow bands, polariscopic work, and meteorological observations.[433] The Joint Permanent Eclipse Committee planned an extensive programme of spectroscopic work, both with slit spectroscopes and with objective prisms of long focus and of dispersion considerably greater than had been previously employed, and also of polariscopic work with Iceland spar prisms, of determinations of the coronal radiation with a bolometer, and of photographs of the corona with instruments of long and short focus.[434]

More precise were the proposals by Charles Dillon Perrine (1867–1951), as summarised at the First International Congress of the Union for Co-operation in Solar Research, held in 1904 at St. Louis.[435] The movements and velocities of coronal matter were to be studied, making use of long-focus cameras to take photographs from different locations at intervals of about one to two hours. Equally, the rotational velocity of the corona relative to that of the Sun's surface was another subject which required further analysis. Whatever was emitting the coronal bright-line radiations had not been identified with any terrestrial element. Solar and coronal hypotheses, such as the theory of the incandescent inner corona, had also to be tested through systematic photometric determinations of the brightness of the corona. The wavelengths of the lines in the spectra of the chromosphere, reversing layer, and corona were also to be determined with the greatest possible accuracy in order to make their identification with terrestrial elements more certain. The use of a moving plate to furnish a continuous record of the spectrum of the Sun's limb was also considered of especial value: it was hoped that it would reveal the heights of the

---

[433] See for instance the article signed by the Superintendent of the U.S. Naval Observatory: Chester, 'The total solar Eelipse of August 29–30, 1905', *Science*, 21 (1905), 635–637.

[434] 'Report of the Council to the Eighty-fifth Annual General Meeting of the Society. Notes on some points connected with the progress of astronomy during the past year', *Monthly Notices of the Royal Astronomical Society*, 65 (1905), 382–391, here p. 391.

[435] Perrine, 'Some Total Solar Eclipse Problems', *Astrophysical Journal*, 20 (1904), 331–337.

elements present and possible changes in the wavelengths of the lines in their spectra, as successive atmospheric layers were covered and uncovered by the advancing Moon. There was also proposed a study of the depths and distribution of the gaseous constituents of the corona throughout one or more sunspot cycles, and of the questionable rotation of the corona. However, the most pressing problem in connection with eclipse work was "that of finding a method for observing the corona in full sunshine" to secure a continuous record, as had already been achieved for prominences.[436] All the methods tried thus far had failed to give positive results.

9.3. American eclipse expeditions to Spain

Those were the main problems and research programmes pursued by astronomers for the total solar eclipse of 1905. It is time now to explore the specific activities of some of the foreign expeditions that visited Spain. As regards to the American expeditions to Spain, it is well known how the U.S. Naval Observatory, with the benefit of an appropriation of $5000 by Congress, undertook the observation of the eclipse of 1905. The party, sent out by the U.S. Government, was under the general direction of Rear-Admiral Colby Mitchell Chester (1844–1932). The main objective of these expeditions was the study of the external corona.[437] For this purpose, the Naval Observatory planned to send two expeditions to Spain. The first, which it was originally intended should go to Burgos, had a stationary camera with an objective of 120 mm and a polar setting apparatus with a camera attached to it with a lens of 210 mm.

The other instruments taken by this first expedition, which planned to visit the central line, were a horizontal photographic telescope of 5 inches (127 mm) aperture and 40 ft (12.2 m) focal length; a photographic telescope of 8½ inches (216 mm) aperture and 12 ft (3.7 m) focal length; a 6-inch (152 mm) Dallmeyer camera of 36 inches (0.91 m) focal length; a portable telescope for contact observations; several grating spectroscopes; spectrophotometric apparatus; and several items of meteorological equipment. The second American station, expected to be set up within the edge of the shadow path, near Valencia, would have a camera 19 m long which produced an image of the Sun 170 mm in diameter, and a polar

---

[436] *Ibid.*

[437] 'El Eclipse de 30 agosto próximo. Expediciones yanquis', *El Mundo Científico*, 7 (1905), 465–466.

setting apparatus with a camera attached to it which had a 150 mm lens.[438] They also had a horizontal photographic telescope of 7½ inches (190 mm) aperture and 65 ft (19.8 m) focal length; a prismatic spectrograph; a photographic telescope of about 9½ inches (240 mm) aperture and 14 ft (4.3 m) focal length with a colour screen; a 6-inch (152 mm) Dallmeyer camera of 40 inches (1 m) focal length; a portable telescope for observing contacts; a grating spectrograph; and several items of meteorological apparatus.[439] Three warships were furnished by the Navy Department for the purposes of the expedition: the U.S.S. *Minneapolis*, the U.S.S. *Dixie*, and the U.S.S. *Caesar*.[440] While the *Dixie* took the African party to Tunis, the other two commissions were sent to Spain. One was destined for Porta Coeli (Valencia), at the edge of the path of totality, and the other set up near the central line at Daroca, about 20 km from Valencia.[441]

When the *Minneapolis* anchored in the harbour of Valencia, it became the first American warship to visit a Spanish port since the 1898 Spanish-American War.[442] Spanish foreign policy had changed since then, particularly after the coronation of Alfonso XIII in 1902.[443] Indeed, Spain now recognised its diminished role as a colonial power in the Americas, and turned its aspirations to North Africa, in particular to Morocco. But France also had colonial interests in that continent. At the same time, Spain was on good terms with Britain, a country very much linked to the United States of America. These two countries had a powerful influence in the final decision on the future of Morocco, which might explain why Spain, a country which aimed to be a colonial power in Africa, tried to

---

[438] *Ibid*. The Spanish source recorded the physical dimensions of the camera and not its focal length.

[439] Chester, 'The total solar eclipse of August 29–30, 1905', *Science*, 21 (1905), 635–637.

[440] Mitchell, *Eclipses of the Sun* (New York, 1923), p. 185.

[441] The former included Lieutenant Commander Hayden, Professor P. B. Littell, G. H. Peters, and G. A. Hill, from the U.S. Naval Observatory, and J. A. Anderson, from the Johns Hopkins University; the latter included Professor Eichelberger and Yowell, from the Naval Observatory, Professor J. H. Bigelow, and from the U.S. Weather Bureau, Hoxton, Olivier, and Mitchell. See: *Ibid*., p. 188.

[442] *Ibid*., p. 186.

[443] On the new image of the country projected by Alfonso XIII, see for example the following primary sources: Juan Pérez de Guzman, 'Alfonso XIII en Londres y las relaciones políticas anglo-españolas', *La Ilustración Española y Americana*, 80 (1905) [Año 49, número 22], 355–359; and Pérez de Guzman, 'El Rey', *La Ilustración Española y Americana*, 80 (1905) [Año 49, número 24], 357–358.

improve its diplomatic relationships with the Americans.[444] The visit of two American warships to Spain is the best example of how these relationships changed in less than ten years. Moreover, all this can also be understood as a part of the civil discourse for science that arose in Spain during these years, discussed in early chapters.

Coronal photographs obtained at the three stations showed not only numerous prominences and great detail, but also the corona extending to two or three solar diameters from the limb of the Moon.[445] But of all the expeditions from the Naval Observatory, the one which obtained more information and interesting results was that located at Daroca, on the central line.[446] On the voyage across the Atlantic, the officers and men of the *Minneapolis* attended astronomical conferences and lectures. Once in Spain, the advance party consisted of four astronomers, eight sailors from the *Minneapolis*, and an interpreter, who arrived at Daroca on 26 July 1905.[447] The greenest field was selected as the observing site and, as Mitchell explained,

> upon application to its owner for permission we found that he was quite satisfied to allow his plot of ground to be used, but thought some compensation should be made for the valuable crop of grass that might possibly be raised during the summer. On receipt of one hundred pesetas, he forthwith proceeded to take a fatherly interest in all of our doings, and explained scientific matters to every one as if he had been chief of the expedition.[448]

---

[444] Not much information is available on the diplomatic relationships between Spain and the United States of America after the Spanish-American War. For an introduction to Spanish foreign policy and international relations during the early years of the twentieth century, see: José María Jover Zamora, *España en la política internacional. Siglos XVIII-XX* (Madrid: Marcial Pons, 1999); Carlos Seco Serrano, *La España de Alfonso XIII. El Estado. La Política. Los Movimientos Sociales* (Madrid: Espasa Calpe, 2002).

[445] Colby M. Chester, 'Preliminary results of the United States Naval Observatory Eclipse Expedition in 1905', *Astrophysical Journal*, 23 (1906), 128–143.

[446] See for example: S. A. Mitchell, 'Preliminary account of flash spectra taken 1905 30 August', *Monthly Notices of the Royal Astronomical Society*, 66 (1906), 382–391.

[447] 'El eclipse de Sol', *El Imparcial*, 27 July 1905, 2. Bigelow arrived at Castellón one day later, having had his gold watch stolen in Valencia. See: 'El eclipse de Sol', *El Imparcial*, 28 July 1905, 1.

[448] Mitchell, *Eclipses of the Sun* (New York, 1923), pp. 189–190.

All hands, astronomers and sailors, worked each day building piers, erecting telescopes with houses to shelter them, mounting spectroscopes, and fixing up a meteorological observatory. In total, about six weeks were spent in Daroca in preparation. The mounting and adjusting of the instruments was almost complete by 25 August, five days before the eclipse, when the party was augmented by the arrival of 35 officers and men from the ship to assist with the observations. In general, officers were placed in charge of instruments with sailors to assist them, and frequent drills were held in order to familiarise each one with his role and ensure that everything would be right on the day.[449] The 40-foot focal length horizontal camera was under the direction of Llewelyn Griffith Hoxton, whilst the time-service and longitude determination was overseen by Everett Irvin Yowell (1870–1959) and various electrical and meteorological instruments under the charge of Mitchell.[450]

On the day of the eclipse, the first contact was observed under clear skies, but clouds obscured the Sun before totality. The citizens of Daroca gathered in great numbers at the eclipse camp chosen by the American astronomers, as closely as they could, and

> during the progress of the partial phase, each Spaniard present seemed bent on telling to his neighbor in a loud tone of voice the full details of what he knew about the subject. The noise of their conversation increased with the progress of the eclipse, and totality was greeted with such a shout that it was impossible to hear the counted seconds which should serve as guides to the exposures. The plan had been for the writer to observe the disappearing crescent of the Sun through a binocular, over one-half of which was arranged a plane mirror and plane grating combination, adjusted in such a way that with one eye through the binocular the crescent Sun could be seen, with the other, its spectrum in the green region … Two unlooked-for features were encountered: one being that the conversation of the Spaniards was rather loud, the second that the total phase occurred some ten seconds or more before it was expected. It was planned to catch the "flash spectrum" with both instruments at the beginning and end of totality, to give two short exposures just before the first flash and, just after

---

[449] S. A. Mitchell, 'Wave-lengths of the chromosphere from spectra obtained at the 1905 eclipse', *Astrophysical Journal*, 38 (1913), 407–495; and Mitchell, *Eclipses of the Sun* (New York, 1923); p. 189. See also: 'El Eclipse de Sol de 30 de Agosto', *El Mundo Científico*, 7 (1905), 544–550.

[450] Mitchell, 'Wave-lengths of the chromosphere … at the 1905 eclipse', *Astrophysical Journal*, 38 (1913), 407–495.

the second flash, one long exposure at mid-totality, and others of varying length between.[451]

Meteorological observations were made by Professor Bigelow. Several photographs were finally secured with different instruments, including some photographs of the corona obtained by Hoxton and Yowell. The spectroscopic work by Mitchell was very successful and his photographs of the spectra were developed in the dark room of the College of Daroca under the supervision of Father Felix Alvarez. They had good definition and normal dispersion, providing excellent images from both parabolic grating photographs from 3300 Å to D3 and plane grating photographs from 3300 Å to 6200 Å. The results of the observations at Daroca are summarised in the four tables in a paper by Mitchell published in 1913.[452] The first table listed each of the 2841 lines identified in the spectrum of the chromosphere, and gave the height to which the chromospheric vapours extend in kilometres, the chromospheric wavelengths compared with the Rowland values, Mitchell's opinion on the identification of the sources of lines in the chromosphere, and the intensities. Many other faint lines were measured, but no lines, not even faint ones, were included in the table unless they had been determined by two or more separate measurements. The second table gave the individual intensities of the different spectral lines. The third was a summary of all lines arranged according to the element producing them and identified where possible with lines in Rowland's tables, or labelled "Unidentified" otherwise. The fourth table was a periodic table of the elements to which had been added the total number of lines from each element identified in the chromospheric spectrum. Finally, the elements were divided into three groups: "Lines strong in the Sun, strong in the chromosphere" (Ca, Mg, Al); "Lines relatively stronger in the chromospheric than in the solar spectrum" (H, He, Ti, Cr, C, V, Zr, Sc, La, Y, Sr, Ba, Nd); and "Lines relatively stronger in the solar than in the chromospheric spectrum" (Fe, Ni, Co, Mn, Na, Nb, Mo, Pd).[453] Mitchell wrote:

> The final conclusion seems to be that the vapors forming the enhanced lines ascend to relatively high altitudes from which results a decrease in

---

[451] Mitchell, *Eclipses of the Sun* (New York, 1923), p. 191.

[452] Mitchell, 'Wave-lengths of the chromosphere … at the 1905 eclipse', *Astrophysical Journal*, 38 (1913), 407–495.

[453] These results were the same as those obtained for the eclipse of 1901. See: *Publications of the Naval Observatory*, Second Series, 4 [App.1], p. 290; *Astrophysical Journal*, 15 (1902), 97.

> pressure and a mixing with hydrogen, and that on account of height, reduced pressure, and the presence of hydrogen, the enhanced lines become relatively strong.[454]

The commissions from the U.S. Naval Observatory were not the only American ones that visited Spain in 1905 to observe the total solar eclipse. The University of California's Lick Observatory also organised three parties, to Labrador, Spain, and Egypt, sponsored by William Henry Crocker (1861–1937). The provisional programme of the expedition headed for Alhama de Aragón in Spain included a photographic intra-mercurial search, photography of the solar corona with a camera of 5 inches (127 mm) aperture and 40 ft (12.2 m) focal length, a polarigraphic study of the polarised light in the corona, recording the general spectrum of the corona with a spectrograph, making continuous records of changes in the spectrum of the Sun's limb at the times of second and third contacts, and the use of spectrographs for determining the wavelengths of the green coronal bright line and, if possible, the accurate wavelengths of the bright and dark lines in the isolated spectrum of the Sun's limb, as close as possible to the time when the dark lines give way to bright ones, and vice versa.[455] The expedition was formed by Campbell, Perrine, MacKiney, Mrs Campbell, and Mrs Perrine. They were later joined by Spanish scientists José Casares Gil, Antonio de Gregorio Rocasolano, Esteban Terradas, Hilarión Gimeno, Felipe Lavill, and Arturo Cuyás, who assisted the American astronomers.[456] Campbell was in charge of taking photographs of the ultraviolet region of the solar spectrum. Perrine was in charge of four long-focus cameras used to search for intra-mercurial planets. MacKiney, together with Mrs Campbell and Mrs Perrine, was in charge of a special device to photograph the polarised light from the corona to study its physical state. However, cloudy weather prevented successful results.

Other American commissions visited Almazán (Soria), including one from the Kirkwood Observatory (Indiana University). Originally formed by members of the Indiana University, John A. Miller (Professor of Mechanics and Astronomy at Kirkwood Observatory), W. A. Cogshall

---

[454] Mitchell, 'Wave-lengths of the chromosphere … at the 1905 eclipse', *Astrophysical Journal*, 38 (1913), 407–495.

[455] Campbell, 'The Crocker eclipse expeditions in 1905', *Astrophysical Journal*, 20 (1904), 235–236.

[456] López Arroyo, *El Real Observatorio Astronómico de Madrid (1785–1975)* (Madrid, 2004).

(Assistant Professor of Astronomy), A. F. Kuersteiner (Professor of Romance Languages), and the young Spaniard J. E. Valdez, the expedition finally included Professor Cogshall, Professor Kuersteiner, John A. Miller, and Mrs Miller, assisted by Mr Charles W. Thompson and Mrs Thompson of California, students E. C. Slipher, F. A. Crull and C. J. Bulleit, and the locals Francisco Jodra, Luis Nebot, Victor Jiménez, and Esteban Milla. The university assumed responsibility for the expedition, but it was aided by generous contributions from *The Indianapolis News* and *The Reader Magazine*, published in Indianapolis.[457] Their main objective was to obtain photographs of the corona, to pursue a photographic search for intra-mercurial planets, and to photograph the flash spectrum and the spectrum of the corona during totality. A photographic telescope of 8 inches (203 mm) aperture and about 70 ft (21 m) focal length was used, fed by a coelostat with a mirror 14 inches (356 mm) in diameter. The programme of observations included five exposures with this telescope to obtain information on the structure of the inner corona. Four other cameras of various focal lengths were planned to be mounted on a polar axis, the principal one of 9 inches (229 mm) diameter and 60 ft (18.3 m) focal length. The equipment for this expedition also included a battery of four cameras of 3½ inches (89 mm) aperture and 11 ft (3.4 m) focal length, with which photographs were to be taken covering the region in which there might be found an intra-mercurial planet. The bad weather conditions did not completely frustrate the observations, and some interesting results were obtained, such as satisfactory short exposures showing prominences. Slightly longer exposures showed considerable coronal detail, the corona extending to about three-quarters of the Sun's diameter.[458]

There were official commissions from other parts of the American continent to Spain, such as that from the Observatory of Tacubaya (Mexico), formed by the astronomers Valentín Gama, Joaquín Gallo, and Antonio Gómez, who also went to Almazán.[459] Finally, the Carnegie Institution of Washington sent Professors Julius Elster (1854–1920) and Hans Geitel (1855–1923), together with F. Harms, to Palma de Majorca to make observations of the electrical conditions of the atmosphere during

---

[457] 'Eclipse expeditions planned for August', *New York Times*, 15 January 1905, SM3.

[458] John A. Miller, 'Solar eclipse of August 30, 1905', *Astrophysical Journal*, 23 (1906), 93–95.

[459] *Archives of the Observatory of Madrid*, 'Eclipse total de sol del 30 de agosto de 1905', Folder 16.

the eclipse.[460] The variation of atmospheric electricity was photographically recorded with a self-registering electrometer and complemented with simultaneous visual observations. The ionisation of the air was studied with a *Zerstreuungsapparat* and an *Ebert's Föncounter*. Besides these observations, exact measurements of the intensity of the solar radiation at short wavelengths were carried out, a special kind of photometer having been prepared for this purpose. Unfortunately these observations, like all others in Majorca, suffered from the bad weather conditions. Continuous measurements of the solar radiation were possible only from first contact to the end of totality. The decrease of illumination, therefore, was determined in a satisfactory manner and without any gaps.

9.4. British eclipse expeditions to Spain

The study of the British eclipse expeditions sent to Spain in 1905 may also highlight the differences and similarities between Spanish and foreign astronomical research for the eclipse. The organisation of the British expeditions was mainly coordinated by the Joint Permanent Eclipse Committee of the Royal Society and the Royal Astronomical Society, a body which arranged to send in 1905 at least five parties to occupy stations in Spain, Algeria, and Egypt.[461] One of these expeditions, led by Alfred Fowler and Hugh Longbourne Callendar (1863–1930), the expenses of which were defrayed from a Government Grant Fund, arrived at Castellón on 3 August 1905. The camp was established in the grounds of the new Provincial Hospital, then under construction, under the supervision of a local architect.[462] The expedition was originally formed by Fowler, the technical assistant T. Banfield, and the volunteers I. Molloy and E. Cahen. Their number rose between 14 and 28 August with the arrival of Callendar, Rayner (National Physical Laboratory[463]) and the volunteer J. J. Steward. The general programme of the expedition,

[460] 'Notes', *Nature*, 72 (1905), 490–494.

[461] 'Report of the Council to the Eighty-fifth Annual General Meeting …', *Monthly Notices of the Royal Astronomical Society*, 65 (1905), 382–391, here p. 391. Most of the observers' reports were presented at a meeting held on 19 October 1905. See for instance: '[Notes:] Reports of the solar eclipse', *The Observatory*, 28 (1905), 430–431.

[462] 'El eclipse de sol', *El Imparcial*, 6 August 1905, 2.

[463] Rayner offered himself as a voluntary assistant to operate the large-scale coronograph and prismatic camera designed by W. Shackleton, whose illness prevented him from going to Spain.

somewhat extended in consequence of the voluntary assistance of local gentlemen, is presented in Table 9-1.

Table 9-1: British eclipse expedition to Spain in 1905

Instrument	Design and purpose	Observers
20-inch reflector	Fitted with equipment to measure solar and coronal radiation. The mirror had an aperture of 20 inches, giving an image of the Sun about 0.4 inches in diameter. A coronal thermopile designed for this instrument, had a receiving surface consisting of ten small blackened rectangles of very thin copper arranged on the circumference of a circle nearly fitting the image of the Sun to receive the greater part of the radiation of the inner corona.	H. L. Callendar
Absolute recording bolometer	Equatorially mounted and designed to obtain by the "electric compensation method" absolute values for the solar radiation when reading the coronal thermopile.	
Horizontal bolometer	For recording vertical component of solar and sky radiation. It consisted of a pair of platinum thermometers wound on a horizontal mica plate fixed in a sealed glass bulb.	
Recording electrical thermometers	For measuring air temperature, etc.	
Slit spectrograph	For photographing the spectra of the corona and chromosphere in the less refracted part of the visible spectrum (adjusted to the region from lines B to F).	A. Fowler Assisted by T. Banfield and Joaquín Garcés (with plate-holders, etc.)
4½-inch equatorial with an Evershed solar spectroscope	For visual observation of the spectrum of the corona during the long exposure with the spectrograph.	A. Fowler

4-inch coronograph	Dallmeyer lens of 38½ ft focal length (4-inch aperture) with a direct-vision prism for coronal and chromospheric spectra (3890 Å), and a 12-inch coelostat and camera.	E. H. Rayner I. Molloy (prism) Assisted by Venancio Soto and José Babiloni (with plate-holders, etc.)
3-inch coronograph	Instrument of 57 inches focal length with green screen for studying the inner corona, as suggested by Shackleton.	T. Banfield
3-inch coronograph	Instrument of 20 inches focal length with a small polar heliostat, to study coronal extensions.	E. Cahen
Spectrograph of small dispersion	Fitted with a small coelostat and image lens for studying the spectrum of the outer corona.	Francisco Betoret
Slitless spectroscopes	Used for drawings of the green ring.	A. Fowler T. Banfield Telmo Vega Francisco Betoret
2-inch telescope	For direct observation of the corona in the region of a previously selected prominence.	J. J. Steward
Thermograph and barograph	—	
Observations of stars during totality	—	I. Molloy
Observations of shadow bands	—	Juan Vilo
Time signals	—	Chronometer: J. J. Steward. Signals during totality: José Badia and Luis Giraudier

In order to prevent any possible interference with the work of the observers, the Military Governor provided a guard of soldiers, stationed on the day of the eclipse in and around the camp. On the day of the eclipse, the owners of the numerous factories in the district closed them so that there was no smoke to ruin the observations. However, though the first and fourth contacts were observed in a perfectly clear sky, not a glimpse of the eclipsed Sun was obtained during totality. As a result, no records were obtained during totality, except those from automatic recordings of solar radiation and temperature.[464]

Special provision was also made for the Government Grant Fund to subsidise the expedition to Pineda de la Sierra led by John Evershed, which also included H. Evershed, Rev. C. D. P. Davies, and R. C. Slater as assistant. The principal objects of the expedition were to obtain large-scale images of the flash spectrum with a prismatic camera of great focal length, for the purpose of studying the actual forms assumed by the different radiating gases giving the enhanced lines; to obtain ultraviolet spectra of the corona with prismatic cameras of glass and of quartz, using the flash-spectrum arcs as reference lines for determining wavelengths; and to study the coronal spectrum in the D to F region at high dispersion, visually and photographically, to ascertain the character of the continuous spectrum and of the radiation at 5303 Å.[465] The members of the expedition were assisted by J. H. White of Burgos, resident manager of the Sierra railway company, and C. Ellis Bevan, an engineer who supervised the setting up of the camp. The budget from the Government Grant Fund was mainly invested in the construction of the essential parts of the large "reflecting prismatic camera" and in a concave mirror of 15 inches (381 mm) aperture and 29 ft (8.8 m) focal length, but also in a dense flint prism, a 16-inch (406 mm) coelostat, and a large revolving plate carrier. The reflecting prismatic camera was designed to make fifteen exposures in rapid succession: seven at the second contact, one at about mid-eclipse, and seven at the third contact. Other instruments included a glass prismatic camera of 2 inches (51 mm) aperture and 46 inches (1.2 m) focal length; an ultraviolet prismatic camera to get good images of the cusp and the flash spectra, which would provide an accurate scale for estimating the wavelengths of the ultraviolet coronal rings; a quartz prismatic camera mounted on a 3-inch (76 mm) equatorial telescope to make one exposure during totality; a direct-vision slit spectrograph where the combined action of three compound prisms caused a deviation of a few degrees of the green part of the spectrum which it was desired to photograph; and a 3¼-inch (83 mm) equatorial telescope together with a high-dispersion three-prism spectroscope for prominence observations. There were also a reflecting

---

[464] H. L. Callendar and A. Fowler, 'Report of the expedition to Castellón de la Plana, Spain', *Proceedings of the Royal Society of London. Series A*, 77 (1906), 1–28. See also: '[Notes:] The following is a summary of work done by various parties during the solar eclipse', *The Observatory*, 28 (1905), 360.

[465] J. Evershed, 'Report of the expedition to Pineda de la Sierra, Spain', *Proceedings of the Royal Society of London. Series A,* 77 (1906), 42–55.

coronograph of focal length 74 inches (1.88 m) and a reflecting prismatic camera of about 60 inches (1.5 m) focal length fed by a coelostat contrived out of an old equatorial stand. Time determination and the correct position of the camp were obtained with the use of a sextant and an artificial horizon. Clouds prevented photography before totality, but some interesting results were obtained during totality:

> The first four exposures with the large prismatic camera yielded well-defined images of the cusp spectrum. Although much obscured by cloud, the Fraunhofer lines are clearly shown, and the hydrogen lines β, γ, δ, and the lines H and K are just visible as bright lines along the edges of the spectrum … The ultra-violet prismatic camera gave two images of the cusp spectrum, at 34 and 30 seconds before the second contact. The spectra are beautifully defined and in perfect focus over the whole length of spectrum photographed, from H_β to λ 3400 … The general falling off in the intensity of the lines close to the limb seems to be simply explained on the supposition that the entire photosphere consists of innumerable vertical columns of incandescent gases partly penetrating the absorbing layer, the lower position of which would be entirely hidden at the limb owing to foreshortening.[466]

Another interesting British eclipse expedition to Spain in 1905 was sent by the Solar Physics Observatory. Located near Palma on Majorca, the party was led by Norman Lockyer and included the astronomers William Lockyer and C. P. Butler, as well as volunteers Howard Payn, Frank McClean, and Lady Lockyer, and a large number of officers and men from H.M.S. *Venus*.[467] The party had to abandon the original plan of observing the eclipse from Algeria because the French authorities objected to the presence of the warship in Philippeville harbour. On 11 August 1905 the erection of the tents loaned by the War Office began in Palma. In the days that followed the various instruments were set up, including a prismatic reflector of 76 ft (23.2 m) focal length, a prismatic camera with three 6-inch (152 mm) 60° prisms, two coronagraphs and an objective grating camera fed by one coelostat, a triple camera of 3 inches (76 mm) aperture and 12 ft (3.7 m) focal length for photographing the eclipsed Sun in colour, and a 3½-inch (89 mm) coronagraph.[468] Groups of observers

[466] *Ibid.* See also: '[…] The following is a summary …', *The Observatory*, 28 (1905), 360.

[467] 'The total eclipse. Accounts from all sources', *Knowledge & Scientific News*, 2 (1905), 246–247.

[468] Lockyer, 'The Forthcoming Total Solar Eclipse', *Nature*, 72 (1905), 399. A slightly different record is found in the 'Report of the Council to the Eighty-sixth

were formed for sketching the corona, observing stars, recording the colours of the corona and the landscape, observing shadow bands and the sweep of the shadow, making meteorological observations, and so on. William Lockyer, in charge of the prismatic camera, secured fourteen photographs. The programme of work was to make four exposures at about the commencement of totality and five near the end. William Lockyer wrote:

> The remaining five plates will be exposed for intervals varying from 5 to 90 seconds, and it is hoped that the two long exposures on each side of mid-totality will add to our knowledge of the wave-lengths of the coronal rings. This prismatic camera is designed to give results suitable for determining accurate wave-lengths of the chromospheric and corona arcs.[469]

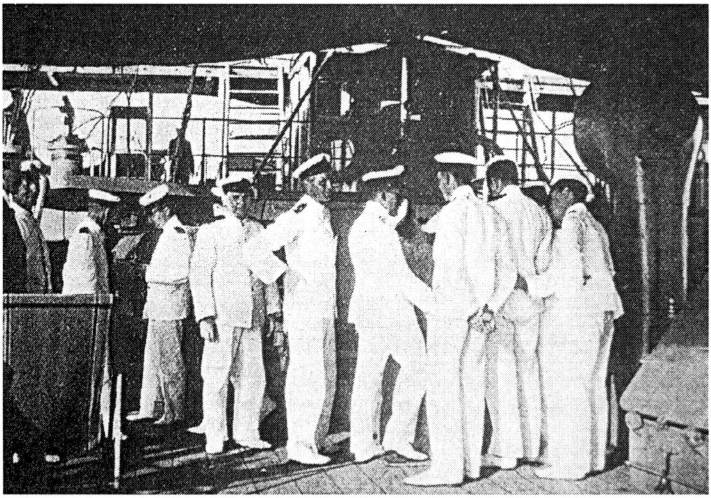

Fig. 9-2: Officers from H.M.S. *Venus* volunteering for eclipse work. Published in *Nature*, 72 (1905).

Annual General Meeting of the Society. Proceedings of Observatories', *Monthly Notices of the Royal Astronomical Society*, 66 (1906), 182–215. There is said here that the two chief instruments to secure large-scale spectrograms of the solar chromosphere and corona were the prismatic camera of six-inch aperture and a prismatic reflector of twelve-inch aperture concave mirror and seventy two feet focus with an objective prism of 10-inch aperture. This source also explains that three coronographs of varying aperture and large-scale *three-colour* camera were employed.

[469] William J. S. Lockyer, 'Total solar eclipse, August 30. The Solar Physics Observatory Expedition', *Nature*, 72 (1905), 457–458.

The 76 ft focal length prismatic reflector was in the charge of Butler. It provided a solar image about 250 mm in diameter and, since the light was made to pass through one prism twice, the dispersion was not excessive. The end of this camera formed part of the darkroom of the camp, and just outside was the siderostat which directed the solar rays onto the long-focus mirror located about 70 ft away from the darkroom.[470] Unfortunately the weather conditions were very unfavourable, and only a small number of photographs were obtained (even these being through thin cloud), including spectra of the corona and prominences, and several fairly good pictures of the corona.[471] Indeed, the prismatic reflector operated by Butler succeeded in securing a picture of the lower corona, the solar diameter being about 200 mm.[472] The 16 ft (4.9 m) coronograph operated by McClean obtained a fine photograph of the corona "with excessively sharp detail and good extension".[473] Lieutenant Trench, in charge of the De la Rue coronograph, secured three negatives. The 3½-inch Newton coronagraph, mounted equatorially and operated by Staff-Surgeon Clift, also obtained two successful exposures. The prismatic camera operated by William Lockyer secured four useful negatives that showed several distinct coronal rings. The objective grating spectroscope worked by Payn produced one successful exposure (of two attempted) showing the spectrum of the larger prominences and the green coronal ring. Finally, the observers of the shadow bands gained a great amount of information on their size, rate of motion, and direction, and the coronal sketchers obtained very concordant results.

In addition to the official parties, a large number of observers sent out under the auspices of the British Astronomical Association and other bodies also observed the eclipse of 1905 from countries such as Algeria, Egypt, and Spain. The ease of the journey to Spain, together with "the fact that the eclipse falls in the summer vacation", tempted astronomers to take

---

470 *Ibid.*

471 'Report of the Council to the Eighty-sixth Annual General Meeting ...', *Monthly Notices of the Royal Astronomical Society*, 66 (1906), 182–215. The paragraph dealing with the Solar Physics Observatory, South Kensington, is: 'Total solar eclipse, 1905 August 30', p. 203. See also: '[...] The following ...', *The Observatory*, 28 (1905), 360.

472 William J. S. Lockyer, 'The Solar Physics Observatory eclipse expedition', *Nature*, 72 (1905), 508–510, here p. 509.

473 *Ibid.*, p. 510.

the opportunity of viewing such an interesting solar phenomenon.[474] For example, four members of the Liverpool Astronomical Society, under the Rev. R. Killip, observed the eclipse from Burgos, as did members of the British Astronomical Association directed by C. Thwaites.[475] Two other members of the British Astronomical Association, Irwin Sharp and Thomas William Backhouse (1842–1920), also observed the eclipse from Peña Corada, near Cistierna (León). Similarly, in the centre of the town of Palma on Majorca, several British astronomers, including Andrew Claude de la Cherois Crommelin (1865–1939), established themselves on the roof of the Grand Hotel. From there, clouds obstructed the view to some extent. Several British observers were aboard large steamers which sailed to arrive at positions well within the limits of totality at the moment of the eclipse. The *Ortona* and the *Arcadia* arranged to anchor and provide passengers with a good view of the phenomenon, and from both ships the atmospheric conditions were fairly good, although no valuable observations, from the purely scientific point of view, were made.[476]

9.5. French eclipse expeditions to Spain

On 14 April 1904 the Bureau des Longitudes informed all the French astronomical observatories of its intention to organise and coordinate several official commissions.[477] It was with this support that the astrophysical Observatory of Meudon sent a party to Spain to observe the

---

[474] 'Report of the Council to the Eighty-fifth Annual General Meeting …', *Monthly Notices of the Royal Astronomical Society*, 65 (1905), 382–391; p. 391

[475] The party included the Rev. T. E. R. Phillips, Mr Raisin, Miss Berry, Mr E. Howarth, Mr H. Kraus Nield, Mr A. Webb, Mr C. Arthur Pearson, Mr W. B. Gibbs, Mr D. E. Benson, Mr T. Thorp, Mr E. Dickson, Mrs H. K. Moore, Miss Woolston, Miss Mary Proctor, Capt. Carpenter, Mr H. Keatley Moore, Mr Skinner, Mr Leetham Jones, Mr Bruford, Mr McClean, Mr John H. Gear, Rev. J. T. W. Claridge, Mr C. L. Brook, Mr and Mrs Johnson, Mr A. E. Eaton, and Mr C. G. J. Dolmage.

[476] 'The total eclipse. Accounts from all sources', *Knowledge & Scientific News*, 2 (1905), 246–247. The following members of the British Astronomical Association observed the eclipse from the steamer *Arcadia*: Mr G. F. Chambers, Mr S. A. Saunder, Mr F. W. Longbottom, Mr J. H. White, Mrs Arthur Brook, and Mr John H. Willis. On the steamer *Ortona*, Professor Joseph Larmor carried out some naked-eye observations of streamers, Venus and Regulus, and shadow bands.

[477] A. Lebeuf and P. Chofardet, 'Observation de l'éclipse totale de Soleil du 29–30 août 1905 à Cistierna, province de Léon (Espagne). Rapport des membres de la mission de l'Observatoire de Besançon', *Annales du Bureau des Longitudes*, 8 (1905), 1–47.

1905 eclipse. Deslandres directed this Bureau des Longitudes mission to Villargamar, a village three kilometres from Burgos, which included the assistant astronomers Kannapell and d'Azambuja. Notable amongst the other members of the expedition were the physicists Charles Fabry (1861–1945) and Jean Becquerel (1878–1953), and the party also included the engineer and volunteer at the Observatory of Meudon, Monsieur Bernard and his wife, as well as Monsieur Blum, member of the Council of the Société Astronomique, Monsieur Fouché, Agregé de l'Université and Vice-President of the Société Astronomique, and Monsieur Sausot, a student.[478] The main instruments carried by this party were a three-mirror coelostat, a large equatorial table, and an 8-inch (203 mm) equatorial. Poor weather conditions, however, prevented any useful spectroscopic results from being obtained. The duration of visible totality was curtailed by the clouds to one minute, which did not include either the second or the third contact. The proposed photography of the spectrum of the chromosphere was therefore impossible.

Nevertheless, photometric observations of the corona were made, and d'Azambuja was able to measure the coronal calorific radiation. Kannapell obtained four photographs of the corona polarised by reflection, including one with a spectrograph which showed the green line at 530 nm.[479] Blum obtained two photographs of the inner corona though coloured screens arranged so as to transmit only the gaseous radiation of the prominences. By comparing these with the ordinary photographs it was hoped to determine whether or not the prominences emit a more intense continuous spectrum than that emitted by the surrounding regions.[480] Finally, Fabry and Bernard were successful in their photometric observations, during the different phases of the eclipse, of the total light of the corona and circumsolar light, respectively. In the case of Fabry's observations, the light intensity measured with a Mascart photometer, modified for

[478] H. Deslandres, 'Note prélimiaire sur l'observation de l'éclipse totale du Soleil du 30 août 1905, à Burgos', *Comptes Rendus*, 141 (1905), 517–519.
[479] Wavelengths were generally given in nanometers rather than angstroms in the French literature.
[480] Deslandres, 'Note prélimiaire sur l'observation de l'éclipse totale du Soleil du 30 août 1905, à Burgos', *Comptes Rendus*, 141 (1905), 517–519. See also: 'Our astronomical column: Further results obtained by the French eclipse expeditions', *Nature*, 72 (1905), 567.

observing the solar corona, five arc minutes from the edge of Sun,[481] fitted perfectly with that obtained empirically by Turner after photographing several regions of the corona during the eclipse of 1893.[482]

Another party from the Observatory of Meudon went to Alcocebre (Castellón).[483] Led by the eminent astronomer Janssen, this expedition included Millochau, Pasteur, and Corroyer, all of them members of the Observatory of Meudon, and Milan Stéfanik (1880–1919) from the University of Prague. As has already been pointed out, this expedition had in Lànderer an enthusiastic collaborator.[484] Amongst the instruments employed were a photographical telescope, two spectroscopes, one theodolite, a meridian circle, two chronometers, and several pieces of meteorological equipment such as thermometers and actinometers. The expedition had several objectives: study of the whole eclipse, making use of photographic cameras; the form and details of the corona and prominences; the spectra of the different layers of the solar atmosphere;

---

481 Charles Fabry, 'Sur l'intensité lumineuse de la couronne solaire pendant l'éclipse totale du 30 août 1905 [présentée par Deslandres]', *Comptes Rendus*, 141 (1905), 870–873.

482 As regards the varying brightness of the corona at different distances from the Sun's centre, Professor Turner had shown that for the corona in 1893 (at solar maximum) and in 1898, the law of luminosity was brightness ~ (distance from the Sun's *centre*)$^{-6}$. This law was thought to apply quite well to the various differents appearances of the corona at different times of the solar cycle, being more satisfactory than that formulated by Professor Harkness in 1878, 'who gave the brightness varying simply as the inverse square from the *limb*, for if the Corona be largely made up of minute particles, other inverse powers of the distance in addition to the ordinary inverse square law of luminosity will enter into the equation'. See: W. Shackleton, 'The Coming Total Eclipse', *Knowledge & Scientific News*, 2 (1905), 45–47; p. 47.

483 See the following works published by Janssen: 'Sur l'éclipse solaire totale du 30 août 1905', *Comptes Rendus*, 141 (1905), 233–234; 'Observation de l'éclipse totale du 30 août 1905 à Alcosèbre (Espagne)', *Comptes Rendus*, 141 (1905), 569–571; and 'Observations de M. J. Janssen, directeur de l'Observatoire de Meudon, à Alcosèbre (Espagne)', *Bulletin de la Société Astronomique de France*, 19 (1905), 491.

484 It was Lànderer who chose the final location for this expedition. See the letter from the President of the Bureau des Longitudes to the Ministre de l'Instruction publique, des Beaux-Arts et des Cultes, signed in Paris on 20 September 1905. Archives Nationales de France—F[17] I [F17/01(B)]), F[17] 13571: Bureau des longitudes. Affaires scientifiques diverses 1881–1922. Folder: *Bureau des Longitudes. Affaires diverses (1901–1914)*. See also: 'Our astronomical column: Further results of the French eclipse expeditions', *Nature*, 72 (1905), 639–640.

and the meteorological changes produced by the eclipse. Janssen was in charge of an equatorial telescope of 6 inches (152 mm) aperture to which spectroscopes were attached. Pasteur had an equatorial telescope of 12 inches (305 mm) aperture to which three cameras were attached, and obtained three photographs of the corona. Millochau was in charge of an equatorial telescope 8 inches (203 mm) in aperture, to which several spectrographs were attached for spectroscopic investigation of the various layers of the solar atmosphere. He obtained several photographs of the flash spectrum and of the corona. For his research, he employed a spectrograph attached to a telescope of 330 mm focal length; a two-prism spectrograph; a camera with a Suter objective of 700 mm focal length; and a direct vision spectroscope by Browning attached to a telescope made by Vion of 1 m focal length.[485]

Millochau obtained several photographs of the lines C to K from the gaseous layers of the corona and of the green coronal line. A second photograph obtained during totality showed that the spectrum of the prominences consisted of lines of hydrogen and helium. The spectrum of the corona included the green line at 530 nm, which was in fact the only line observed in the coronal spectrum between lines C and H. Finally, the continuum spectrum showed an intense region from 470 nm to 600 nm, with the maximum around 550 nm.[486] Finally, Corroyer employed a photographic telescope from Meudon to secure photographs of the solar surface, and Stéfanik undertook the visual study of the spectrum during the eclipse through a fixed horizontal telescope, which received light from a Silbermann heliostat. To record the coronal spectrum he employed a Duboscq spectroscope attached to a telescope made by Prazmowsky of 0.8 m focal length. Stéfanik later reported that he observed the line D3 and the green line at 530 nm. No lines, however, were observed at the extreme red end of the spectrum.[487]

The Bureau des Longitudes also subsidised an expedition mounted by members of the Observatory of Besançon, together with some members of the Observatory of Paris, to Cistierna. The expedition was formed by

---

[485] G. Millochau, 'Sur l'observation de l'éclipse totale du 30 août 1905 à Alcosebre (Espagne)' [presentée par J. Janssen], *Comptes Rendus*, 141 (1905), 586–588.

[486] Millochau, 'Sur l'observation de l'éclipse totale du 30 août 1905 à Alcosebre (Espagne) [présentée par Janssen]', *Comptes Rendus*, 141 (1905), 815–818.

[487] See: Milan Stefanik, 'Recherches spectroscopiques pendant l'éclipse du 30 août 1905 à Alcosebre (Espagne)' [présentée par J. Janssen], *Comptes Rendus*, 141 (1905), 585–586.

Pierre Henri Puiseux (1855–1928), Maurice Hamy, Jules Baillaud, and Le Morvan, from the Observatory of Paris; Benjamin Baillaud (1848–1934), Director of the Observatory of Toulouse; the astronomers P. Chofardet and A. Lebeuf (who travelled with his wife), both from the Observatory of Besançon; as well as G. Prin, de Sailly, Mme Voisin, P. Gautier, Bouty, Vion, and the Spaniard Professeur M. D. Berrueta, from León.[488] As Lebeuf and Chofardet reported,

> Le voyage … se fit sans incident, suivant le trajet Besançon, Paris, Bordeaux, Irun, Burgos, Venta de Baños et Léon [sic] … l'arrivée eut lieu à Cistierna le 5 août … au bruit des acclamations enthousiastes des habitants de la petite cité.[489]

Puiseux and Hamy were in charge of the astrophysical work, while Lebeuf and Chofardet focused their efforts on positional astronomy, visual and photographic study of the corona, and observation of shadow bands, as well as some meteorological and magnetic observations.[490] Clouds prevented successful results during totality, although Lebeuf obtained several photographs of the partial phases of the eclipse.[491]

The Observatory of Bordeaux also sent an expedition to Burgos under the auspices of the Bureau des Longitudes to photograph the phases of the eclipse. The instruments brought by the Director of the Observatory, Georges Antoine Pons Rayet (1839–1906), and the astronomers Courty and Jimeno, included a photographical Grubb equatorial of 120 mm diameter and 1.1 m focal length, and a visual equatorial with an objective of 8 inches (203 mm) and about 3 m focal length with a spectroscope with three 60° prisms. Although clouds prevented some of the planned images from being obtained, Courty did get two photographs of the corona which showed the structure at a sunspot maximum.[492] Several measurements of the polarised light of the corona were attempted by Georges Meslin

---

488 Lebeuf and Chofardet, 'Observation de l'éclipse totale de Soleil du 29–30 août 1905 …', *Annales du Bureau des Longitude*s, 8 (1905), 1–47.
489 *Ibid.*
490 *Ibid.*
491 See: Lebeuf and Chofardet, 'Résultats des observations faites, pendant l'éclipse totale de soleil du 30 août 1905, à Cistierna, Espagne', *Astronomische Nachrichten*, 176 (1907), 215–218; and Observatoire de Paris. Ms 1060 II-A-3: *Les Activites scientifiques de l'observatoire—éclipses 1850–1908 (1858, 1860, 1864, 1867, 1868, 1875, 1882, 1905, 1908).*
492 G. Rayet, 'L'Éclipse totale de Soleil du 30 août 1905', *Comptes Rendus*, 141 (1905), 490–491.

(1862–1918), from Montpellier, who also joined the party.[493] Similarly, the Observatory of Lyon sent to Tortosa an expedition led by Charles Louis François André (1842–1912).[494] The members of this expedition included Guillaume, who was in charge of the astronomical studies to solve the problems of the lunar theory, and Le Cadet, in charge of investigating whether the eclipse had any influence on the electrical behaviour of the atmosphere. The main instrument for the former was a 6-inch (152 mm) equatorial telescope through which the Sun's image was projected onto a screen.[495] The latter investigation was performed with an Exner transportable apparatus and a Gerdien *aspirateur*. Although clouds interfered with observations, the French expedition to Tortosa managed to measure the intensity of the Earth's electric field and the atmospheric ionisation throughout the eclipse.[496] The intensity of the electric field was measured in volts per metre with a device consisting of a small acetylene lamp about 2.50 m above the ground, linked to the electroscope through a fine wire 5 m in length; it was measured 98 times during five hours, but uncertain conclusions were obtained.[497]

Another French institution to send an expedition to Spain in 1905 was the University of Montpellier. This expedition, led by Marcel Moye

---

[493] See: Observatoire de Paris. Ms 1060 II-A-3: *Les Activités scientifiques de l'observatoire—éclipses 1850–1908 (1858, 1860, 1864, 1867, 1868, 1875, 1882, 1905, 1908).*

[494] C. André, 'Sur l'éclipse totale du Soleil du 30 août 1905 à Tortosa', *Comptes Rendus*, 141 (1905), 867–869. See also: Misión del Observatorio de Lyon, comandada por su director André [Letter from the President of the Bureau des Longitudes to the Ministre de l'Instruction publique, des Beaux-Arts et des Cultes, signed in Paris on 20 September 1905. Archives Nationales de France—F[17] I [F17/01(B)]), F[17] 13571: Bureau des longitudes. Affaires scientifiques diverses 1881–1922. Folder: *Bureau des Longitudes. Affaires diverses (1901–1914)*. André's interest in total solar eclipses led him to design apparatus with which to practise eclipse observations; see: André, 'Appareil à éclipses artificielles de Soleil', *Comptes Rendus*, 141 (1905), 168–169.

[495] These results were reported together with those obtained by Stéphan and Borelly in Guelma, Algeria, in: André, 'Quelques remarques sur les observations des contact dans les éclipses totales de Soleil', *Comptes Rendus*, 143 (1906), 806–809.

[496] 'Our astronomical column: French observations of the total solar eclipse', *Nature*, 72 (1905), 518.

[497] G. Le Cadet, 'Mesures de l'intensité du champ électrique terrestre et de l'ionisation de l'atmosphère pendant l'éclipse totale de Soleil du 30 août 1905 [présentée par Loewy]', *Comptes Rendus*, 141 (1905), 925–928.

(1873–1939), was located at the village of Alcalá de Chisvert. Its programme was to observe shadow bands and the corona (with an opera glass), and to provide a general description of the main aspects of the eclipse. The good weather conditions allowed coronal streamers and prominences to be sketched, and the green coronal line to be observed.[498] Also at this village was a party of amateur astronomers from the Société Astronomique Flammarion de Montpellier, who observed the corona and the chromosphere with portable telescopes, spectroscopes, and cameras, and also with the naked eye.[499] The eclipse was observed under good conditions; the corona was very brilliant, the longest streamers being observed to the south and the north-east, and the green line was very apparent. Shadow bands were well seen both before and after totality, but were invisible during totality.[500]

Fig. 9-3: French expedition set up at Alcalá de Chisvert. From *El Mundo Científico*, 7 (1905).

---

[498] See: Marcel Moye, 'The total eclipse of 1905. At Alcala de Chisvert', *Knowledge & Scientific News*, 2 (1905), 271–272; Moye, 'Observation de l'éclipse du 30 août à Alcalá de Chisbert (Espagne)', *Comptes Rendus*, 141 (1905), 458.

[499] Moye, *A la poursuite d'une ombre: l'éclipse totale de soleil du 30 août 1905* (Montpellier: G. Firmin, Montane et Sicardi, 1905). See also: 'Our book shelf: *A la Poursuite d'une Ombre* by Prof. Moye', *Nature*, 73 (1905–1906), 606.

[500] 'Our astronomical column: French observations of the total solar eclipse', *Nature*, 72 (1905), 518.

9.6. Other foreign expeditions

Expeditions from all over the world came to Spain to study the total solar eclipse of 1905. As well as the Americans, British, and French, there was, for example, the expedition organised by the Academy of Science of Amsterdam and led by Hendricus Gerardus Van de Sande Bakhuyzen (1838–1923), who observed the eclipse from Burgos together with Willem Henri Julius (1860–1925), A. A. Nyland, and J. Wilterdink. They were assisted by W. J. J. Moll; by the Spanish Fathers Eleuterio Martínez and Antonio de Madariaga, lecturers in physics and theology, respectively, in Burgos; and by J. Kaplan, from St Petersburg. Amongst the scientific instruments they employed were a coelostat, two spectrographs, a prismatic camera, a radiometer, a pyrheliometer, a sextant, and three chronometers. Their main objectives were the study of the corona and limb phenomena of the solar surface, and of the radiation of the corona. First contact was not observed, but the sky had cleared by second contact. Julius, who was in charge of the study of the radiation from the corona and from the solar surface, concluded that the minimal radiation was 1/200,000 of the regular solar radiation. On the other hand, Nyland, who was in charge of the prismatic camera, and Wilterdink, in charge of the spectrographs, were unsuccessful in their measurements at the time of the eclipse.[501]

The Imperial Academy of Science of St Petersburg organised an eclipse expedition to Spain led by Nicolas Donitch to study the chromosphere and the corona. The programme of observations included the study of the spectrum of the solar atmosphere with a direct vision spectroscope, and the observation of dark lines in the coronal spectrum with a high-dispersion spectrograph. It also incorporated the study of the rotation of the corona through the displacement detected in its monochromatic (blue) radiation, and obtaining two spectrograms of the chromosphere and the corona during the second and third contacts. The expedition was divided into two: one party went to Assouan (Egypt) and the other to Alcalà de Chisvert (Castellón, Spain). The latter consisted of Donitch, who made spectral observations, and his assistant M. von der

[501] 'Académie des Sciences d'Amsterdam', *Revue générale des Sciences pures et appliquées*, 17 (1906), 55–56.

Pahlen, who took photographs of the corona and was in charge of the meteorological observations.[502]

Unfortunately, only the first 40 seconds of totality were observed from Alcalà without any clouds. Nonetheless, wavelengths and origins of 110 chromospheric lines were determined, and the altitudes of the various elements in the chromosphere were calculated.[503] As has been pointed out, even worse was the luck of expeditions sent to Majorca, for example the Swiss party at Santa Ponza (Majorca), formed by the Director of the Observatory of Geneva, Adolph Raoul Gautier (1854–1931), together with J. Pidoux, astronomer from the observatory, and François Alphonse Forel (1841–1912), an honorary professor at Lausanne University.[504] As Gautier wrote:

> [At Palma de Majorca] nous y avons joui de la société à la fois aimable et intéressante de Sir Norman Lockyer, de sa famille et des membres de l'expédition qu'il dirigeait, puis de la mission écossaise dirigée par M. le Dr Sutherland Black, de celle dirigeait par M. H.P. Hollis et de la mission allemande de MM. Elster et Geitel pour l'étude des phénomènes d'électricité atmosphérique accompagnant l'éclipse.[505]

The main objective of the Swiss expedition was to photograph the corona and the prominences, and to search for comets. Gautier and Pidoux were both in charge of taking the photographs, whilst the Spaniard Anckermann, who assisted the party, made the spectroscopic observations. Forel, together with the Spaniard Fontserè, dealt with the observations

---

502 Nicolas Donitch, 'Observations de l'éclipse totale du Soleil du 29–30 août 1905', *Bulletin de l'Académie Impériale des Sciences de St Pétersbourg*, 23 (1905), 283–288.

503 The highest altitudes were given by Hydrogen (49,300 km) and Calcium (44,000 km). See: 'Astronomical Publications: Eclipse Observations, August 1905', *Journal of the British Astronomical Association*, 18 (1907–1908), 185.

504 Also in Palma de Majorca were several relevant observers such as Dr. Black and Dr. Hunter, from Edinburgh, some members of the British Astronomical Association (A. C. D. Crommelin and his wife, Major Baden-Powell, Prof. Leahy, Mr. Gavin Burns, Mr. Allen, Miss Alice Everett, Mr. R. Hodge), Mr. and Mrs. Hollis as representing the journal The Observatory, and a party of Jesuit Fathers led by the Director of the Manila Observatory, Father Algué. See: 'The total solar eclipse of 1905, August 30', *The Observatory*, 28 (1905), 355–356.

505 Raoul Gautier, 'L'Éclipse totale de soleil du 30 aout 1905: observations de la Mission Astronomique Suisse a Santa Ponza (Ile de Majorque)', *Archives des Sciences Physiques et Naturelles*, 20 (1905), 650–674.

related to terrestrial physics, temperature, shadow bands, and so on. Table 9-2 lists some other foreign and local astronomers who also observed the total solar eclipse of 1905 from Spain.[506]

Table 9-2: Some foreign and local scholars who observed the total solar eclipse of 1905 from Spain.

<table>
<tr><th>Observers</th><th>Location</th><th>Observations and results</th></tr>
<tr><td>Observatory of Nice: Martial Simonin, Henri Chretien, and S. Javelle[507]</td><td rowspan="3">Alcalá de Chisvert</td><td>Cloudy weather. Several photographs were obtained.</td></tr>
<tr><td>J. Esquirol</td><td>Reported the existence of five protuberances with double coloration (white and pinkish) at the eastern limb of the Sun.</td></tr>
<tr><td>G. Tramblay</td><td>Shadow bands, prominences, corona, chromosphere</td></tr>
<tr><td>Swedish commission led by Svante August Arrhenius, assisted by Kobb</td><td>Alhama de Aragón</td><td>Continuous photography of the solar spectrum</td></tr>
</table>

[506] See: 'El Eclipse de Sol de 30 de Agosto', *El Mundo Científico*, 7 (1905), 544–550, here pp. 548–549; J. Esquirol, 'Éclipse totale de Soleil du 30 août 1905. Protubérances solaires à deux couleurs [présentée par Deslandres]', *Comptes Rendus*, 142 (1906), 757; 'Observations de M. G. Tramblay, à Alcala de Chisvert (Espagne)', *Bulletin de la Société Astronomique de France*, 19 (1905), 499–500; 'Observaciones del eclipse de Sol de 30 de Agosto', *El Mundo Científico*, 7 (1905), 559–562; C. Nielsen, 'The total eclipse of 1905. At Campo de la Isla, Burgos', *Knowledge & Scientific News*, 2 (1905), 272; 'Observations de M. G. A. Quignon, à Burgos', *Bulletin de la Société Astronomique de France*, 19 (1905), 498; 'Observations de MM. Georges, Henri et André Duval à Calatayud (Espagne)', *Bulletin de la Société Astronomique de France*, 19 (1905), 498–499; 'The total solar eclipse of August 30, 1905', *The Observatory*, 28 (1905), 377–382; Augustin Morford, 'The total eclipse of 1905. At Vinaroz, Spain', *Knowledge & Scientific News*, 2 (1905), 273; and 'B.B.', 'El eclipse en Burgos', *La Ilustración Española y Americana*, 81 (1905) [Año 49, número 33], 131–134.

[507] Letter from the President of the Bureau des Longitudes to the Ministre de l'Instruction publique, des Beaux-Arts et des Cultes, signed in Paris on 20 September 1905. Archives Nationales de France—F^{17} I [F17/01(B)]), F^{17} 13571: *Bureau des longitudes. Affaires scientifiques diverses 1881–1922*. Folder: *Bureau des Longitudes. Affaires diverses (1901–1914)*.

Dungan and Hatmann (Potsdam Astrophysical Observatory)		Hatmann was in charge of photographing the solar spectrum. Dungan employed a photographic telescope of 13 m focal length to study the internal region of the corona
Esteban Terradas and Casares (Central University)		Photography of the corona
C. Nielsen	Burgos	Sketch of the corona, study of prominences, colours of the Earth and sky, Baily's beads, shadow bands, and effects of the eclipse on animals
G. A. Quignon		Shadow bands
German commission led by M. F. S. Archenholg		Photography of the eclipse
Nippoldt and Ludeling (Postdam Astrophysical Observatory)		Influence of the eclipse on terrestrial and atmospheric magnetism
Georges Duval, Henri Duval, and André Duval	Calatayud (Aragón)	Contact times, prominences, description of the corona, meteorological observations, study of the light intensity
Professor Alexander Herschel	Castellón de la Plana	Totality was entirely obscured by clouds, but the first and last contacts were observed in a clear sky. Good records of radiation and temperature were obtained.
Jesuit Fathers under the orders of Father Algué (Director of the Manila Observatory)	Palma de Majorca	Photographs and drawings
Father Cortie, assisted by A. Liddell, L. Cafferata, and G. de Aquilera. The British Father Morford, of Saltash, was also in the party	Vinaroz	Father Cortie had three instruments to photograph the spectra and a camera, for the corona, of about 20 ft focal length. Morford brought a refractor of 4¼ inches, to make visual observations of the prominences. He also made naked-eye observations and studied changes in light levels and the behaviour of birds.

It is worth pointing out here how many amateur astronomers paid attention to the study of shadow bands. The shadow bands were considered a subsidiary phenomenon, and one that was usually "neglected or treated in an unscientific manner".[508] It was a field in which mainly amateur rather than professional astronomers concentrated their efforts. Although there were several theories of shadow bands, it was not clear whether they were a scintillatory effect of the atmosphere or a diffraction effect. During the eclipses of 1900 and 1905 many amateur astronomers observed the phenomenon and tried to establish its origin.

The earlier theory that the bands were a diffraction effect was challenged in 1887 by Cleveland Abbe (1838–1916), who proposed that the bands were caused by the interference of rays of light which were slightly inclined to one another as a result of irregular refraction in an inhomogeneous atmosphere.[509] Authors such as Catharine O. Stevens, who observed the eclipse of 1905 from Majorca, argued that the shadow bands were another aspect of the boiling distortions on the Sun's limb, always present in any solar observation.[510] After this eclipse, many amateur astronomers agreed that the shadow bands were a phenomenon of the Earth's atmosphere.[511] Others, however, explained through mathematical and geometrical methods such as Fresnel's theory of diffraction, the impossibility of these bands being diffraction phenomena.[512]

---

[508] Shackleton, 'The coming total eclipse', *Knowledge & Scientific News*, 2 (1905), 71–73, here p. 72. An attempt to approach the problem in a more systematic and scientific way is found in: A. Lawrence Rotch, 'Eclipse shadow bands', *Nature*, 72 (1905), 307–308, in which the author, from the Blue Hill Observatory (Mass., USA), suggested what information about shadow bands was needed and how to make the necessary observations. As he pointed out, 'the report should consist of a statement of the methods employed by the observer or observers in making the observations, and the results obtained', including the orientation of the bands, their direction of motion, velocity, width, and distance apart, and the direction of the wind—all before and after totality.

[509] See, for instance: 'The total solar eclipse of August 30, 1905', *The Observatory*, 28 (1905), 377–382.

[510] Catharine O. Stevens, 'Letters to the Editor: The troblem of "shadow-bands"', *Nature*, 72 (1905), 631.

[511] See, for example: Wilhelm Krebs, 'Fliegende Schatten bei der Sonnenfinsternis vom 30. August 1905', *Astronomische Nachrichten*, 170 (1906), 283–284; 'Our astronomical column: Atmospheric origin of "shadow bands"', *Nature*, 72 (1905), 611; and 'Les Bandes d'ombre pendant les éclipses totales', *Revue générale des Sciences pures et appliquées*, 16 (1905), 1018.

[512] F. Biske, 'Are diffraction phenomena possible at solar eclipses?', *Astrophysical Journal*, 38 (1913), 192–196.

9.7. Spanish amateur research

In Spain, amateur astronomy continued to play an important role in the development and support of astronomy during the early twentieth century. As in 1900, many Spanish amateur astronomers observed the eclipse of 1905 with the aim of contributing to the progress of astronomy. Indeed, their astronomical work is comparable to that pursued during these eclipses by foreign amateur astronomers who visited Spain, as Table 9-3 shows.[513]

Table 9-3: Spanish amateur astronomers at the total solar eclipse of 1905

Observers	Location	Observation and results
Gonzalo Brañas Fernández (Secondary school professor of physics and chemistry)	A Coruña	Magnetic observations
Luis Tortosa	Alboraya	Shadow bands
Gerardo Solís and Oscar Pulido	Avilés	Sketching the corona, contact times, meteorological observations, and effects on animals and plants during totality
Father Joaquín M. Barnola	Barcelona	Visual and meteorological observations
J. Inglés, P. Rius, and Gatell		Meteorological observations, photography of the partial phases
Engineers from the Instituto Geográfico	Burgos	Positive photographs of the corona (sent by Vicente López Puigcerver)
Capt. Juan López Soler		Photometrical observations
Fernando Estévez	Cádiz	Meteorological observations
Domingo Audía	Coll de Moro	Contact times
Pascual y Cantó	Desierto de las Palmas	Unsuccessful observations with his 108 mm aperture telescope
Francisco González Prieto	Gijón	General observations and description
Justino Guitart and Luis Oms	Moncayo	General and meteorological observations

---

[513] Many others whose research dealt mainly with contact times and meteorological observations could also be found in the *Archives of the Observatory of Madrid*, ‘Eclipse total de sol del 30 de agosto de 1905’, Folder 16. See also: ‘Observaciones del eclipse de Sol de 30 de Agosto’, *El Mundo Científico*, 7 (1905), 559–562.

"Second astronomical brigade": Ubaldo de Azpiazu, Carmelo Sanchos, and José Vila[514]	Narón (A Coruña)	General observations, contact times, photography, and effects on animals and plants during totality
Expedition from the College of San Rafael, led by Rafaél Vilás	Tortosa	General observations
Military engineers, led by Lieut. Luis Gómez de Barreda y Salvador	Valencia	Shadow bands at the limits of totality
Meteorological Observatory of the University of Valencia: F. Palencia and S. Martínez Perales		Meteorological observations, contact times, naked-eye observation of stars, study of the light intensity, and shadow bands
Father Manuel Cazador (Meteorological Observatory of the Secondary School of San Julián)[515]	San Julián de Vilatorta	Contact times, sunspots, shadow bands, naked-eye observation of stars, study of the light intensity, meteorological observations, and effects on animals and plants during totality (instruments: Bardou telescope of 61 mm aperture, Negretti & Zambra sextant, Porthouse chronometer)
A. Cruz	Zaragoza	Sketch of the corona and effects on animals and plants during totality

As has already been pointed out, perhaps the most outstanding Spanish amateur astronomer during the late nineteenth and the early twentieth century was Lànderer. During the solar eclipse of 1905, Lànderer and Meslin continued the studies on the polarisation of coronal light that they had begun five years before. Both Lànderer and Meslin had observed the eclipse of 28 May 1900 from Elche (Spain). For the eclipse of 1905, Meslin used a method similar to one used by Lànderer in 1900.[516] This original method was based on a Cornu photopolarimeter and a telescope of

[514] This expedition was completed by: Luis Ribera, Father Plácido Angel Rey, and Ramón María Aller (Observatory of Lalín), Santiago Casares, and José Rodríguez and his son.

[515] See: Manuel Cazador, 'El eclipse de Sol de 30 de Agosto de 1905 en el Colegio de San Julián de Vilatorta (Vich)', *El Mundo Científico*, 7 (1905), 696–697.

[516] J. J. Lànderer, 'Sur la lumière polarisée de la couronne solaire', *Bulletin de la Société Astronomique de France*, 19 (1905), 495–497, here p. 497. This paper is very similar to one that had appeared a month before under the same title in *Comptes Rendus*, 141 (1905), 589–590.

100 mm aperture. Meslin, assisted by Aroles (a physics lecturer at the Lycée de Montauban) and Chaudier (*préparateur* at the Faculty of Science in Montpellier), also employed in 1905 a more efficient Cornu photopolarimeter to measure the degree of polarisation of light.[517] He also studied the elliptical polarisation of light with a Bravais polariscope positioned inside the telescope's eyepiece. Although no elliptical polarisation was detected, Meslin found that the degree of polarisation of light from both the polar and equatorial regions was about 50%. These results agreed with Lànderer's observations for a region in the ecliptic a few arc seconds from the Sun's limb.[518]

Because of the characteristics of the eclipse of 1905, Lànderer decided to study the polarisation of the light from the corona using a different method than the one he employed in 1900. He set up an optical system which consisted of an astrophotographic objective 75 mm in diameter and a 62 mm Wollaston prism attached to a camera mounted on an equatorial axis. Six photographs were obtained during totality. On a scale of 0 for natural light and 10 for totally polarised light, Lànderer found that the polarisation of light from the corona was between 0.50 and 0.60, a very similar value to what he obtained in 1900.[519] This was an important and relevant scientific finding, and was often cited as a comparison for subsequent results. The 1905 expedition led to Guelma (Algeria) by Newall also worked on the atmospheric polarisation and the polarisation of the corona.[520] And at Mount Hamilton several experiments were undertaken in order to increase, by polariscopic methods, the theoretical contrast between the strongly polarised portions of the corona and the non-

---

517 See: J. J. Lànderer, 'En vísperas del eclipse', *La Ilustración Española y Americana*, 81 (1905) [Año 49, número 31], 93.

518 Geroge Meslin, 'Sur l'eclipse du 30 août 1905 et sur la polarisation de la couronne solaire', *Comptes Rendus*, 141 (1905), 493–496, here p. 494.

519 J. J. Lànderer, 'Sur la lumière polarisée de la couronne solaire' [presentée par Janssen], *Comptes Rendus*, 141 (1905), 589–590; Lànderer, 'Sur la lumière polarisée de la couronne solaire', *Bulletin de la Société Astronomique de France*, 19 (1905), 495–497. His measurement of 1900, adopting Cornu's notation, led to an average result of R = 0.52 for the degree of polarisation of light from the corona. (The results of these observations were: $-2\omega = 58°$, $(\omega_2 - \omega_1) = 30°$, which led to the values R = 0.53 and R = 0.50.) See: Lànderer, 'Sur la proportion de lumière polarisée de la couronne solaire', *Comptes Rendus*, 130 (1900), 1524–1525.

520 H. F. Newall, 'Note on polarisation phenomena in the solar corona, 1905 August 30', *Monthly Notices of the Royal Astronomical Society*, 66 (1906), 475–481.

polarised sky illumination.[521] As has been already shown, the Lick Observatory commission also studied the polarised light from the corona during the eclipse of 1905.

The professional interaction and respect between Lànderer and Meslin shows how Spanish astronomers not only hosted foreign scholars and assisted them, but also how they took active part in research of general scientific interest. Indeed, Meslin mentioned Lànderer's results and used them to contrast and confirm his own observations.[522] Other Spanish authors who published papers on the eclipse of 1905 in recognised international journals were José Comas i Solà, Ricardo Cirera, and Mario Roso de Luna (1872–1931). The latter was a theosophist and amateur astronomer interested in the origin of the shadow bands, for which he and others employed in Soria, on 30 August 1905, a device to solve it from a geometrical point of view.[523]

9.8. Jesuit expeditions for the eclipse of 1905

Ricardo Cirera, as we have seen, was the Jesuit Father who founded and directed the Ebro Observatory. As part of his aim of solving the problem of the influence of eclipses on the terrestrial magnetic field, he coordinated the study of magnetic variations at Ebro Observatory during the 1905 eclipse. Any attempt to analyse the astronomical observations undertaken during the eclipse of 1905 needs to pay attention to the work of the Jesuit Fathers. As a note in *The Observatory* on June 1905 explained:

> The track of totality passes over no less than fifteen Jesuit colleges or houses of the Society in Spain, and this religious body, which seems to us much associated with eclipse work and solar physics generally, will participate largely in the observation of the eclipse. Father Cortie will possibly go to Tortosa, near the mouth of the Ebro, where Father Cirera has established a first-class astrophysical observatory: totality here lasts 2.6 minutes. If Father Cortie does not go here he may go to Palma, in Majorca, where the Spanish Jesuits have a residence: totality 3 minutes. A party

---

[521] Perrine, 'Some total solar eclipse problems', *Astrophysical Journal*, 20 (1904), 331–337.

[522] Meslin, 'Sur l'eclipse du 30 août 1905 …', *Comptes Rendus*, 141 (1905), 493–496, here p. 495.

[523] Mario Roso de Luna, 'Eclipses totales de Sol [boceto de un aparato para investigar la posición de los elementos productores de las "shadow bands" (bandas oscilantes) en el espacio]', *Astronomische Nachrichten*, 171 (1906), 87–90.

> from the Jesuit College of Grenada [sic] will place themselves at a college at Carrion de los Condé [sic].[524]

Indeed, several Jesuits moved to within the path of totality in Spain to observe the eclipse of 1905. To coordinate these observations, the Jesuits from Tortosa published a leaflet with instructions on the most valuable observations for Jesuit individuals and institutions to undertake, whatever the instruments that they had available.[525] An example is the work of the Society of Jesus of Oña, whose observations of the eclipse were later published. Their programme is summarised in Table 9-4.[526]

Table 9-4: Programme of observations of the Society of Jesus of Oña in 1905

Type of observation		**Observer (if known)**
Naked-eye observations	Light intensity	
	Visible stars	
	Baily's beads	
	Prominences	
	Observations of the corona and coronal streamers	
	Shadow bands	Fermín Chanal
	Sketches of the corona	Enrique Ascunce, Luciano Ascunce, Florencio Legarreta, Manuel Serapio, Eliseo de la Torre
	Meteorological observations	Pedro Fernández (director of the observatory and lecturer in physics)
Observations with scientific instruments (including several quadrants, a Molteni's solar chronometer, and a sextant)	Time-keeping	Enrique Tavernier
	Photographs of the landscape	Pablo Simón
	Photographs of the corona (the problem of the short focal length of the cameras being used was solved by using two telescopes, of 100 mm and 200 mm aperture)	Mariano Ayala, Antonino Oraa

524 'The total solar eclipse of August 30', *The Observatory*, 28 (1905), 260–261.

525 Observatorio de Física Cósmica del Ebro, *Instrucciones para la observación del eclipse total de sol del 30 de agosto de 1905 (Colegio de Jesús, Tortosa)* (Tortosa: Imprenta de José L. Foguet, 1904).

526 *Observaciones del eclipse ... en el Colegio de Oña* (Oña, 1906).

	Lunar polariscopic work (with a Nicol prism)	Pedro Echeverría
	Spectroscopic work on the lunar limb (with a direct-vision prism attached to binoculars)	
	Green coronal line	
	Flash spectrum	Fernando Gutiérrez del Olmo
	Other telescopic observations: contact times, observations of sunspots, prominences, etc.	

The main findings of this expedition by the Society of Jesus of Oña were that the observed contact times did not match the predicted ones because the lunar diameter used for the calculations was too large, and that the structure of the prominences seemed to contradict Bigelow's electromagnetic hypothesis for the solar corona.[527] The comparison of the two coronal photographs that they obtained with the five photographs from the Jesuit College of Gijón, where totality was experienced 10 minutes before it was in Oña, showed no changes in the corona and confirmed that both the corona and its extensions are real and belong to the Sun, as the prevalent theories suggested.[528] Indeed, one of the main advantages of the several expeditions organised by the Jesuits in Spain was that it was possible to coordinate observations and compare results. In fact, Jesuits organised in Spain more than ten expeditions to observe the solar eclipse of 1905. Table 9-5 summarises some of them.[529]

[527] As Agnes Clerke stated in her *A Popular History of Astronomy During the Nineteenth Century* (Edinburgh: A. & C. Black, 1885), the most promising electrical theory of the solar corona developed by the late nineteenth century was that of the American Professor Frank Bigelow, from Washington.

[528] *Observaciones del eclipse ... en el Colegio de Oña* (Oña, 1906), pp. 42–48.

[529] 'Capítulo IV. Otras observaciones fuera del Observatorio', *Memorias del Observatorio del Ebro*, 1 (1906), 43–51.

Table 9-5: Jesuit expeditions in Spain to observe the total solar eclipse of 1905

Location	Observers	Observations and results
Espina (organised by the Ebro Observatory)	Expedition led by Father Ubach	Observation of contact times and shadow bands. Sketch of the corona by Father Maina. Meteorological observations and naked-eye observation of stars.
Montsiá (organised by the Ebro Observatory)	Included Fathers Pujadas, Vidal, Vilarrubias, and Esturí	Sketch of the corona and study of prominences with binoculars
Alcosebre (organised by the Ebro Observatory)	Fathers Audí, Willot, and Zurbitu	This party operated the Ebro Observatory's Dover magnetometer, a Tonnelot mercury barometer, three Richard graphic barometers, one thermometer, one hygrometer, one psychrometer, and a Richard portable anemometer. Their main purpose was to complete the meteorological and magnetic observations carried out at the Ebro Observatory.
Vinaroz	Father Cortie and A. Liddell, assisted by G. Aguilera and L. Cafferata. Liddell was also assisted by some locals: Roca, Coas, and Ribera	Instruments: a 6 m focal length photographic camera (100 mm objective) fed horizontally by a coelostat; three spectroscopic cameras attached to a camera (or coronograph). Several photographs of the corona, prominences, and filaments and streamers were obtained.
Gijón	Expedition led by Father Jambrina	Observation of contact times and the flash spectrum (Nicolás E. Ozalla was in charge of the spectroscope), photography of the corona during totality (with a camera of about 1.5 m focal length and 88 millimetres aperture, in charge of which was Francisco Menéndez Rua), several sketches of the corona, shadow bands, meteorological observations, and naked-eye star observations.
Comillas	Three expeditions led by Father José M. Ibero were set up at this location	Observation of prominences and the corona, as well as meteorological observations. Bad weather conditions.
Burgos	German Fathers Esch and Baur, from the Observatory of Valkenberg (Holland)	This party used an equatorial telescope to search for intra-mercurial planets, and obtained a photograph of 3 minutes exposure. They also took a photograph of the corona.

	Eight Jesuit Fathers from the College of Burgos assisted the Belgian astronomer Spée; four from the College of Burgos assisted Francisco Iñiguez; Fathers Martínez and Madiraga also helped the Dutch expedition	Sketches of the corona and assistance to several astronomers.
Veruela	Expedition led by Father Luis Puiggrós (Rector of the College of Veruela)	Meteorological observations, contact times, and study of the colour of the landscape.
Zaragoza	Expedition led by Father Albiñana	Contact times, photographs and sketches of the corona, observations of shadow bands and meteorological observations.
Palma	Expedition led by Father Algué, Director of the Manila Observatory, assisted by several Fathers of the College of Barcelona	Contact times, shadow bands, photographs and sketches of the corona and the prominences, meteorological and magnetic observations (with the Dover magnetometer, in charge of which was Father Sallaberry)

Table 9-5 shows how most of the astronomical observations made by Spanish Jesuits were comparable to those made by amateur astronomers. Only Jesuit observers from established astronomical institutions, such as Father Cortie at Vinaroz, attempted observations relevant to the development of astronomy. Amongst the astronomical research groups formed to study the eclipse, there stands out the one located at the Ebro Observatory, which pursued an extensive programme of astrophysical work.[530] Dozens of observers took part: three of them were in charge of the study of the shadow bands; the magnetic and meteorological sections were under Fathers Merveille and Dressel, respectively; Father Balcells attempted to measure the wavelength of the coronium line; Father Vitoria studied the intensity of the light during the eclipse; Fathers Kramers and

---

[530] See: Cirera, *Notice sur l'observatorie ...* (Barcelona, 1906), Chapter 3, 'El Eclipse en el Observatorio', pp. 31–42. For a summary of the observations, see: Puig, *El Observatorio del Ebro* (Tortosa, 1927); pp. 145–148.

Gravalosa were placed in charge of the operation of several photographic cameras; Father Prósper attempted to photograph the spectrum of the corona; and Fathers Schaeffers and Balasch employed a Cornu photopolarimeter to obtain data on the polarisation of light and its rotation. Many other Jesuits, Fathers and Brothers, took part in the astronomical and astrophysical activities on 30 August 1905 at the Ebro Observatory.[531] However, the most original contribution at the Ebro Observatory came from the work of the German Father Theodor Wulf (1868–1946) and the Belgian J. D. Lucas. As the *Journal of the British Astronomical Association* recorded:

> Fathers Wulf and Lucas employed selenium cells for the purpose of estimating the variation of sunlight during the eclipse, and also for determining exactly the beginning and end of totality … It is probable that very accurate determinations may be made by this method.[532]

The results were published in several international journals.[533] The use of a selenium cell for an accurate determination of the exact instants of the beginning and end of totality was a practical way of avoiding the subjectivity of observers. The effect was based on the way in which the electrical conductivity of selenium is affected by the light of the Sun, as explained in an article published in *Nature* in April 1906:

> The electrical conductivity of selenium increases on exposure to sunlight, being especially sensitive to the less refrangible end of the spectrum; therefore, by placing the cell in series with a battery and a delicate reflecting galvanometer, the amount of light falling on the selenium may be registered by registering the movements of the galvanometer beam of light.
>
> Whilst the decrease of light, during an eclipse, is not *visible* until the eclipse is well advanced, the galvanometer needle at Tortosa was seen to move immediately after first contact, and for nearly an hour showed a

---

[531] The previous list of names has been obtained from: J. Soler de Morell, 'El día 30 de Agosto en El Observatorio del Ebro', *Memorias del Observatorio del Ebro* (1906), 53–56.

[532] Whitmell, 'Contact times of the eclipse on 30 August', *Journal of the British Astronomical Association*, 16 (1905–1906), 220–226.

[533] An extract of the original work is in: 'Observations made with selenium cells during the total solar eclipse of August 30, 1905', *Astrophysical Journal*, 23 (1906), 153–161 [Translated, condensed and diagrams redrawn by William F. Rigge from Th. Wulf and J. D. Lucas, 'Zwei Beobachtungen mittels Selenzellen bei der totalen Sonnenfinsterniss am 30 August 1905', *Physikalische Zeitschrift*, 6 (1905), 838–847].

uniformly increasing resistance. Assuming that the light during totality was of the same quality as that obtaining at dawn, the results derived from the observations show that its brightness was about equal to that of the sky some thirty or forty-five minutes before sunrise.

The results obtained regarding the instants at which totality began and ended were very satisfactory.[534]

The French Académie des Sciences delayed the presentation of the main conclusion from the magnetic observations at the Ebro Observatory during the eclipse until 26 December 1905.[535] The Académie was initially very reluctant to publish the results about a general magnetic "cosmic perturbation", independent of the eclipse and detected on that same day, which had affected the results of the observations. However, the results obtained by the magnetic section of the Ebro Observatory with the Dover magnetometer were later confirmed by the German Johannes Christoph Alfred Nippoldt (1874–1936) and the French Charles Nordmann.[536] There are no further details about the problems with and the negotiations about publishing the text, but this incident could be seen as an episode after which both Cirera and the Spanish astronomical community were reinforced. Indeed, once published in the *Comptes Rendus*, Cirera's magnetic records obtained during the total solar eclipse, showing extraordinary deviations from the normal diurnal curves, were broadly publicised and appeared in other notable journals such as *Nature*.[537]

[534] 'Our astronomical column: Employment of selenium cells during total solar eclipse', *Nature*, 73 (1905–1906), 617.

[535] Cirera, 'Observations magnétiques de l'Observatoire de l'Ebre …', *Comptes Rendus*, 141 (1905), 1270–1271.

[536] The work of Nippoldt, 'Zum Einfluβ der totalen Sonnenfinsternis vom 30. August 1905 auf die erdmagnetischen Variationen', *Physikalische Zeitschrift*, 7 (1906) 242–248, was based on the observations made by the German expedition to Burgos. Nordmann's article, 'Recherches sur les effets magnétiques de l'éclipse totale du 30 août 1905', *Bulletin Astronomique*, 24 (1907), 93–106, compared his observations at Philippeville with those of Burgos, Tortosa, Poissy and Stonyhurst. Both cases are mentioned in Cirera, 'Prólogo', in E. Merveille, *La Section magnétique. Memorias del Observatorio del Ebro*, 3 (1908); pp. 5–8. See also: 'Société Française de Physique', *Revue générale des Sciences pures et appliquées*, 18 (1907), 336.

[537] See the note published in *Nature* in February 1906: 'Our astronomical column: Magnetic observations during the total eclipse', *Nature*, 73 (1905–1906), 400–401.

9.9. Other Spanish research

For the eclipse of 1905, the Royal Academy of Sciences and Arts of Barcelona and the Fabra Observatory sent to Vinaroz a commission led by José Comas Solá. This time, Comas i Solà had several assistants: not only his wife, as in 1900, but also Salvador Raurich (1869–1945), and Laur Clariana i Ricart (1842–1916) and his son, as well as the photographer Riba, the electrician Antoni García, and Sebastià Bover. They obtained three pictures of the corona with a Grubb equatorial telescope of aperture 6 inches (152 mm, owned by Comas i Solà), mounted equatorially, which showed the filaments and prominences. Salvador Raurich also made a sketch of the corona, and García observed the flash spectrum. As part of their research programme, they took photographs of the chromospheric spectrum, observed the spectral lines of the prominences, and made several meteorological observations. The photographs of the chromospheric spectrum were taken by Comas himself with a Bardou objective of 110 mm together with a 60° flint prism. Particularly interesting is the cinematographic method to observe the spectrum during the totality of the eclipse, one of the earliest instances of this technique ever reported.[538] The cinematographic method of taking spectroscopic photographs, which had similarities to some techniques used by foreign scientists at previous eclipses, was expected to illuminate the study of the inversion of lines and the flash spectrum. Each exposure was one-fifth of a second. The developed images yielded very interesting results. However, the 25 metres of film prepared to record the crucial moments was not enough, and only one minute's worth of images was recorded before totality began. Consequently, the inversion of lines was successfully observed but not the flash spectrum.

Spanish astronomical activity included a large number of individuals, and several institutional parties went to various locations to observe and

---

[538] This cinematograph was lent by the French firm Gaumont and included a Maihlat prism located in front of the objective, made by Goerz. On the work of Comas i Solà during the total solar eclipse of 1905, see following publications of his: 'Observations de MM. José Comas Sola, directeur de l'Observatoire Fabra et Salvador Raurich, à Vinaroz (Espagne)', *Bulletin de la Société Astronomique de France*, 19 (1905), 495; 'Observations sur l'eclipse totale du Soleil du 30 aôut 1905 [présentée par M. Loewy]', *Comptes Rendus*, 141 (1905), 616–618; 'La Corona solar del eclipse de 1905', *El Mundo Científico*, 7 (1905), 818–819; and 'El eclipse total de sol del 30 de agosto de 1905', *Memorias de la Real Academia de Ciencias y Artes de Barcelona*, 5 [24] (1906), 479–512.

study the eclipse. For example, the Professors González Martí and Cabrera from the Central University of Madrid, assisted by Gil and the volunteers Alberto Inclán, Suárez Inclán, and José Cabrera, planned to go to Rasillo de Cameros (Logroño) to confirm the relationship between the frequency and strength of terrestrial magnetic storms and the level of solar activity.[539] The Observatory of Madrid sent a party to Lilaila (Burgos), led by its Director, Francisco Iñiguez. During March and April 1904, Francisco Iñiguez had exchanged letters with his superiors about the necessity to buy new instruments for the study of the forthcoming eclipse, and to set up a photographic laboratory at the Observatory.[540] The bureaucratic problems encountered by the Observatory in obtaining the desired new equipment were, however, evident. The frustration that these problems induced is well illustrated in one of Iñiguez's letters to his superiors, in which he described how humiliating it would be for the Observatory to have to study the solar eclipse of 1905, surrounded by so many well-equipped foreign astronomers, without a long-focus camera.[541] It was not until October 1904 that the Spanish Government granted an additional 10,000 pesetas for the acquisition of new equipment.[542]

The work carried out by the Observatory of Madrid during the eclipse of 1905 included several meteorological, magnetic, and actinometrical observations, and also some astronomical work such as the determination of times of contact, photography of the interior and outer corona, and photography of the spectrum of the corona and the chromosphere.[543] The programme of observations set out by the Observatory of Madrid was

---

[539] 'El mundo al día. El próximo eclipse y las borrascas magnéticas', *El Imparcial*, 8 July 1905, 1.

[540] The budget for the laboratory was finally agreed on 14 April 1904. See: *Archives of the Observatory of Madrid*, 'Eclipse total de sol del 30 de agosto de 1905', Folder 16.

[541] See the letter signed by Iñiguez on 5 April 1904 in the *Archives of the Observatory of Madrid*, 'Eclipse total de sol del 30 de agosto de 1905', Folder 16. Iñiguez was aware of the interest that the official Spanish party had aroused in several foreign observers, such as Mary Proctor, who wrote to the Director of the Observatory of Madrid before the eclipse asking to join the Spanish expedition.

[542] See the letter written to Iñiguez by the Director del Instituto Geográfico y Estadístico and signed on 5 April 1904, in the *Archives of the Observatory of Madrid*, 'Eclipse total de sol del 30 de agosto de 1905', Folder 16.

[543] Dirección General del Instituto Geográfico y Estadístico, *Eclipse total de sol de 30 de Agosto de 1905. Reseña de los trabajos efectuados para su observación* (Madrid: Imprenta de la Dirección General del Instituto Geográfico y Estadístico, 1906).

Fig. 9-4: Photograph of the eclipse of 30 August 1905 by Comas i Solà. Published in *El Mundo Científico*, 7 (1905).

divided into four main areas: meteorological observations (temperature, pressure, wind, humidity, etc.), magnetic observations (variation of the magnetic declination during the eclipse and measurement of other magnetic elements), actinometric observations (change in the intensity of the solar radiation during the eclipse), and astronomical observations (determination of contact times; photography of the different parts of the corona, of the surroundings of the Sun, of the chromospheric spectrum, and of the visible and ultraviolet spectra; naked-eye observations, study of shadow bands, etc). The magnetic observations were overseen by the engineer Luis Cubillo, and the actinometric measurements, with an Angstrom pyrheliometer owned by the Observatory of Madrid, were coordinated by the engineer Aurelio Capilla, assisted by José Fernández Villalta and José López Morcillo. The study of the corona and of the surroundings of the Sun was a task for the Professor of the Secondary School of Pamplona, Modesto Díez del Collar, while the photographs of the inner corona were left to the engineer José Galbis, who used a coronograph of 150 mm aperture and 8 m focal length which received the light from a Grubb coelostat—the same one that was used in 1900. The Director of the Meteorological Station of Oviedo, Professor Arturo Pérez Martín, was in charge of the meteorological observations. Some direct observations of the corona, and photography that aimed to determine the light intensity variation during the eclipse, were left to Captain Juan López Soler. The official commission from the Observatory of Madrid was also assisted by some students of the Jesuit College of La Merced, who observed the shadow bands.

Several members of the Observatory of Madrid travelled to Burgos to carry out the astronomical programme of observations. In total, five astronomical instruments were set up by the official expedition to Burgos to work during totality: two photographic cameras to study the corona, one prismatic camera, and two spectrographs. A thatched house was built in Burgos for the astronomers, with bedrooms, living-room and kitchen, at a cost of 1250 pesetas. Francisco Cos photographed the chromosphere and the corona with a prismatic camera of 200 mm aperture and 2 m focal length; Antonio Vela photographed the ultraviolet spectrum with a two-quartz-prism spectrograph which received the light from a two-mirror Grubb coelostat; the assistant Pedro Jiménez employed a camera of 150 mm aperture and 1 m focal length mounted on an equatorial Steinheil telescope to photograph the corona and the Sun's surroundings; and the assistants Victoriano Fernández Ascarza and Gonzalo Reig photographed the visible spectrum with a three-flint-prism spectrograph which received the light from the same coelostat employed by Vela.[544] Francisco Iñiguez also observed the contact times with a Merz telescope of 108 mm aperture. However, their work was poor in that it was not original, and the results were merely qualitative. Other institutions, such as the Observatory of San Fernando, observed contact times, searched for intra-mercurial planets, and made meteorological observations. The main instruments they employed were a Grubb photographic camera, a theodolite, a chronograph, and a siderostat (or coelostat) with cameras.[545]

9.10. Solar eclipses and the development of aeronautics

There is no doubt that astrophysics experienced a continuous development as a result of the solar physics studies at total solar eclipses, like those made in 1905. Solar eclipses were also very important for more traditional astronomical research. They were particularly useful, for instance, in attempts to develop a proper lunar theory of the Moon's

---

[544] *Ibid.*

[545] See: *Archives of the Observatory of Madri*d, 'Eclipse total de sol del 30 de agosto de 1905', Folder 16.

motion.[546] At the same time, the eclipse of 1905 was a boost to the development of other sciences such as meteorology and aeronautics. With the former it should not be surprising: as has been already mentioned, by the end of the nineteenth century many observatories were devoted to both astronomical and meteorological studies. In fact, the observation of meteorological conditions during the progress of the eclipse was considered by the scientific community to be the most important of the secondary work. Although considered less crucial than studies of the corona and the spectrum, or of polariscopic work and the photographic search for intra-mercurial planets, it is significant that it was considered even more important than the determination of the relative position of the Sun and the Moon at the time of the eclipse, the photometric work, or the study of shadow bands.[547]

The case of aeronautics is, on the other hand, quite remarkable. It was at the meeting of the International Scientific Aeronautical Commission of 1904 in St Petersburg[548] that the Spanish Colonel Pedro Vives y Vich (1868–1938) presented a project to observe the eclipse of 30 August 1905 from the skies.[549] The Servicio de Aerostación Militar,[550] encouraged by the international scientific community, sent from Burgos three airborne expeditions on board dirigibles, with the intention of comparing their observations with ground-based ones. Several local and foreign scholars took part in this pioneering project, which attracted some public attention in Spain[551] and was a model for subsequent attempts made in other parts of

[546] To help with this study, Charles Louis François André constructed an artificial model that mimicked total and partial eclipses. See: André, 'Appareil à [sic] eclipses artificielles de Soleil', *Comptes Rendus*, 141 (1905), 168–169.
[547] Chester, 'The total solar eclipse of August 29–30, 1905', *Science*, 21 (1905), 635–637.
[548] It was held from 29 August to 3 September. A programme of the congress can be read in *The Aëronautical Journal*, 8 (1904), 58–59.
549. Pedro Vives y Vich, *Observaciones del eclipse total de sol de 30 de agosto de 1905 por medios de globos* (Madrid: Imprenta del Memorial de Ingenieros, 1905).
[550] The 'Military Aeronautical Park', as translated in: 'The Spanish provisional programme of balloon observations in the coming total eclipse', *The Aëronautical Journal*, 9 (1905), 47–49.
[551] See for instance: 'El próximo eclipse total de Sol. Observaciones por medio de globos', *El Mundo Científico*, 7 (1905), 210–211; 'B. B.', 'El eclipse en Burgos', *La Ilustración Española y Americana*, 81 (1905) [Año 49, número 33], 131–134. This ballooning enterprise, however, was eclipsed by the coverage the Spanish press gave to the visit of the Spanish royal family to Burgos.

the world.[552] The first balloon, *Júpiter*, under the command of Vives y Vich, also carried the meteorologist and aeronaut of the Prussian Aeronautical Observatory Arthur Berson (1859–1942) and the Spanish spectroscopy specialist Octaviano Romeo. Their mission was to make meteorological and spectroscopic observations as well as studying the shadow bands. *Urano*, the second balloon, was piloted by Captain Alfredo Kindelán Duany (1879–1962), and included in its crew Augusto Arcimis y Werle (1844–1910), astronomer and Director of the Instituto Central Meteorológico of Madrid. They were in charge of taking photographs and making meteorological observations. In the third balloon, *Marte*, were the aeronaut Jesús Fernández Duro and General Emilio Herrera y Linares (1879–1967). They had to draw the corona, make meteorological observations, and study the shadow bands.

From his observations, Herrera concluded that the shadow bands were a merely atmospheric phenomenon, as the higher one was, the thinner the bands were.[553] Shadow bands were also studied from Burgos, both from terrestrial stations and from balloons. At the same time, several members of the army coordinated their observations made from the edges of the path of totality, from Sagunto (Valencia), Sigüenza (Guadalajara), Valladolid, Peña de la Población (Navarra), and Albeada (Huesca); and from Benifayó (Valencia), Catarrosa (Valencia), Tabernas Blanques (Valencia), Guadalajara, Junquera (Guadalajara), Viana (Valladolid), Posadles (Valladolid), Matapozuelos (Valladolid), Losarcos (Navarra), and Monterruja (Navarra), respectively. Also, the meteorological observations made by the Spanish army from localities such as Burgos, Sigüenza, and Guadalajara were coordinated with the ones made by the Jesuits in Alcocebre, Gijón, and the Ebro Observatory.[554] At the same time, several non-piloted balloons were sent up from Burgos to an altitude of about 2 km to complement the projected study of air currents, which

[552] Balloon ascents were also expected to be made from Paris, Prague, and Algeria in order to study the variation in air temperature and solar radiation at various altitudes. See: 'Astronomical publications: The eclipse of 30 August 1905', *Journal of the British Astronomical Association*, 15 (1904–1905), 346.
[553] Emilio Herrera y Linares, 'Observación aerostática de las sombras volantes en el eclipse de sol del 30 de agosto de 1905', *Asociación Española para el Progreso de las Ciencias, Tercer Congreso (Granada del 20 al 25 de Junio de 1911). Tomo III* (Madrid: Eduardo Arias, 1911), pp. 65–71.
[554] Pedro Vives y Vich, *Avance de los resultados obtenidos en las observaciones del eclipse total de sol de 30 de agosto de 1905* (Madrid: Imprenta del Memorial de Ingenieros, 1906).

was expected to confirm the ideas of the American meteorologist Henry Helm Clayton (1861–1946) about how solar eclipses produce cold air cyclones.[555] Finally, coordinated observations and photographs of the solar corona were planned from different stations. The more successful results, from aerial and ground-based observations, were the spectroscopic observations of the corona made by Romeo (from the *Jupiter*) and Captain Juan López Soler, who detected in the bright (emission) spectrum of the chromosphere the lines C, F, and G' of hydrogen, and in the emission spectrum of the corona the D3 helium line and lines of sodium, magnesium, and coronium.[556]

The development of this programme of observations shows how Spanish institutions such as the Instituto Geográfico y Estadístico (which included the Observatory of Madrid and the Instituto Central Meteorológic*o*) and the Servicio de Aerostación Militar coordinated their efforts around a particular scientific research programme. As with the Ebro Observatory's programme,[557] one can see a division of labour and a degree of organisation which are characteristics of the new astronomy. This new way of pursuing astronomical research can be understood as being very much related to the development of astrophysics and its innovative techniques, as well as to the progressive specialisation of science and the need to organise new and bigger research programmes. During these years, even the official Spanish observatories of Madrid and San Fernando transformed their activities by placing a greater emphasis on

---

[555] Published in: *The Annals of the Astronomical Observatory of Harvard College*, 43 [Part 1] (1901) and cited in: 'El próximo eclipse total de Sol. Observaciones por medio de globos', *El Mundo Científico*, 7 (1905), 210–211. As Sean Clarke explained to the BBC in an article published on 21 June 1999 (http://news.bbc.co.uk/1/hi/sci/tech/specials/eclipse99/355056.stm): 'While the winds during a total eclipse drop, they can rise before and after and also in areas on the edges of totality. An eclipse cyclone is set up because the cooling causes a slight variation in atmospheric pressure, which in turn sets up a weak system of winds. Winds with definite cyclonic circulation have been observed on some occasions. American meteorologist H. Helm Clayton concluded in 1901 that the Moon's shadow is accompanied by a weak, cold-centred cyclone. Eclipses where totality is of short duration or over large expanses of water will not give the air and surface enough time to cool, and this will result in a much weaker eclipse cyclone.'

[556] Vives y Vich, *Avance de los resultados ... del eclipse total de sol de 30 de agosto de 1905* (Madrid, 1906).

[557] A description of how the observation of the eclipse was planned and completed at the Ebro Observatory can be found in: Soler de Morell, 'El día 30 de Agosto en El Observatorio del Ebro', *Memorias del Observatorio del Ebro* (1906), 53–56.

astrophysical research, as had happened during the late nineteenth century at observatories such as Paris and Greenwich.

9.11. Local versus foreign astronomy

How do the local and foreign expeditions that observed the 1905 total solar eclipse from Spain compare? First of all, one has to recognise that the observational programmes and resources of Spanish amateur astronomers were very similar to those of foreign amateurs. Also, as in 1900, some interesting and original research was also pursued by Spanish institutions and individuals, for example the studies of the polarisation of light pursued by several scientists such as the Jesuits from Ebro Observatory and the American commission from the Lick Observatory. Outstanding was the research by Lànderer, whose method of observation employed in 1900 was copied and improved by Meslin in 1905. Another good example is the cinematographic method employed by Comas i Solà to record the eclipse. Furthermore, the most important magnetic observations were those pursued by Jesuits from various countries who saw in the Ebro Observatory a reference point for their research.

As regards the search for intra-mercurial planets, it is interesting that the Observatory of San Fernando was the only Spanish institution to participate, along with the foreign commissions from the Valkenberg, Lick, and Kirkwood Observatories. I think this can be understood in a more general context. During the late nineteenth and early twentieth centuries the Observatory of San Fernando became an astronomical centre very much engaged with important international projects dealing with positional astronomy, such as the *Carte du Ciel*. One can regard the search for intra-mercurial planets as another of these international enterprises.

Other Spanish institutions, such as the Observatory of Madrid and the Fabra Observatory, had a stronger interest in the development and consolidation of their own astrophysical research programmes. That explains their attention to eclipse photography and the study of the spectra of the corona and the chromosphere, in common with a large number of foreign commissions as described in this chapter. In fact, the astronomical instruments employed by these Spanish institutions were very similar to those employed by other commissions who followed similar programmes of research. For example, the photographic study of the ultraviolet spectrum was attempted not only by the commission led by Evershed and the Lick Observatory expedition, but also by the Observatory of Madrid making use of a two-quartz-prism spectrograph receiving light from a two-mirror Grubb coelostat—a set-up resembling that used on Evershed's expedition, which consisted of a glass prismatic camera and a quartz prismatic camera mounted on a 3-inch (76 mm) equatorial telescope.

Similarly, the Grubb 6-inch (152 mm) equatorial employed by the Fabra Observatory to photograph the corona was comparable to instruments such as the camera of 5-inch (127 mm) aperture used by the Lick Observatory for the same purpose, and to the 120 mm Grubb photographical equatorial employed by the expedition from the Observatories of Besançon and Paris. Also, to photograph the corona the Observatory of Madrid employed a coronograph of 150 mm aperture receiving light from a Grubb coelostat and a camera of the same aperture set up on a Stenheil telescope. These instruments were to some extent poorer than the equatorial telescope from the Observatory of Meudon of about 305 mm aperture to which three cameras were attached.

Equally, the Fabra Observatory employed a Bardou telescope of 110 mm aperture to photograph the chromospheric spectrum, with slightly less resolution than the equatorial of about 200 mm aperture used by the commission led by Deslandres. Likewise, to photograph the chromosphere and the corona the Observatory of Madrid used a prismatic camera of 200 mm aperture which was comparable to the coronograph of about 100 mm aperture and the coelostat and camera of about 305 mm aperture employed by the expedition led by Fowler and Callendar for that same purpose. The instruments employed by these Spanish institutions for astronomical research were therefore appropriate ones and, although inferior in some cases, were very similar to these employed by the more prestigious foreign astronomers.

At the same time, at the conference on solar research held on 23 September 1904 in St Louis (United States of America) were represented various associations such as the Royal Society of London, the Royal Astronomical Society, the Société Astronomique de France, the French Physical Society, the German Physical Society, the Amsterdam Academy of Sciences, the Society of Italian Spectroscopy, the Vienna Academy of Sciences, the St Petersburg Academy of Sciences, the Subcommittee of the International Meteorological Committee, the Stockholm Academy of Sciences, the Astronomical and Astrophysical Society of America, the

American Physical Society, and the National Academy of Sciences.[558] Significantly, although no Spanish official commission was present at the second meeting of the International Unit for Co-operation in Solar Research held at Oxford from 27 to 29 September 1905, the Spanish Jesuit astronomer Cirera was invited to attend the meeting as a member of the Société Astronomique de France.[559] This indicates how foreign scholars now recognised the important progress being made in astronomy in Spain through the astronomical work by Spanish astronomers at the total solar eclipses of 1900 and 1905. Even the American members of the U.S. Naval Observatory expedition pointed out in 1906 that their especially complete meteorological observations made on the day of the eclipse were helped by "the co-operation of the Spanish officials in charge of the observatories at Tortosa, Zaragoza, Guadalajara, and Castellón".[560]

The eclipse of 1905 saw the consolidation of a process of institutionalisation of astronomy begun in Spain after the eclipse of 1900 which laid the groundwork for the further development of this discipline. For the 1905 eclipse Spanish expeditions were more complete and better organised, not only logistically but also in the number and quantity of the instruments employed and because of the appropriate training of most of the Spanish astronomers who took part in the different astronomical research programmes. Indeed, the new Spanish astronomical institutions founded during the early years of the twentieth century saw in the eclipse of 1905 the opportunity to become involved in very important astronomical research. That was also the case with some older organisations and individuals that, in many cases, benefited from the presence of the foreign scientists who visited the country. Consequently, the eclipses of 1900 and 1905 opened a period in which Spanish contributions to international projects and programmes of collaboration became regular. After these eclipses, the activity of Spanish astronomers was definitively well regarded amongst the international scientific community. And Spanish scholars were striving to join the international astronomical and astrophysical community just when that community was being formalised and organised. The foundation of the International Union for Cooperation in Solar Research in 1904 illustrates the institutionalisation

---

[558] Hale and Perrine, 'International co-operation in solar research', *Astrophysical Journal*, 20 (1904), 301–305.

[559] 'International Unit for Co-operation in Solar Research: Meeting at Oxford, 1905 September 27–29', *The Observatory*, 28 (1905), 382–387.

[560] Chester, 'Preliminary results of the United States Naval Observatory eclipse expedition in 1905', *Astrophysical Journal*, 23 (1906), 128–143.

of international collaboration in astronomy and astrophysics in a period characterised by the promotion of cooperative scientific research. Indeed, the word "international" was beginning to take on connotations of partnership, collaboration, and alliance. Spanish scholars reaped the benefit as they emerged into the international scientific community during this transformation.

CHAPTER TEN

SOCIAL ASPECTS OF ECLIPSE EXPEDITIONS

10.1. Introduction

The recent book by Alex Soojung-Kim Pang on eclipse expeditions has improved our understanding of the values, organisation, and politics of Victorian science.[561] As Pang pointed out, eclipse expeditions had not only a scientific but also a cultural and a social dimension, particularly in the case of amateur eclipse expeditions. In the new and vast literary market of the second half of the nineteenth century and the early twentieth, eclipses were publicised and popularised by observers in newspapers, magazines, books, scientific journals, and reviews. According to Pang's thesis, eclipse expeditions also reflect the imperialistic attitudes of various European countries. That was certainly the case of British eclipse expeditions organised to India in 1871 and 1898.[562] Similarly, condescending (although not strictly imperialistic) attitudes are apparent in some of the eclipse expeditions organised by the most powerful European countries to Spain. Several anecdotes recorded in Spanish newspapers tell of the concern expressed by the British commissions about the existence of wild people in the hills near Santa Pola (Alicante) in 1900.[563] More significantly, the local press reported how the fenced field where Norman Lockyer installed his party to observe the 1900 eclipse was regarded almost as British territory, as the running-up of the English flag illustrated.[564] At the same time, as Pang has identified, eclipse expeditions during the second half of the nineteenth century absorbed much from the

---

[561] Pang, *Empire and the Sun* (Stanford, 2002).
[562] *Ibid.*
[563] P. Ibarra, 'Desde Elche', *Las Provincias*, 27 May 1900. From: Vicent F. Soler Selva (coord.), *L'eclipsi total de sol de 1900 al Baix Vinalopó* (Elx: Ajuntament d'Elx, 2000).
[564] Ibarra, 'Desde Elche', *Las Provincias*, 23 May 1900. From: Soler Selva (coord.), *L'eclipsi total de sol de 1900 ...* (Elx, 2000).

novel European cultural phenomenon of tourism.[565] Not only were eclipse expeditions very appealing from a scientific point of view, but they also (as will be pointed out in this chapter) had an important exotic element, as is apparent from the records of amateur and official eclipse expeditions organised by British and French observers. Moreover, many of the scientific expeditions of these years may be considered an archetype of privileged bourgeois travel.

Planning these eclipse expeditions involved several months of work. There was a suitable location to select, itineraries and observing programmes to create, decisions to be made on instruments and means of recording observations, and logistical, technical, and political problems to solve. However, the tourism element also played an important role, particularly in expeditions such as those organised by the British Astronomical Association. For example, when the British Astronomical Association took the first steps towards the organisation of several expeditions to observe the eclipse of 28 May 1900, George Frederick Chambers (1841–1915), as early as June 1898, presented in a meeting to the Association some practical aspects to bear in mind.[566] In his presentation, Chambers described not only the path of totality through Portugal and Spain and the relative ease of reaching different observing stations, but also some potential tourist halts on the way. Equally, as early as 1902, Chambers took similar steps towards the organisation of expeditions to Spain to observe the eclipse of 1905.

At the British Astronomical Association, the organisation and coordination of expeditions was in the hands of an Eclipse Committee,[567] which had several tasks. One of the most important was the selection of an appropriate location. Where to observe the eclipse would depend mainly on weather prospects, and the Committee studied and presented to the Association any relevant weather reports that were published. For the eclipse of 1900, Andrew Crommelin read to the Association on 22 February 1899 a paper on the weather prospects for Spain, Portugal, and

---

[565] Pang, *Empire and the Sun* (Stanford, 2002), pp. 49–82. See also: Alex Soojung-Kim Pang, 'The social event of the season: Solar eclipse expeditions and Victorian culture', *Isis*, 84 (1993), 252–277.

[566] 'Report of the meeting of the Association, held on June, 29, 1898', *Journal of the British Astronomical Association*, 8 (1897–1898), 363–373, here pp. 366–371.

[567] See: '[Notes:] The total solar eclipse of August 30 next', *The Observatory*, 28 (1905), 70.

Algiers[568] which presented weather statistics based on the annual publications of the Observatories of Madrid, Coimbra, Lisbon, and San Fernando, and daily reports from the *Bulletin of International Meteorological Observations of Washington*. He also presented predictions of weather conditions for the eclipse, based on an analysis of weather records published in the *Washington International Weather Bulletins*.[569]

But were there any differences between the organisation of eclipse expeditions by amateurs and professionals? As regards the organisation of official expeditions, the French case is a paradigmatic example. In France, the organisation of official commissions was mainly centralised in the hands of the Bureau des Longitudes. This institution had a global financial plan to help fund the organisation of the most interesting proposed eclipse expeditions.[570] For example, in 1905 the Bureau's budget for all the eclipse expeditions that year was 97,000 francs.[571] The organisation of official eclipse expeditions involved quite a lot of bureaucracy. For example, members of French official expeditions required diplomatic passports, and for those who went on the eclipse expeditions sent to Spain in 1905 it took a long time to obtain them.[572] The organisation of these

---

[568] A. C. D. Crommelin, 'The weather prospects for the total solar eclipse of May 28, 1900, in Spain, Portugal, and Algiers', *Journal of the British Astronomical Association*, 9 (1898–1899), 204–207.

[569] Crommelin, 'The coming Spanish eclipse. Probable weather conditions', *Journal of the British Astronomical Association*, 15 (1904–1905), 138–141.

[570] See for instance the letter of April 1905 dealing with the expedition led by Henri Alexandre Deslandres and Pierre Puiseux from Paris to Venta de Baños via Madrid. Observatoire de Paris. Ms 1060 II-A-3: *Les Activités scientifiques de l'observatoire—éclipses 1850–1908 (1858, 1860, 1864, 1867, 1868, 1875, 1882, 1905, 1908)*.

[571] However, the distribution of the funding was unequal, and some astronomers, such as Georges Rayet, Director of the Observatory of Bordeaux, argued that the budget for their parties was insufficient. See: Letter from the Observatoire de Bordeaux signed on 18 June 1905. Observatoire de Paris. Ms 1060 II-A-3. Note that the exchange rate by 1905 was 32.9 pesetas to 1 sterling pound and 1 peseta to 1.3 French francs. Consequently, 1 sterling pound was 42.8 French francs. Therefore, 97,000 French francs were about 2270 sterling pounds (or 74,615 pesetas).

[572] The President of the Bureau des Longitudes had to write to the Ministre de l'Instruction Publique, des Beaux-Arts et des Cultes to obtain a diplomatic passport for some of the expedition members, including Georges Rayet and Fernand Courty, both from the Observatory of Bordeaux. See the letter signed at Paris on 12 July 1905. *Archives Nationales de France* (*ANF*)—F[17] I [F17/01(B)], F[17]

expeditions also entailed a large amount of administrative work. For example, in March 1905, when De la Baume Pluvinel joined the official expedition from the Observatory of Meudon led by Janssen,[573] the concession of the "mission gratuite" implied a large amount of bureaucracy.[574] The Ministère de l'Instruction Publique encouraged the Ministère des Affaires Etrangères to write a letter of reference for De la Baume Pluvinel so that he might receive all possible assistance in Spain.[575] The reference letter, addressed to the Agents Diplomatiques et Consulaires en Espagne, made him an official representative of France during his time in Spain.[576]

The Ministère de l'Instruction Publique et des Beaux Arts also had to negotiate with steam or railway companies, such as the Compagnie Générale Transatlantique and the Compagnie des Chemins de Fer du Midi, for the transport of scientific material and to obtain a reduction of the tariff.[577] The payment of travel expenses for the members of the different expeditions also entailed much work for the organisers of official

---

13571: *Bureau des Longitudes. Affaires scientifiques diverses 1881–1922*. Folder: *Bureau des Longitudes. Affaires diverses (1901–1914)*. The formal request for diplomatic passports was forwarded to the Ministry of Foreign Affairs, which agreed to supply them. See the letters from the Ministère de l'Instruction Publique to the Ministère des Affaires Etrangères (18 July 1905), from the Ministère des Affaires Etrangères to the Ministère de l'Instruction Publique (21 July 1905), and from the Ministère de l'Instruction Publique to the Bureau des Longitudes (24 July 1905). ANF—F[17] 13571.

[573] See the letter to the Ministère de l'Instruction Publique et des Beaux Arts et des Cultes signed on 15 March 1905. ANF—F[17] II [F17/02(A)], F[17] 17274, LA BAUME-PLUVINEL : Missions gratuites en Espagne (éclipses du 1900 et 1905), Senegal (16 avril 1893), Sumatra (mai 1901), Égypte (11 Novembre 1901), Crimée (Théodosie, 21 août 1914); *Mission gratuite en Espagne pour y observer l'éclipse de soleil du 30 août 1905*—document 1.

[574] Letters signed by the Ministère de l'Instruction Publique et des Beaux Arts on 20 March 1905 to Janssen, De la Baume Pluvinel, and the Ministère des Affaires Etrangères. ANF—F[17] 17274—documents 2, 3, and 4. The official communication is to be found in document 5.

[575] Letter signed by the Ministère de l'Instruction Publique et des Beaux Arts on 18 March 1905 to the Ministère des Affaires Etrangères. ANF—F[17] 17274—document 7.

[576] Letter signed by the Ministère de l'Instruction Publique et des Beaux Arts on 28 March 1905 to the Ministère des Affaires Etrangères. ANF—F[17] 17274—document 9. See also documents 10 to 12.

[577] ANF –F[17] 17274—documents 14 to 17.

expeditions.[578] Consultations on this issue had to be redone each time new members joined an expedition, as happened with the late addition of L. Ancel and M. Juéry to the eclipse expedition of 1905 led by De la Baume Pluvinel.[579] Finally, it had to be arranged for scientific equipment not to be inspected at the Spanish and French custom houses in order to avoid any risk of damage.[580] All these aspects were regular issues in the organisation of any official expedition. In the French case, one has the examples of the eclipse expedition of De la Baume Pluvinel to Spain in 1900,[581] the expedition to Algeria in 1905 led by Henri de la Vaulx (1870–1930),[582] the visit of Jean Mascart (1872–1935) to Tenerife (Spain) to observe Halley's comet in 1910,[583] and the expedition to the Iberian Peninsula by Fred Vlès (1885–1944) to study the eclipse of 1912,[584] all of them under the auspices of the Bureau des Longitudes.

In the British case, the last years of the nineteenth century also saw the formation of a body comparable to the Bureau des Longitudes in its efficient organisation of eclipse expeditions. In 1894 the Royal Astronomical Society and the Royal Society agreed to form a small, permanent planning committee which could not only apply directly to the Government Grant Committee of the Royal Society for funds, but could also purchase and manage instruments and control the publication of expedition reports. The remit of this Joint Permanent Eclipse Committee illustrates the various tasks involved in planning an expedition, including: drafting proposals for funding and convincing the government to support

---

578 For the expedition to observe the eclipse of 1905 from Paris to Alcalá de Chisvert (through Toulouse, Port-Bou, and Tarragona) conducted by De la Baume Pluvinel, and which originally included J. Peridier, Albert Senonque, Fernand Baldet, and Baptiste Longuet, see: *ANF*—F[17] 17274—documents 13, 17, 20.

579 See for instance: *ANF*—F[17] 17274—document 22.

580 See for instance: *ANF*—F[17] 17274—documents 35, 36.

581 *ANF*—F[17] II [F17/02(A)], F[17] 17274, LA BAUME-PLUVINEL: Missions gratuites en Espagne (éclipses du 1900 et 1905), Senegal (16 avril 1893), Sumatra (mai 1901), Égypte (11 Novembre 1901), Crimée (Théodosie, 21 août 1914); *Mission gratuite en Espagne à Elché (eclipse 1900).*

582 *ANF*—F[17] II [F17/02(A)], LA VAULX : Mission aéronautique à Constantine à l'effet d'observer à bord de son aérostat l'éclipse totale de soleil du 30 août 1905.

583 *ANF*—F[17] II [F17/02(A)], F[17] 17278, MASCART : Mission gratuite à l'île de Ténériffe à l'effet d'y poursuivre des observations astronomiques relatives à la comète Halley, 1910.

584 *ANF*—F[17] II [F17/02(A)], F[17] 17292, VLÈS : Demande de mission gratuite en Espagne et au Portugal pour etudier et cinématographier l'éclipse de soleil du 17 avril 1912.

the expedition; the logistical work of researching observation sites, making travel plans, finding local volunteers, and arranging accommodation; and the technical work of choosing instruments, deciding what kind of observations should be made, and who should go into the field.[585]

Local contacts were very important to properly organised expeditions, both amateur and institutional, as they could provide first-hand information. In particular, the comments and suggestions of good local astronomers and meteorologists played a crucial role in the final decision of where eclipse expeditions should be sent. For the eclipses of 1900 and 1905, this interaction proved very fruitful for Spanish astronomers as well. Spanish scholars built strong links with several foreign individuals and institutions that opened the doors to their integration into the international astronomical community. There are many good examples: the correspondence between the organisers of the official Swiss mission, commanded by Raoul Gautier, Director of the Observatory of Geneva, and the Spaniard Eduard Fontserè, Professor of Mechanics at the University of Barcelona[586]; the collaboration between the French astronomer Auguste Victor Lebeuf (1859–1929) and Francisco Iñiguez, Director of the Observatory of Madrid[587]; and the helpful advice given by Lànderer to several foreign parties, including those from the U.S. Naval Observatory, the one led by Janssen,[588] and from the expedition mounted by the Société

[585] The British case has been studied in depth by Pang in *Empire and the Sun* (Stanford, 2002). Particularly relevant is Chapter 2. 'Planning eclipse expeditions', pp. 11–48.

[586] Gautier, 'L'Éclipse totale de soleil du 30 aout 1905 …', *Archives des Sciences Physiques et Naturelles*, 20 (1905), 650–674.

[587] Lebeuf and Chofardet, 'Observation de l'éclipse totale de Soleil du 29–30 août 1905 …', *Annales du Bureau des Longitudes*, 8 (1911), 1–47.

[588] These expeditions followed Lànderer's suggestions, published in several journals during the years leading up to the eclipse: 'L'Éclipse totale de soleil des 29–30 aôut 1905', *Astronomische Nachrichten*, 159 (1902), 125–130 [an extract from this paper was presented by A. M. W. Downing under the title 'The total solar eclipse of August 1905', *Journal of the British Astronomical Association*, 14 (1903–1904), 367–368]; 'L'Éclipse totale de soleil des 29–30 aôut 1905 (Duexième note)', *Astronomische Nachrichten*, 160 (1903), 309–312. See also his publications 'L'Éclipse totale de soleil du 30 août 1905', *Bulletin de la Société Astronomique de France*, 17 (1903), 105–108; and 'El eclipse total de sol del 30 de agosto de 1905', *La Ilustración Española y Americana*, 22 (1902), 375–378. The American expedition went to Mont Colibre, in the Columbretes islands, while the French one went to Alcossebre. Lànderer stayed with the latter and organised the setting up of the instruments. As Janssen wrote in his paper 'Sur l'éclipse solaire

Astronomique Flammarion of Montpellier.[589] In the case of the British Astronomical Association, Chambers also had correspondence with the Director of the Observatory of Madrid, Francisco Iñiguez, whose help and comments greatly assisted the organisation of the eclipse expedition to Spain in 1905. As Chambers reported to the Association:

> With respect to the climate of the mainland, Señor Iñigues [sic] evidently thinks that an undercurrent alarm ran through my original remarks, which is without sufficient foundation so far as regards any excessive inconvenience to be feared from great heat. He says: "I can assure you that by August 30 the great heat of summer is quite over, and that the temperature is below that of the continent generally at that time, and this is especially the case in the north of Spain and in the Balearic Isles; indeed, you would do well to caution your friends who might come to Burgos, Soria, &c., to provide themselves with overcoats, for they will certainly require them in the evenings." The Director has sent me a book of Meteorological Observations for the years 1895 and 1896 (the latest published).[590]

All these contacts were well maintained, and correspondence flowed back and forth on a regular basis. Local contacts were very important for foreigners both for scientific information and for the purposes of tourism. As has been said above, the Director of the Observatory of Madrid was in touch with several individuals and institutions, such as members of the British Astronomical Association, to provide them with as much information as was needed for the organisation of eclipse expeditions in 1900 and 1905. In some cases the information provided transcended merely scientific aspects. For example, in connection with the eclipse of 1905, Iñiguez sent a letter to the British Astronomical Association on 13

---

totale du 30 août 1905', *Comptes Rendus*, 141 (1905), 233–234, the destination of the expedition was chosen 'par mon éminent ami, l'astronome espagnol, M. Lànderer, dont le nom est bien connu de tous les membres de l'Académie'—see also Janssen, 'Observation de l'éclipse totale du 30 août 1905 à Alcosebre (Espagne)', *Comptes Rendus*, 141 (1905), 569–571; and in the Archives Nationales de France, F[17] 13571, Bureau des longitudes. Affaires scientifiques diverses 1881–1922.

[589] This last expedition had also the help of the Catalan Félix Via, from Amposta. See: Moye, *À la poursuite d'une ombre* ... (Montpellier, 1905).

[590] Chambers, 'The total eclipse of the Sun of 1905', *Journal of the British Astronomical Association*, 12 (1901–1902), 280–281. The Association also benefited from Iñiguez's comments on weather conditions for the eclipse of 1912. See: Crommelin, 'Further note on the eclipse of 1912', *Journal of the British Astronomical Association*, 22 (1911–1912), 226.

February 1904 in which his comments on the weather prospects also included the following advice:

> Undoubtedly, the temperature of the northern coast of Spain will be lower than that of the southern coast at the date of the eclipse; but I do not think that there need be any doubt as to the temperature of the south coast being quite supportable and tolerable for English people.[591]

10.2. Eclipses and tourism

In February 1899, Chambers presented to the British Astronomical Association some of the results of his research on travelling arrangements, steamer and rail routes, fares, hotels, inland travelling, and practical points "of a very essential character, as regards the comfort and convenience of those who would travel" to the path of totality in 1900.[592] He suggested that the length of the journey (the Spanish rail network was not very extensive) should be profited from to see some of the beauties and sights of the country, including cities such as Granada, Seville, Cordova, and Burgos. In his opinion, the more they travel by sea, covering only short distances inland from the coast, the more they would enjoy the trip. Some of these arguments, based on maximising the travellers' comforts, conflicted considerably with scientific requirements. Chambers himself recognised the strong touristic character of expeditions such as the ones organised by the British Astronomical Association, pointing out that

> earnest astronomers would have to rough it, because they would wish to go far inland, while those who went for pleasure would enjoy themselves most by keeping near the coast.[593]

The Story of the Eclipses, a book by Chambers, had an appendix for travellers wishing to visit Spain and Portugal to observe the eclipse of 1900 which dealt solely with geographical and transport matters.[594] The

---

[591] 'Report of the Meeting of the Association, held on May 25, 1904, at Sion College, Victoria Embankment', *Journal of the British Astronomical Association*, 14 (1903–1904), 303–315, here pp. 305–306.

[592] 'Report of the Meeting of the Association held on February 22, 1899', *Journal of the British Astronomical Association*, 9 (1898–1899), 193–200, here pp. 193–197.

[593] *Ibid.*, pp. 193–197.

[594] George F. Chambers, *The Story of the Eclipses* (London: George Newnes, 1899), pp. 239–252.

study and analysis of the routes to be followed and the methods of travel was a very important non-scientific part of the organisation of any eclipse expedition. In many cases, the organisers of expeditions had to negotiate with steamship companies to arrange special trips, with travel agents, with foreign governments (about relaxing customs regulations and other formalities), and even with the British Government (to obtain facilities such as free transport in one of Her Majesty's ships).[595] The administrative arrangements for unofficial eclipse observers were just as much of a problem as they were for official eclipse expeditions. For example, the British Astronomical Association planned to travel to Majorca to observe the eclipse of 1905, but found out that their ocean-going steamer could not enter the island's harbour, and that astronomers would have to make the steamer their hotel because no answer had been received from inquiries about hotel accommodation.[596] And the 1905 eclipse expedition organised by Norman Lockyer intended to anchor a gunboat that was at their disposal in the harbour of Philippeville (Algeria), but the French objected to the presence of a warship of a foreign power in their waters, and the South Kensington expedition had to go instead to Palma de Majorca.[597]

The Eclipse Committee of the British Astronomical Association was in charge of researching and providing information to the members of the Association about the variety of options and costs for travelling to the path of totality to observe the solar eclipse. Many reports were sent to the Association pointing out the most suitable stations from where the total eclipse could be viewed and itineraries to be followed from Britain.[598] With this information obtained, the Eclipse Committee began to organise expeditions. For instance, for the eclipse expedition of 1900, the members of the Association were informed on June 1899 about the following

---

[595] See for instance: 'Report of the meeting of the Association held on March 29, 1899', *Journal of the British Astronomical Association*, 9 (1898–1899), 241–251, here p. 247.

[596] G. F. Chambers, 'The coming total eclipse of the Sun', *Journal of the British Astronomical Association*, 15 (1904–1905), 164–166. For some of the research on accommodation, see for instance: Chambers, 'The total eclipse of the Sun of April 17, 1912', *Journal of the British Astronomical Association*, 22 (1911–1912), 92–99.

[597] '[Notes:] The total solar eclipse', *The Observatory*, 28 (1905), 328–330.

[598] See for instance: Alexander Burton-Brown, 'Notes on stations for viewing the total eclipse of the Sun. 29–30 August 1905', *Journal of the British Astronomical Association*, 15 (1904–1905), 333–334; and 'Astronomical publications: The August total eclipse of the Sun', *Journal of the British Astronomical Association*, 15 (1904–1905), 211.

passage rates: return to Lisbon or Oporto, £10; to Cadiz, giving access to the centre of Spain and back, £14; to Alicante, £16; and to Algiers, £22 10s.[599] In February 1900, the Association received from the Royal Mail Company an offer to use the steamer *Tagus* for £3600. It was reported to the Association that

> of this sum half must be paid down on signing the contract. The steamer will carry 189 passengers as we propose to arrange the berths. About 80 have already engaged berths and paid half fares as deposit, giving about 1,000 [pounds] in all ... We have now to give a final answer to the Royal Mail Company immediately ... If only two-thirds of the berths now vacant be finally taken, the guarantors would have their money returned in full. The Committee would ask the Members to give their special attention to the great increase of usefulness and reputation which would accrue to the Association if the proposed expedition were successfully carried out, and to the corresponding loss of prestige which is to be feared if the enterprise has to be dropped from lack of support.[600]

So, eclipse expeditions were clearly more than scientific endeavours. In some cases it was simply a way for the wealthy to gain prestige—these expeditions were, indeed, cultural expressions. The touristic dimension of these eclipse expeditions is well illustrated by the plans for the tour as initially conceived by Chambers for the eclipse of 1905. As Chambers himself pointed out, it was "one week for the outward voyage, nearly two weeks for business and sightseeing, partly on land and partly at sea in Spanish waters; and the inside of a week for the homeward voyage"[601]:

Aug. 21 Monday	Leave —. (Tilbury, *or* Liverpool?)
" 22 Tuesday	Call at Havre to pick up passengers from Southampton and the Continent
" 23 Wednesday	At sea

599 Friends of members would pay in guineas instead of pounds (1 guinea = £1 1s). See: 'Report of the meeting of the Association held on June 28, 1899', *Journal of the British Astronomical Association*, 9 (1898–1899), 357–365, here pp. 363–364. Note that the exchange ratein 1900 was 32.6 pesetas to 1 sterling pound and 1 peseta to 1.3 French francs.

600 Circular from the British Astronomical Association, 'The total solar eclipse of May 28th, 1900: Proposed expedition to Spain and Algeria', attached to *Journal of the British Astronomical Association*, 10 (1899–1900), attached to issue No. 4.

601 G. F. Chambers, 'The total eclipse of the Sun, 30th August 1905', *Journal of the British Astronomical Association*, 15 (1904–1905), 93–98. In order to convert sterling pounds into pesetas, the exchange rate in 1905 was 32.9 pesetas to 1 sterling pound.

" 24 Thursday	Call at Lisbon to pick up passengers from Portugal
" 25 Friday	Pass Gibraltar
" 26 Saturday	Arrive at Valencia
" 27 Sunday	At Valencia
" 28 Monday	At Valencia. Day trip to Alicante and go ashore to see the palm trees at Elche …
" 29 Tuesday	At Valencia
" 30 Wednesday	At Valencia. Day trip to central line of eclipse, 30 miles N. of Valencia
" 31 Thursday	At Valencia
Sept. 1 Friday	Leave Valencia
" 2 Saturday	Call at Malaga to land passengers for Granada, &c.
" 3 Sunday	Arrive at Gibraltar
" 4 Monday	At Gibraltar. Inland day trip to Ronda by rail
" 5 Tuesday	Gibraltar to African coast, Tangiers, &c.
" 6 Wednesday	At Gibraltar
" 7 Thursday	At sea
" 8 Friday	Call at Cadiz to take up the Granada, &c., Section
" 9 Saturday	Call at Lisbon, and wait a few hours
" 10 Sunday	At sea
" 11 Monday	Call at Havre
" 12 Tuesday	Arrive —.- (Tilbury, *or* Liverpool?)

Programme of the Land Journey viâ *Granada, and Seville*

Sept. 2 Saturday	Land at *Malaga* and sleep there
" 3 Sunday	At *Malaga*
" 4 Monday	Malaga to *Granada* (119 miles)
" 5 Tuesday	At *Granada*
" 6 Wednesday	At *Granada*
" 7 Thursday	Granada to *Seville* (172 miles)
" 8 Friday	Seville to *Jeres* (64 miles)
" 9 Saturday	Jeres to Cadiz (31 miles), and return to the steamer

Special circular tickets are issued by the Spanish railways in books capable of being split up into coupons of 5 kilometres and upwards. The charge for batches of such tickets, beginning with 3000 kilometres (say, 1860 miles) is 247 pesetas 1st class; 185 pesetas 2nd class; equal to about £7 5s. and £5 7s. respectively; which portioned out for the use of parties of six comes to about 24s. or 18s. per passenger for the round from Malaga to Cadiz. To this must be added board and lodging, say, 10s. each per day, or £3 10s. for the 7 days. The cost of sight-seeing and local travel must of course be added, *ad. lib.*

As the President of the British Astronomical Association himself said, a visit to Spain need not to be confined to astronomical work: there was a great deal of pleasure to be derived from the scenery and the architectural monuments; Spain was

> a very attractive country to all who were not too fastidious as regards their bodily comforts. The people were very pleasant but exasperatingly dilatory; and no language but Spanish was of any use, even in the larger towns. Travelling, whether by rail or diligence, was very tedious; and though comfortable hotels might be found in such places as Oviedo, León, Paléncia, Búrgos, and Teruel, yet in the remote villages the inns were still as in Don Quixote's day.[602]

Eclipse expeditions organised by institutions such as the British Astronomical Association have to be understood in terms of tourism, scientific curiosity, and upper-class leisure.

10.3. Eclipses and business

Making eclipses commercially and intellectually accessible helped to establish the cultural legitimacy of astronomy and astrophysics. As Alice Walters has explained, not only were the astronomical broadsides (publications which predicted or explained different astronomical phenomena to the general public) produced in Britain during the eighteenth century an important medium for the consumption of scientific knowledge, but they also mobilised public participation in the observation of an event such as an eclipse: "these prints emphasized the public, open character of the new science, increasing popular awareness of, and interest in, astronomy and the astronomical community".[603] At the same time the producers of these broadsides had an opportunity for both personal promotion and the advertisement of related goods. Similarly, the new social and cultural context of astronomy in the late nineteenth and early twentieth centuries led to a market for astronomical material, catalysed by the interest that solar eclipses excited in the population, particularly in the upper classes. During the months leading up to the eclipse, manufacturers advertised in different journals and magazines the most suitable products

---

[602] 'Report of the Meeting of the Association, held on January 25, 1905, at Sion College, Victoria Embankment', *Journal of the British Astronomical Association*, 15 (1904–1905), 155–161, here p.159.

[603] Alice N. Walters, 'English broadsides of solar eclipses', *History of Science*, 37 (1999), 1–43.

to best observe the event. Figure 21 is a good example: an advertisement of Sandell's “perfect” plates for astronomical and microscopic photography. Firms such as T. Thorp & Co., of Whitefield, Manchester, advertised themselves as experts in supplying instruments for observing eclipses. Another example is the map of the whole of Europe, showing the path of totality and lines of constant phase, published by the French Bureau des Longitudes and on sale for two and a half French francs .[604]

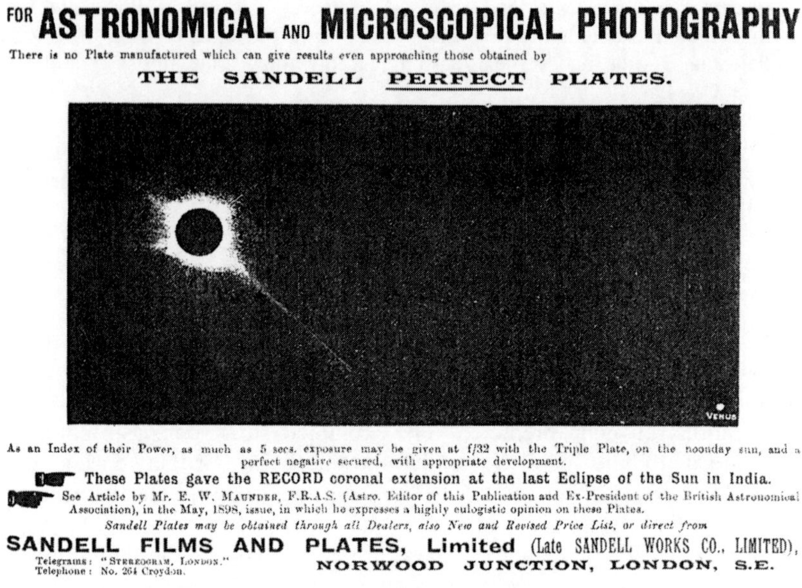

Fig. 10-1: Advertisement in the journal *Knowledge* on 1 March 1900, p. vi.

Travel agents also developed tours in 1900 and 1905 based around observing the eclipse from the path of totality. Figure 22 shows an advertisement by Thomas Cook & Son for a tour from London to Talavera, visiting Paris, Bordeaux, Biarritz, and Madrid, for £30, including tickets, accommodation, and all fees.[605] These tours were widely advertised in popular scientific journals. Some steamship companies advertised that steamers on regular services which brought them to the

[604] ‘[Notes:] The total solar eclipse’, *The Observatory*, 28 (1905), 328–330. By 1905 one sterling pound was about 42.8 French francs.

[605] For this particular example, see also: ‘[Notes:] The coming solar eclipse of May 28’, *The Observatory*, 23 (1900), 147.

central line at the time of the eclipse would heave to for a few hours to give passengers the opportunity to observe the event.[606]

For the eclipse of 1905 three different possible itineraries were presented by Walter Maunder to the British Astronomical Association: a twenty-strong party expedition organised by the Eclipse Committee, following the programme arranged by Thomas Cook, for 30 guineas; a route via Barcelona and Valencia to the east coast; and a tour to Majorca, also arranged by Cook, for 33 guineas.[607] A tour to Burgos to observe the 1905 eclipse was also organised, to leave London on 25 August 1905; the cost of 30 guineas would cover all necessary expenses from London and back, "with accommodation at the best hotel in Burgos".[608] After some negotiations, an advantageous arrangement was made for twenty members of the Association to observe the eclipse from Burgos, including hotel accommodation and full board at Burgos for ten days, and a first-class rail travel from London to Burgos and back, through Boulogne, Paris, and Bordeaux, for the sum of 18 guineas.[609]

The eclipse expedition business was not a British monopoly. Firms such as the Italian Steamship Company, of Genoa, also arranged excursions to the path of totality to observe the 1905 eclipse. The programme that the Italian Company advertised began with departure from Genoa on the evening of 23 August and arrival at Barcelona on 25 August. After a day or two in Barcelona for sightseeing, the party was expected to reach Palma on 28 August. No hotel accommodation was arranged, and passengers had to remain on the vessel overnight. After leaving the island on 31 August, the ship would return to Genoa on 2 September 1905. The charges included food and wine for the trip and ranged from 200 to 300 Italian lire (between about £8 and £12) according to the class of accommodation on board. The Company also offered another route via Marseilles, including some nights of hotel accommodation, at slightly

---

606 See for instance: G. F. Chambers, 'Proposed eclipse expedition to the Mediterranean', *Journal of the British Astronomical Association*, 15 (1904–1905), 384–385.

607 E. Walter Maunder, 'The total solar eclipse of August 30, 1905', *Journal of the British Astronomical Association*, 15 (1904–1905), 240–242.

608 '[Notes:] The total solar eclipse of August 30 next', *The Observatory*, 28 (1905), 70.

609 See the "Important" note from the British Astronomical Association about the 'Total Solar Eclipse of 1905, August 30' attached to the *Journal of the British Astronomical Association*, 15 (1904–1905), issue No. 5.

higher cost.[610] Similarly, the Peninsular and Oriental Company arranged to send a vessel by way of Gibraltar and thence to Marseilles.[611]

Fig. 10-2: Advertisement in the *Journal of the British Astronomical Association*, 15 (1904–1905) [issue number 7].

10.4. The social impact of eclipse expeditions in Spain

In Spain, passion for eclipses exploded in 1900. A good example is the success of the special trains sent to the path of totality by the various train companies. In this initiative, which had no precedents, either in the United States and the rest of Europe or elsewhere, cheap one-day return tickets to the path of totality were offered to ordinary Spaniards. Some routes and fares are given in Table 10-1.[612]

---

[610] '[Notes:] The total solar eclipse of August 30', *The Observatory*, 28 (1905), 260–261.

[611] '[Notes:] The total solar eclipse', *The Observatory*, 28 (1905), 328–330.

[612] This information is illustrative and not complete. See: 'El eclipse de Sol', *El Imparcial*, 22 May 1900, 1; and 'El eclipse de Sol', *El Imparcial*, 23 May 1900, 1.

Table 10-1: Some train routes and fares to the path of totality on 28 May 1900

Departure	Arrival	Price
Madrid	Navalmoral de la Mata	3 pesetas (for groups of more than three people)
	Argamasilla	Luxury train: 25 pesetas
	Alcázar	Luxury train: 20 pesetas
		1st class: 12 pesetas 2nd class: 8 pesetas 3rd class: 4 pesetas
All the stations from Cartagena to Beniaján	Hellín	1st class: 12 pesetas 2nd class: 8 pesetas 3rd class: 5 pesetas
All the stations from Murcia to Cieza		1st class: 10 pesetas 2nd class: 7 pesetas 3rd class: 4 pesetas
Linares, Vadollano, Vilches, Santa Elena, and Cárdenas	Manzanares	1st class: 10 pesetas 2nd class: 7 pesetas 3rd class: 4 pesetas
Valdepeñas		1st class: 4 pesetas 2nd class: 3 pesetas 3rd class: 2 pesetas

By 26 May, more than 800 tickets had been sold to travel from Madrid to Navalmoral de la Mata (Cáceres) in first, second, and third class trains. One day later, the day before the eclipse, more than 4000 tickets from Madrid to Navalmoral had already been sold.[613] Even tickets for the luxury train rapidly sold out and more had to be put on sale; in all about 600 luxury train tickets were sold for Alcázar de San Juan (Ciudad Real) and Argamasilla de Alba (Ciudad Real).[614] But all expectations were exceeded on the day of the eclipse. Some tickets were put up for sale with an extra charge of 25 per cent of their original cost.[615] On the morning of 28 May 1900, starting at 5.00 a.m., several trains left Madrid for the path of totality. Three trains left early in the morning from Delicias for Navalmoral, with more than a thousand passengers each. Five trains left from Madrid for Navalmoral, along the Madrid–Cáceres–Portugal railway, on the day of the eclipse.[616] After the first three left, a fourth luxury train

---

[613] 'No hay billetes', *El Heraldo de Madrid*, 27 May 1900, [año 11, nº 3484], 2.
[614] 'Para el eclipse: Servicio de trenes', *El Heraldo de Madrid*, 26 May 1900 [año 11; nº 3483], 2.
[615] 'El eclipse', *El Imparcial*, 29 May 1900, 1.
[616] I am very grateful to Juan Pedro Esteve García for information about the train departures from Madrid.

with more than 400 passengers, and a fifth train carrying about a thousand, also departed for the path of totality.[617] Three more trains left from the station at Atocha for Alcázar and Argamasilla, along the Madrid–Zaragoza–Alicante railway, with 1900 passengers between them, and later a luxury train departed with about 700 people.[618] In all, about 10,000 to 12,000 people left Madrid by train on the day of the eclipse for various locations in the path of totality, such as Alcázar, Argamasilla, and Navalmoral.[619] Similar initiatives were also taken by several train companies for the eclipse of 1905 to encourage these "one-day expeditions".[620]

On the days leading up to the eclipse of 28 May 1900, inns were full in cities located in the path of totality, such as Plasencia.[621] Some Spanish innkeepers had special offers during the day of the eclipse. For example, the innkeeper of Alcázar offered for one and a half pesetas a menu which included soup, one fried dish, hake, chicken and rice, bread, wine, and desserts.[622] Opticians also made good money out of the eclipse. For example, the firm Casa de Oliva sold more than ten thousand "observe-eclipses" devices at a price of one peseta each.[623] And more than a hundred thousand pieces of smoked glass were sold for one and two *perras gordas* (ten-cent coins) by hawkers in Madrid on the day of the eclipse.[624] Several books and leaflets in Spanish related to the solar eclipses of 1900 and 1905 were also on sale.[625]

---

[617] 'El eclipse total', *El Imparcial*, 29 May 1900, 1–2.

[618] 'El Eclipse de Sol', *El Heraldo de Madrid*, 28 May 1900, [año 11, nº 3485], 1.

[619] 'Día de tregua', *El Liberal*, 29 May 1900, 1. See also: 'Trenes baratos', *El Heraldo de Madrid*, 23 May 1900, [año 11, nº 3480], 3.

[620] See for example: 'El eclipse en Burgos', *El Imparcial*, 27 June 1905, 2; 'Las fiestas del eclipse', *El Imparcial*, 29 June 1905, 2.

[621] 'El eclipse', *El Imparcial*, 27 May 1900, 2.

[622] 'El eclipse de Sol', *El Imparcial*, 22 May 1900, 1.

[623] 'El Eclipse de Sol', *El Heraldo de Madrid*, 28 May 1900, [año 11, nº 3485], 1.

[624] 'Telescopios a perra gorda', *El Liberal*, 29 May 1900, 2.

[625] For example, the leaflet *El eclipse solar del 28 de Mayo de 1900* was on sale for ten cents at the La Irradación library. See: 'El eclipse de Sol', *El Imparcial*, 22 May 1900, 1. For the 1905 eclipse, José Ruiz Castizo and Gabriel Galán wrote *El eclipse de Sol de 1905* (Madrid: Adrian Romo, 1905). Comas i Solà also had on sale a leaflet in which he condensed all the relevant information on the eclipse of 1905 for amateur observers. See: 'El Próximo eclipse total de Sol de 30 de Agosto. Recuerdos interesantes', *El Mundo Científico*, 7 (1905), 517–520, here p. 518. A brief note on the most useful texts published before the eclipse of 1905, including those produced by the Ebro Observatory and the observatories of Madrid and San

So, the eclipses were good business not only for instrument-makers and travel agents. As reported in *The Observatory*, the night before the eclipse of 30 August 1905 the price for bed and breakfast in Burgos was £4.[626] The business generated by the eclipse was also described in a Spanish newspaper:

> A large number of *professionals* and an even larger number of learned amateurs and curious people of all kinds are arriving at Burgos to observe the solar eclipse, and some natives of Burgos, trying to make the occasion as successful and profitable as possible, aim to strip the very skin from the Spaniards and foreigners in their grip.
>
> They have taken from a Yankee—from what we have been told—the unbelievable quantity of 1800 pesetas for *three days'* accommodation …
>
> Considering the immense publicity that this eclipse has afforded Burgos, the natives should be pleased to provide hospitality, for their own dignity and for that of Spain, and not take too much advantage of the occasion.[627]

All in all, the eclipses of 28 May 1900 and 30 August 1905 had a large impact on Spanish society. The expected visits of famous foreign astronomers were recorded in all Spanish newspapers during the months before the eclipses. The facilities and resources provided by the Spanish government to foreign astronomers wishing to observe eclipses were expected to project an image of modernity of great value to the country.[628] Astronomy became a topical subject with Spaniards, many of whom became interested in science and particularly in astronomy. As Flammarion wrote after his visit of 1900:

> Bornons-nous à remarquer aujourd'hui le mouvement considérable causé par ce grand phénomène. Pendant quinze jours, l'Espagne n'a été occupée tout entière que de l'éclipse. Du plus haut de l'échelle sociale jusqu'aux classes les plus humbles, on a voulu assister à cette grande leçon de cosmographie … Le progrès est lent dans l'humanité, assurément, mais il existe, et … il y a une différence immense entre ce qui s'est passé en Espagne en 1860, lors d'une éclipse analogue, et ce qui vient d'arriver sous

---

Fernando, as well as by individuals such as Lànderer, Comas i Solà, and Ruiz Castizo and Galán, is found in: *Observaciones del eclipse … en el Colegio de Oña* (Oña, 1906), pp. 6–7.

626 'The total solar eclipse of August 30, 1905', *The Observatory*, 28 (1905), 377–382.

627 'Mirando al cielo. El Eclipse de Sol', *El Universo*, 29 August 1905.

628 See for instance the Royal Order published by the Finance Ministry: 'Ministerio de Hacienda. Real Orden', *Gaceta de Madrid*, 11 May 1905 [num. 131], 554.

> nos yeux. Les intelligences s'élèvent, s'instruisent avec curiosité, apprécient la valeur de la science, sont altérées de vérité et de lumière.[629]

Restaurant

DE LA

estación de Alcázar de San Juan

Menú del 28 de Mayo

El fondista de la estación de Alcázar tiene el gusto de ofrecer á los señores viajeros que acudan á dicho punto á presenciar el **ECLIPSE TOTAL DE SOL,** un almuerzo que costará únicamente

Pesetas : 1,50 cubierto,

y se compondrá de los siguientes platos:

Sopa
Un frito
Merluza en salsa
Pollo con arroz á la valenciana
Pan, vino, postres.

Alcázar, 18 de Mayo de 1900.

Fig. 10-3: Menu offered by the innkeeper of Alcázar with occasion of the eclipse of 28 May 1900.

What Flammarion is suggesting here is the approach taken in the second part of this book: to study the progress of science in Spain, particularly of astronomy, through the history of total solar eclipses visible from the country. Similarly, the French astronomer, meteorologist, and populariser of science, Abbot Théophile Moreux (1867–1954), wrote the following words about his experience in Spain in 1900:

> Bercé par la lenteur proverbiale des chemins de fer espagnols, j'avais eu le temps de lire les journaux et de constater que l'éclipse occupait toutes les colonnes. "Mieux vaut cela qu'un récit de course de taureaux, me disais-je. Décidément l'Espagne se passionne pour la science. Là est peut-être le signe de son relèvement social." Les journaux illustrés firent comme leurs

---

[629] Flammarion, 'L'Éclipse totale de soleil', *Bulletin de la Société Astronomique de France*, 14 (1900), 289–297, here p. 297.

confrères, et les caricaturistes, pendant plusieurs semaines, durent aiguiser leur verve et leur talent sur l'éclipse.[630]

The region that attracted the most observers for the total solar eclipse of 1900 was the east coast of Spain, in particular the villages of Elche and Santa Pola, where both Flammarion and Moreux travelled to. Here were also located the expeditions from the observatories of San Fernando, the Vatican, Paris, and the universities of Montpellier and Toulouse. Such a large concentration of foreign astronomers attracted many Spaniards to these locations. The best example is that of Comas i Solà, who passed by Elche on his way to Hellín to observe the eclipse from there; but, seeing so many prestigious astronomers in the area, he decided to stay in Elche.[631] In total, some ten thousand visitors—twice its population—descended on Elche for the eclipse.

Almost every Spanish journal and newspaper wrote enthusiastically about the solar eclipses, including suggestions and advice for amateur observers.[632] In fact, many of them sent correspondents to several towns to report on the event. Others contacted local amateur journalists to write articles and notes for them.[633] For example, *El Correo*, a newspaper in Valencia, sent Eduard de Bonet to Elche.[634] Similarly, Pere Ibarra i Ruiz (1858–1934) wrote some articles for the Valencia-based *Las Provincias*.[635] Other newspapers and periodicals, including *Catalònia*, *El Imparcial*, *Heraldo de Madrid*, *La Unión Democrática*, *Las Provincias de Levante*, *La Correspondencia de España*, *Nuevo Mundo*, *Ilustración Española*, *Blanco y Negro*, *El Noticiero*, *Mar y Tierra*, *El Orbe Católico*, and *El*

---

630 T. Moreux, 'Autour de l'éclipse. Notes de voyage', *Bulletin de la Société Astronomique de France*, 14 (1900), 308–324; p. 308.

631 'En visperas del eclipse', *El Heraldo de Madrid*, 26 May 1900, [año 11, nº 3483], 3.

632 See for instance: 'Para el eclipse de mañana. Guía práctica del observador', *El Heraldo de Madrid*, 27 May 1900, [año 11, nº 3484], 1.

633 Ibarra, 'Desde Elche', *Las Provincias*, 23 May 1900. From: Soler Selva (coord.), *L'eclipsi total de sol de 1900 ...* (Elx, 2000).

634 Moreux, 'Autour de l'éclipse. Notes de voyage', *Bulletin de la Société Astronomique de France*, 14 (1900), 308–324, here p. 311.

635 Ibarra, 'Desde Elche. Multitud de sucesos y escasez de cronistas ...', *Las Provincias*, 23 May 1900; and Ibarra, 'Desde Elche. La fotografía y el eclipse ...', *Las Provincias*, 27 May 1900. The director of this newspaper, Teodor Llorente, finally went to Elche himself to observe the eclipse, and wrote 'El eclipse en Elche', *Las Provincias*, 30 May 1900. All these articles are to be found in: Soler Selva (coord.), *L'eclipsi total de sol de 1900 ...* (Elx, 2000).

Mundo Científico published notes and articles. Table 10-2 illustrates the large impact of the eclipses of 1900 and 1905 in the Spanish press. It shows the unusually large number of contributors that the Madrid-based newspaper *El Liberal* had all around the path of totality for the eclipse of 1900.[636]

Table 10-2: Contributors to *El Liberal* for the eclipse of 28 May 1900

Contributor	Location
Bernardino Torres González	Manzanares
Jose María López Campello	Elche
Joaquín Dicenta	
Orozco	Santa Pola
Miguel Aguilar	Plasencia
Carlos Puente	
Carlos del Río	Argamasilla de Alba
Mariano Martín Fernández	
Carlos Miranda	Alcázar de San Juan
Niño	Daimiel
Eduardo Rosón	Navalmoral de la Mata
Antonio R. Lázaro	

Spanish periodicals reported almost everything related to the solar eclipses of 1900 and 1905 during the days before these events. The Spanish press, indeed, played a crucial role in the catalysing effect of the total solar eclipses of 1900 and 1905 on the development of astronomy and astrophysics in Spain during the early twentieth century:

> The press, the opportunism of publishers, and commercial temptation have spread leaflets, descriptions, prints, and optical devices all over the country to enable the progress of the eclipse to be observed in all its detail—all its rays, protuberances, and coroniums … Interesting popular texts such as those by Comas Solá, good versions of the very valuable works by Flammarion, and other similar ones are purchasable for a few centimes … Some of this will remain in the people's souls, and may ferment into higher instruction. Even if it does not, it will always exalt and strengthen us, and brighten our national pessimism, like an aurora, to see people, even if just circumstantially, becoming involved with Science.[637]

[636] 'Antes de la batalla', *El Liberal*, 28 May 1900, 1.

[637] 'A. A.', 'El eclipse de Sol de 30 de Agosto', *El Mundo Científico*, 7 (1905), 544–547, here p. 547.

At the same time, the Spanish press asked the most prestigious scientists and astronomers, both local and foreign, to explain the importance of their astronomical researches.[638] The solar eclipses of 1900 and 1905 marked a turning point not only for Spanish scientists, but also for Spanish journalists, who began to take interest in scientific subjects and to report scientific news in general periodicals.

The *regenerationist* movement promoted not only scientific research in Spain but also interaction between Spanish and foreign scholars. The visit of well-known astronomers to Spain from all over the world in 1900 and 1905 helped to establish a network which encouraged Spanish scholars to develop their own scientific programme in their country. Consequently, progress in astronomical research in Spain accelerated after the total solar eclipses of 1900 and 1905, particularly as a result of the efforts made to emphasise the significance of these events to the Spanish population.

Foreign scientists were eagerly received and hosted by the Spanish middle and upper classes—to do so was understood by many as a national obligation.[639] The middle and upper classes and Spanish scientists helped them in every possible way. For example, for the official party of English and Irish located in Plasencia in 1900, being next to the Spanish official party proved of great assistance, as

> it enabled them to have a reserve ground kept which the Spanish party had secured for the occasion; and also to have a detachment of the civil guard on the spot to keep off the "madding crowd", so that they could pursue their observations without molestation … The Marquis of Mirabel, a Spanish nobleman, who had a palace in Plasencia … as soon as he heard that Mrs. Downing was to accompany him (the speaker) to Plasencia, he, with the greatest hospitality and courtesy, offered to place his palace in

---

[638] The 26 May 1900 edition of the newspaper *El Heraldo de Madrid* published on is a great example, containing in its pages: José Echegaray, 'El Sol y el eclipse', *El Heraldo de Madrid*, 26 May 1900 [Año 11, nº 3483], 1; Deslandres, 'Origen eléctrico de las protuberancias, de la cromosfera y de la corona solar', *El Heraldo de Madrid*, 26 May 1900 [Año 11, nº 3483], 5; De la Baume Pluvinel, 'Fotografías del eclipse', *El Heraldo de Madrid*, 26 May 1900 [Año 11, nº 3483]; Victoriano F. Ascarza, 'El análisis espectral', *El Heraldo de Madrid*, 26 May 1900 [Año 11, nº 3483], 6. The contributions of other important foreign astronomers such as Flammarion will be presented later in this chapter.

[639] See the Royal Order published in: 'Real Orden Circular', *Gaceta de Madrid*, 11 August 1905 [nº 22], 548.

Plasencia at their disposal, as he realised that there were no hotels or inns in the town fit for a lady to go.[640]

The Spanish authorities received foreign observers with the utmost courtesy and assured them of all possible assistance. A good example is that of the British expedition to Spain led by Callendar in 1905. After the first members of this expedition arrived at Castellón,

> the work of clearing the ground, building piers for the instruments, and erecting huts was commenced without delay, under the supervision of Don Francisco Tomás, a local architect, who generously performed his own part of the work gratuitously. During the preparations the Military Governor (His Excellency General Don Juan Manrique de Lara), the Civil Governor (His Excellency Señor Don Sanchez Ortiz), and the Mayor of Castellón made constant inquiries, either personally or by messenger, as to the needs of the expedition; and, in addition, Señor Don José Marza, Town Councillor, was almost constantly in attendance at the camp to ensure that nothing was wanting to facilitate the work.[641]

The attention paid to everything related to these solar eclipses boosted astronomy in Spain. In the words of Fontserè, writing in 1900, these eclipses were the "signal" that helped to create a "compact group" of people interested in astronomy and astronomical research in Spain.[642] Science began to acquire a higher status in Spanish society. Hosting foreign scholars became a privilege for many Spaniards, and the Spanish middle and upper classes were happy to make facilities available to prestigious observers, particularly to those from abroad. At the same time, many astronomers installed themselves and their equipment for free at houses offered for that purpose by wealthy Spaniards. Finally, the huge expectation raised by the eclipses of 1900 and 1905 spread throughout Spanish society:

---

[640] 'Report of the Meeting … held on June 27, 1900 …', *Journal of the British Astronomical Association*, 10 (1899–1900), 341–351, here p. 343. The party was formed by Dr Downing, Howard Grubb and his son, Dr Rambaut, Prof. Joly, Mr Wilson, Mr Geoghegan, Mr Burgon, and Mrs Downing.

[641] Callendar and Fowler, 'Report of the expedition to Castellón …', *Proceedings of the Royal Society of London*, 77 (1906), 1–28, here p. 2. Many examples are found in other reports. See for instance: Copeland, 'Preliminary note on … the total solar eclipse of 1900 …', *Proceedings of the Royal Society of London*, 67 (1900), 385–391.

[642] Fontserè, 'Eclipse total de 28 de mayo de 1900', *El Mundo Científico*, 2 (1900), 403–410, here pp. 404–405.

> The solar eclipse has awoken the public interest, even more than the general elections in the country, with which it has coincided.[643]

Even the royal family was involved in amateur expeditions: Princess Isabel travelled to Argamasilla in 1900, and King Alfonso XIII led an expedition to Burgos in 1905. But the eclipses of 1900 and 1905 were also of interest to the working classes. For example, about five thousand people from the working classes congregated at El campo de las Vistillas in Madrid to observe the solar eclipse of 28 May 1900.[644] For Spaniards, the solar eclipses of 1900 and 1905 were "a national *fiesta*",[645] as experienced in full at Burgos for the eclipse of 1905. Burgos not only received the King of Spain and the President of the Spanish Government, Eugenio Montero Ríos (1832–1914), but also organised several celebrations during the days immediately before and after the eclipse, including the unveiling of monuments, a young-pigeon target practice, a photographic competition, and a first-class bullfight.[646] But the 1900 and 1905 eclipses also proved to be of great importance for the development of science in Spain. The *regenerationists* had called for the total involvement of the government in the scientific development of the country. As *El Liberal* commented on the day after the eclipse of 1900:

> There was a curiosity to see astronomical instruments and to admire the work of both foreign and local scholars … People … have faith in science … The government has now to appreciate this spirit and proceed accordingly … honouring the scholars who worked on these very important astronomical observations.[647]

These eclipses had also important consequences for the image of Spain abroad. The facilities that the Spanish society provided for foreign scientists were crucial for the projection of an international image of modernity. Moreover, there was wide international press coverage, not only in scientific journals and magazines, but also in daily newspapers and

---

[643] 'A. A.', 'El eclipse de Sol de 30 de Agosto', *El Mundo Científico*, 7 (1905), 544–547, here p. 546.

[644] 'El Eclipse', *El Imparcial*, 29 May 1900, 1. See also: 'El eclipse de Sol', *El Heraldo de Madrid*, 28 May 1900, [año 11, nº 3485].

[645] 'La fiesta de la ciencia', *El Imparcial*, 29 May 1900, 1.

[646] See, for example: 'El eclipse en Burgos', *El Imparcial*, 27 June 1905, 2; and 'Perfiles del día. El eclipse en Burgos', *El Imparcial*, 30 June 1905, 2. The various activities were compiled in a fifteen-sheet brochure entitled *Burgos. Eclipse total de sol. 30 Agosto 1905*.

[647] 'La fiesta de la ciencia', *El Imparcial*, 29 May 1900, 1.

general periodicals such as the German *Hamburger Nachrichten.*[648] The most prestigious ones, like *The Times* of London, published several notes on the eclipse of 1900.[649] Particularly relevant to the topic discussed here are those articles in which British astronomers such as Norman Lockyer showed how pleased they were with the reception they were given in Spain and the facilities that were provided to the different expeditions.[650] At the same time, the relationships established with foreign scientists who visited Spain in 1900 and 1905 provided a unique opportunity to improve the poor image that Spain had abroad. Things were slowly changing in the country, as Chambers pointed out at a meeting of the British Astronomical Association on 29 January 1900:

> Though it may seem a little early to begin talking about it, I find that though the Spaniards are not very go-ahead people, yet they have made a very vigorous start in the matter. The Madrid Observatory has already published a very fine, large-scale chart of the path of the eclipse across Spain, a copy of which I now hold, and a summary of the facts of which I beg now to submit to the Association.[651]

The eclipses of 1900 and 1905 were a sort of "international exhibition organised by the Nature"[652] which produced excellent results for Spain, as they modified favourably the opinion of the country abroad.[653]

---

[648] See for instance the contributions to this German periodical by Enrique Sivera from Argamasilla, mentioned in: 'El eclipse de Sol', *El Heraldo de Madrid*, 28 May 1900 [Año 11, nº 3485].

[649] See for instance: 'The coming eclipse of the Sun', *The Times*, 24 May 1900.

[650] See for example: 'The solar eclipse', *The Times*, 26 May 1900.

[651] G. F. Chambers, 'The total eclipse of the Sun of 1905', *Journal of the British Astronomical Association*, 12 (1901–1902), 167–169 [Read at the meeting of the Association held on 29 January 1902].

[652] Antonio Vela, 'El eclipse de sol del 30 de agosto', *El Imparcial*, 10 July 1905, 1.

[653] 'Después del eclipse', *El Heraldo de Madrid*, 30 May 1900, [año 11, nº 3487], 1.

Fig. 10-4: Front page of the Spanish periodical *El Mundo Científico*, 7 (1905) [issue 284]

10.5. Astronomy and gender in eclipse expeditions

The role of women in the eclipse expeditions of 1900 and 1905 can be examined in the context of these expeditions as cultural expressions of the period. Movements claiming more rights for women had become active during the second half of the nineteenth century in all developed countries. These movements, which can be regarded as a heritage of the Enlightenment and the French Revolution, became particularly strong in

Great Britain and the United States of America, mainly as a consequence of the social ideas of Protestantism. Following the example of the American National Female Antislavery Association, founded in 1837, women began to organise themselves into associations all over Europe during the second half of the nineteenth century.[654] Well-known examples are the German Allgemeinen Deutscher Frauenverein and the British National Society for Women's Suffrage. Their leaders were, in general terms, reformist, cultured, and refined women from the middle classes. The publication in 1869 of the book *The Subjection of Women* by the British philosopher and economist John Stuart Mill (1806–1873) brought the British feminist movement to wide public attention. The boost that feminism received during the last years of the nineteenth and early twentieth century is personified in the figure of Emmeline Goulden Pankhurst (1858–1928), whose suffragist offensive included acts of physical violence and destruction. Feminist strategies of alliance with democratic, socialist, and anarchist movements during the nineteenth century brought women's rights to the forefront of public debate.

In the natural sciences, including astronomy, these new attitudes had an impact not only in popular science but also in the practice of science itself.[655] As early as the seventeenth and eighteenth centuries, astronomy had been a field in which a small number of women had been active, more so than in most other sciences. That was the case in Germany, where several women were active in astronomy between 1650 and 1710, including Maria Cunitz (1610–1664), Elisabetha Hevelius (1647–1693), Maria Klara Eimmart (1676–1707), Maria Winkelmann (1670–1720), and Christine Kirch (1696–1782).[656] However, as feminist movements came more to the fore during the nineteenth century, new and more effective initiatives were taken. Colleges opened their doors to women. In Britain, the careers of Caroline Herschel (1750–1848) and Mary Sommerville (1780–1872) offered two early models for women interested in astronomy. In 1835 they were the first two women to be elected honorary fellows of the Royal Astronomical Society. However, women were not allowed to

---

[654] Anne-Marie Käppeli, 'Escenarios del feminismo', in Georges Duby and Michelle Perrot (dirs.), *Historia de las mujeres en Occidente*. [Volumen 4, *El siglo XIX*] (Madrid: Taurus, 1993), pp. 497–531.

[655] A good compilation of several interesting texts on the role of women in science is: Sally Gregory Kohlstedt (ed.), *History of Women in the Sciences. Readings from* Isis (Chicago and London: The University of Chicago Press, 1999).

[656] See for instance: Londa Schiebinger, 'Maria Winkelmann at the Berlin Academy. A turning point for women in science', *Isis*, 78 (1987), 174–200.

join the Society as regular members: those who were nominated as ordinary fellows of the Society, like Mrs Maunder, the former Annie Scott Dill Russell (1868–1947), who worked at Greenwich and Cambridge, did not get the necessary support from the membership to be elected.

As has been mentioned, one of the reasons for the foundation in 1890 of the British Astronomical Association was to fully integrate women in astronomy. Between the election of Caroline Herschel and Mary Sommerville and the foundation of the British Astronomical Association, only Anne Sheepshanks (1789–1876) had been elected an honorary fellow of the Royal Astronomical Society, in recognition of her donation of several instruments.[657] However, the new British astronomical organisations, in particular those devoted to physical astronomy, welcomed women. At the turn of the century, women were also included in eclipse expeditions planned by these institutions, such as those organised by the British Astronomical Association. Things were slowly changing, and by the beginning of the twentieth century many astronomers had a high opinion of women astronomers and popularisers of astronomy such as Agnes Clerke.[658]

The study of the eclipse expeditions sent to Spain to observe the eclipses of 1900 and 1905 provides some insights into the increasingly public role of women in western societies. Several women travelled to Spain together with male colleagues to observe these total eclipses, and some of their contributions to astronomy were outstanding. The "most remarkable" result on the extent of the corona before the eclipse of 1905 was the photograph obtained by Annie Maunder at the total solar eclipse of 1898 "with lenses of 1½ inches [38 mm] in diameter, and 9 inches [229 mm] focal length ... showing one streamer extending to a distance of 12.9 lunar radii from the Moon's limb".[659] Amongst the women who took an

---

657 See: Peggy Aldrich Kidwell, 'Women astronomers in Britain, 1780–1930', *Isis*, 75 (1984), 534–546.

658 On the figure of Agnes Clerke, see for example: Bernard Lightman, 'Constructing Victorian heavens: Agnes Clerke and the "new astronomy"', in Barbara T. Gates and Ann B. Shteir (eds.), *Natural Eloquence: Women Reinscribe Science* (Madison: University of Wisconsin Press, 1997), pp. 61–75; and M. T. Brück, *Agnes Mary Clerke and the Rise of Astrophysics* (Cambridge: Cambridge University Press, 2002).

659 Lockyer, 'Thorp gratings and small cameras in eclipse work', *Knowledge & Scientific News*, 2 (1905), 117–119, here p. 117. See also: Shackleton, 'The coming total eclipse', *Knowledge & Scientific News*, 2 (1905), 45–47, here p. 47;

active part in the expeditions to Spain were the Frenchwomen Mme Flammarion, Mme de la Baume Pluvinel, and Mme Bourget, who travelled to Elche in 1900 with their husbands. Mme de la Baume Pluvinel took an active part in the scientific research programme, being in charge of the observation of visible stars and planets.[660] Equally, for the eclipse of 1905 two of the eleven members of the expedition sent to Spain by the Société Astronomique Flammarion of Montpellier were Mlle Fanny Domerc and Mme Estève.[661]

The compositions of the various eclipse expeditions to Spain suggest that it was in Great Britain that the movement for women's rights was most active, for most of the women who travelled to Spain to observe the eclipses of 1900 and 1905 were British. They included Mrs Downing, who accompanied her husband and the rest of the Irish party to Plasencia in 1900, and was engaged in looking for visible stars during totality;[662] Lady McClure, who in 1900 took a photograph of the corona from Elche;[663] and Catharine O. Stevens, who after observing the eclipse of 1900 from Algiers visited Majorca in 1905 to study the origin and nature of the shadow bands.[664] From Palma of Majorca the 1905 eclipse was also observed by Mrs Crommelin, Miss Alice Everett, who worked at Greenwich for a few years in the 1890s, and Mrs Hollis, who represented the journal *The Observatory*.[665] In this location was also Lady Lockyer, who was in charge of a novel attempt to photograph the eclipsed Sun with a three-colour camera.[666] The long exposures required for this instrument,

---

and 'El eclipse total de sol del próximo agosto', *El Mundo Científico*, 7 (1905), 289–291, here pp. 290–291.

[660] Flammarion, 'L'Éclipse totale de soleil', *Bulletin de la Société Astronomique de France*, 14 (1900), 289–297.

[661] Moye, *À la poursuite d'une ombre ...* (Montpellier, 1905).

[662] 'Report of the meeting ... held on June 27, 1900 ...', *Journal of the British Astronomical Association*, 10 (1899–1900), 341–351, here pp. 344–345.

[663] *Ibid.*, p. 350.

[664] She concluded that the shadow bands were of atmospheric origin. In fact, she identified shadow bands with 'boiling' distortions found at the Sun's limb in ordinary observations. See: Catharine O. Stevens, 'Letters to the Editor: The Problem of "Shadow-bands"', *Nature*, 72 (1905), 631; and Wilhelm Krebs, 'Fliegende Schatten bei der Sonnenfnsternis vom 30. August 1905', *Astronomische Nachrichten*, 170 (1906), 283–284.

[665] 'The total solar eclipse of 1905, August 30', *The Observatory*, 28 (1905), 355–356.

[666] Ono this novel method, see: Lockyer, 'The approaching total solar eclipse of August 30', *Nature*, 71 (1905), 393–395, here p. 395.

however, could not be secured because there was too much cloud.[667] Other female members of the British Astronomical Association who observed the eclipse of 1905 in Spain were Mrs Tabor, from Espiña; Mrs Arthur Brook, on the steamer *Arcadia* in the Mediterranean; and Miss Berry, Miss Woolston, Mrs Moore, Mrs Johnson, and Miss Mary Proctor (1862–1944), the successful populariser of astronomy, from Burgos.

In the United States of America, the movement for higher education, which began in the 1870s and 1880s, rapidly succeeded and led during the following decades to the gradual emergence of what we would now call a sex-blind practice of science. In astronomy, a number of women found permanent positions at various astronomical institutions (for example, about twenty women assistants worked at Harvard College Observatory between 1885 and 1900), and some of them carried out original research.[668]

No official party was organised in the United States of America to observe the eclipse of 1900 from Spain: the Spanish-American War of 1898 prevented it. In fact, as noted in chapter 9, the first American warship to visit Spain after the war was the one that anchored off Valencia in 1905 with astronomers on board to observe the eclipse of that year. There is no doubt that the long distance that had to be travelled was another reason to discourage an expedition to Spain from the United States. However, the Americans did organise an eclipse expedition to Spain in 1905, and two American women were included. Together with their husbands, who were commissioned by the Kirkwood Observatory (Indiana University), Mrs Miller and Mrs Thompson of California travelled to Almazan (Soria).[669]

The situation was slightly different in Spain, where in the second half of the nineteenth century women could express their cultural identity only in the private sphere. Women were regarded as having only a domestic role, and their work in any other context was discouraged. It was not until the end of the nineteenth century and the first decades of the twentieth that women workers began to organise themselves in Spain and to seek social

---

[667] Lockyer, 'The solar physics observatory eclipse expedition', *Nature*, 72 (1905), 508–510, here p. 510.

[668] Margaret W. Rossiter, '"Women's work" in science, 1880–1910', *Isis*, 71 (1980), 258–275.

[669] Miller, 'Solar eclipse of August 30, 1905', *Astrophysical Journal*, 23 (1906), 93–95.

acceptance for their work.[670] The education of women in Spain during the first two-thirds of the nineteenth century had been focused on their domestic role. Only after the Glorious Revolution of 1868 were some initiatives taken to change the traditional education for women, encouraged by krausist ideologies. The public debate around this issue continued into the twentieth century and, in many cases, other minority movements such as masonry and spiritualism supported women's claims for further rights—as they saw defending women's claims as part of a strategy to legitimate their own beliefs.[671] Some women, such as Amalia Domingo Soler (1835–1909), made use of spiritualism to present some of their claims as the opinions of supernatural entities.[672] One should not forget that spiritualism was then considered by its proponents to be a scientific subject. Science was a successful vehicle for validating women's claims as it was understood by the ruling classes of Western societies as synonymous with progress.

Women's movements were, however, less active in Spain than in other European countries. That explains why only a few women took an active part in observing the eclipses of 1900 and 1905. The practice of science was understood as an activity mainly for men. Nevertheless, there were a few women, such as the wives of Lànderer and Comas i Solà, who observed the eclipse of 1900 from Elche together with their husbands. Alice Arcimis observed the eclipse with her father, Agusto Arcimis, from Argamasilla, making a sketch of the corona that was published in the *Bulletin de la Société Astronomique de France*.[673] Likewise, Rosa Sensat saw her sketch of the corona, also made in 1900, published in the Spanish scientific magazine *El Mundo Científico*.[674] In sum, the eclipses of 1900 and 1905 not only meant a regeneration of astronomy in Spain, but also illustrate a new attitude towards women and science in the country.

---

[670] Mary Nash, 'Identidad cultural de género, discurso de la domesticidad y la definición del trabajo de las mujeres en la España del siglo XIX', in Duby and Perrot (dirs.), *Historia de las mujeres en Occidente* (Madrid, 1993), pp. 497–597.

[671] Pilar Ballarín, 'La construcción de un modelo educativo de "utilidad doméstica"', in Duby and Perrot (dirs.), *Historia de las mujeres en Occidente* (Madrid, 1993), pp. 599–611.

[672] María del Carmen Simón Palmer, 'Mujeres rebeldes', in Duby and Perrot (dirs.), *Historia de las mujeres en Occidente* (Madrid, 1993), pp. 629–641.

[673] See: Flammarion, '[...] Notes complémentaires', *Bulletin de la Société Astronomique de France*, 14 (1900), 298–307, here pp. 300–303.

[674] Fontserè, 'Eclipse total de 28 de mayo de 1900', *El Mundo Científico*, 2 (1900), 403–410, here pp. 404–405.

However, the sex-blind practice of science in Spain had to wait longer for its consolidation.

10.6. The visit of Camille Flammarion in 1900

Many of the astronomers who visited Spain to observe the total solar eclipses of 1900 and 1905 stayed in private houses owned by the higher classes of Spanish society. Indeed, hosting foreign scholars was seen both as a national obligation and as a way of gaining social prestige. This section looks briefly at the facilities enjoyed by the French in 1900, as France was the country that sent the most official expeditions to Spain that year. For example, Hamy and Lagarde, astronomers from the Observatory of Paris and members of the expedition organised by the Bureau des Longitudes, were installed in 1900 in "El Pino", a house three kilometres south of Elche owned by banker "and owner of the city" Jaime Brotons Pastor.[675] The Comte De la Baume Pluvinel, sent by the Ministry of Public Instruction, and his wife were put up at the house of the young mayor of Elche, Sebastián Canales, just one kilometer west of the city. The members of the expedition from the Universities of Toulouse and Montpellier, Lebeuf (lecturer in science at the University of Montpellier), Meslin (lecturer in chemistry at the University of Montpellier), Carrère (Observatory of Toulouse), and Bourget (lecturer in astronomy at the university of Toulouse) and his wife, were installed at "San Antonio", a house three kilometres from Elche owned by Gervasio Torregrosa.[676] North of there, in "El Toscar", the house owned by the physician José María López Campello, were José J. Lànderer, helped by the physicians and amateur astronomers Llopis and Pomares.[677] In Elche, only the Spanish party from the Observatory of San Fernando had to pay any rent for their stay, and this was because the owner of the house they were using, the businessmen Jaime Beltran, had to interrupt the production of brick at his factory to make his workshops available for the astronomers and the students of the Escuela de Ampliación who were with them.[678] The

---

[675] Ibarra, 'Desde Elche', *Las Provincias*, 27 May 1900. From: Soler Selva (coord.), *L'eclipsi total de sol de 1900 ...* (Elx, 2000).

[676] It seems that Torregrosa moved to a different house to make room for the observers. See: J. M. Grau Gadea, 'Desde Elche', *El Heraldo de Madrid*, 13 May 1900 [Año 11, nº 3470], 1.

[677] Other sources state that Hamy and his party stayed at 'El Toscar'. See: *Ibid.*

[678] The commission was formed by Juan Bautista Viniegra and thirteen other members: Cheriguini, Graiño, Somoza, Guardia, Sunnyer, Sáiz, Castellanos,

Spanish Father Angel Rodríguez de Prada (1859–1935), Director of the Vatican Observatory from 1898 to 1905, stayed at the same place as the San Fernando party together with some other Augustinian fathers including Antonio Tonna, Manuel Díez, and Cándido de la Puente.[679] Finally, the Irish party, in Plasencia, were stationed in a palace owned by the Marquis of Mirabel.[680]

However, the most widely reported visit to Spain for the 1900 eclipse was that of the prestigious French astronomer and populariser of astronomy Camille Flammarion. Flammarion, "whose name has such a deserved and great popularity, probably not even unknown in the smallest town of the civilised world",[681] was highly respected all over Europe by the end of the nineteenth century. For the eclipse of 1900, the French astronomer was asked by one of the leading Spanish newspapers, *El Liberal*, to write a pair of articles. *El Liberal* was a very important Spanish newspaper, mainly read by a liberal (non-conservative) bourgeoisie, cultivated and open to ideas from outside Spain. Consequently, these articles had a large impact in Spanish society and elevated the figure of Flammarion above other men of science.[682] The first of these articles was published on 18 May 1900 and dealt with the description of an eclipse and how and with what to observe it.[683] He also contributed to this newspaper for the solar eclipse of 1905.[684]

After a four-day journey, Flammarion arrived at Barcelona on 24 May 1900, where he had lunch in the "Gran Hotel" and left immediately for Tarragona.[685] He continued from Tarragona to Castellón, where despite the

---

Vélez, Sotelo, Lobo, Quijano, Torres, and Pujazón, as reported in: '[El eclipse de Sol] En Elche', *El Heraldo de Madrid*, 26 May 190 [Año 11, nº 3483], 7.

679 'El eclipse de Sol', *El Heraldo de Madrid*, 21 May 1900 [Año 11, nº 3478], 1.

680 'Report of the meeting ... held on June 27, 1900 ...', *Journal of the British Astronomical Association*, 10 (1899–1900), 341–351, here p. 343.

681 'Flammarion en España', *El Imparcial*, 18 May 1900, 1.

682 Moreux, 'Autour de l'éclipse. Notes de voyage', *Bulletin de la Société Astronomique de France*, 14 (1900), 308–324, here pp. 309–310. articles in tribute to Flammarion were also published in several journals and magazines. See for example: 'Los héroes de la ciencia', *El Mundo Científico*, 2 (1900), 417–418; 'Flammarion', *El Heraldo de Madrid*, 2 June 1900 [año 11, nº 3490], 1.

683. Flammarion, 'El eclipse total de Sol del 28 de Mayo', *El Liberal*, 18 May 1900, 6.

684 See for instance: Flammarion, 'El eclipse total de Sol del 30 de agosto', *El Imparcial*, 14 August, 1905, 1.

685 'Flammarion en España', *El Imparcial*, 24 May 1900, 3.

bad weather conditions more than a hundred people, representatives of the city, welcomed him. After Castellón he stopped in Valencia, where no official representatives met him, as he arrived earlier than expected.[686] However, several events there were held in his honour, including a banquet for seventy people in the Botanic Garden of the University of Valencia on 25 May 1900, organised by the University and the Ateneo Científico of the city.[687] On his train journey from Valencia to Alicante, Flammarion was welcomed at every stop he made. The interest he awoke in the general public was expressed on banners at several stations.[688] For example, at the train station of Villena there was written: "Reconnaissance à celui qui nous a fait connaître l'Univers et a élevé nos âmes dans la philosophie de l'infini".[689] Flammarion arrived in Alicante on 26 May 1900 and continued his trip to Elche, where a multitude welcomed him.[690] On that day in Elche he attended a banquet in which several local dignitaries as well as other astronomers such as Viniegra, Lànderer, De la Baume Pluvinel, Donitch, Rodríguez Prada, and Lockyer were present.[691]

It was claimed that in total about a hundred thousand Spaniards welcomed Flammarion on his trip to Elche at the many different train stations through which he passed or stopped.[692] His visit was recorded in almost every single Spanish newspaper. When he arrived at Elche,

> the rich and the poor, the wise and the ignorant, the workers from the city and the labourers from the fields, all walks of life that constitute the totality of Elche were represented.[693]

---

[686] 'Flammarion en España', *El Imparcial*, 25 May 1900, 1.
[687] 'En honor de Flammarion', *El Liberal*, 26 May 1900, 2. See also: Zelt, 'Flammarion en Valencia', *El Heraldo de Madrid*, 24 May 1900 [año 11, nº 3481], 1; and Zelt, 'Flammarion', *El Heraldo de Madrid*, 25 May 1900 [año 11, nº 3482], 2.
[688] See for example : 'Flammarion en España', *Blanco y Negro*, 2 June 1900, 4.
[689] Moreux, 'Autour de l'éclipse. Notes de voyage', *Bulletin de la Société Astronomique de France*, 14 (1900), 308–324; p. 311. The enthusiastic reception in Villena (and later in Alicante) was also recorded in: 'Flammarion', *El Heraldo de Madrid*, 26 May 1900 [año 11, nº 3483], 2.
[690] 'El eclipse', *El Imparcial*, 27 May 1900, 2.
[691] 'El eclipse', *El Imparcial*, 28 May 1900, 1–2, here p. 2.
692. Moreux, 'Autour de l'éclipse. Notes de voyage', *Bulletin de la Société Astronomique de France*, 14 (1900), 308–324, here p. 310.
[693] Joaquín Dicenta, 'Flammarion. Instantánea', *El Liberal*, 29 May 1900, 1.

Indeed, of the 30,000 inhabitants of the city of Elche, 25,000 went to receive Flammarion.[694] He was courted and attended to by weighty figures. As Flammarion himself wrote,

> une trentaine de personnes nous accompagnaient, gouverneur, amiraux, généraux, professeurs, amis de la Science, s'offrant généreusement à nous pour être utilisés à une observation quelconque.[695]

All the bibliographical sources consulted agree on the large number of people that Flammarion attracted during his visit. His departure from Alicante to Madrid on 31 May 1900 is another good example.[696] Once in Madrid, Flammarion and his wife were hosted by Ernesto de la Guardia.[697] De la Guardia, on behalf of the Astronomical Society of Madrid, awarded Flammarion a commemorative medal of his visit to Spain.[698] Flammarion was also honoured in Madrid with a banquet organised by the Spanish newspaper *El Imparcial* and a reception hosted by the Spanish publication *Blanco y Negro*.[699] Many figures from the political, cultural, and scientific sectors of Spanish society were present at these events.[700] Flammarion's party made a trip from Madrid to Toledo and then spent a couple more days in Madrid, where they visited the capital's observatory. The absence of the Director of the Observatory of Madrid, Francisco Iñiguez, meant that Ventosa had to receive and entertain them. He even gave them a guided tour of the museum of art, El Prado.[701] They were also received by the royal family, whose members had also studied the solar event. Princess

---

[694] Moreux, 'Autour de l'éclipse. Notes de voyage', *Bulletin de la Société Astronomique de France*, 14 (1900), 308–324, here p. 314.

[695] Flammarion, 'L'éclipse totale de soleil', *Bulletin de la Société Astronomique de France*, 14 (1900), 289–297; p. 290.

[696] 'Después del eclipse', *El Imparcial*, 30 May 1900, 2.

[697] Moreux, 'Autour de l'éclipse. Notes de voyage', *Bulletin de la Société Astronomique de France*, 14 (1900), 308–324, here p. 322. See also: 'Flammarion en Madrid', *El Liberal*, 1 June, 1900, 2.

[698] 'Flammarion', *El Imparcial*, 2 June 1900, 2; and 'Visita de Mr. y Mme. Flammarion a "Blanco y Negro"', *Blanco y Negro*, 9 June 1900, 7–8.

[699] 'Flammarion', *El Imparcial*, 3 June 1900, 1–2.

[700] 'En honor de Flammarion', *El Liberal*, 3 June 1900, 1. See also, for example: 'Flammarion en Madrid', *El Heraldo de Madrid*, 31 May 1900, [año 11, nº 3488], 2; 'En honor de Flammarion', *El Heraldo de Madrid*, 2 June 1900, [año 11, nº 3490], 3; 'Flammarion en Madrid', *El Heraldo de Madrid*, 3 June 1900, [año 11, nº 3491], 1.

[701] Moreux, 'Autour de l'éclipse. Notes de voyage', *Bulletin de la Société Astronomique de France*, 15 (1900), 308–324, here pp. 323–324.

Isabel had observed the eclipse from Argamasilla, while observations from the Royal Palace of Madrid had been made by the Queen (who made some drawings of the phases of the eclipse, notes on the effect of the eclipse on flowers, and meteorological observations) and by her son, the future King Alfonso XIII, who took some photographs of the corona at different phases of the eclipse.[702]

[702] Flammarion, '[…] Notes complémentaires', *Bulletin de la Société Astronomique de France*, 14 (1900), 298–307, here p. 300. See also, for example: 'Flammarion en Palacio', *El Heraldo de Madrid*, 2 June 1900 [año 11, nº 3490], 1.

CHAPTER ELEVEN

THE SOLAR ECLIPSES OF 1912 AND 1914

11.1. Introduction

It is clear from previous chapters that the eclipse of 1900 saw the "re-emergence of astronomy" in Spain.[703] Spanish astronomy experienced an important boost in 1900 and in 1905, and at the international astrophysics conferences of the first decade of the twentieth century the prestige of both the country and its representatives grew rapidly. That was the case at the Fourth Conference of the International Union for Cooperation in Solar Research, held from 29 August to 6 September 1910, which the Spanish Jesuits Cirera and Balcells attended as representatives of the Reial Academia de Ciències i Arts de Barcelona and the Sociedad Española de Física y Química, respectively.[704] The frequency with which total solar eclipses visited Spain during these years led to an increasing interest in astronomical and astrophysical studies in the country. The larger number of individuals and institutions involved in these fields was reflected in the foundation of a new section for Astronomy and Geophysics at the Asociación Española para el Progreso de las Ciencias.[705] Moreover, prestigious astronomers such as De la Baume Pluvinel expressed an interest in cooperating with Spanish astronomers and scientific institutions to pursue an international and coordinated programme of observations for the total solar eclipses of 9 May 1910, 28 April 1911, and 17 April

---

703 R. Cirera, 'Recientes progresos de las ciencias astronómicas en España', in *Asociación Española para el Progreso de las Ciencias, Tercer Congreso (Granada, del 20 al 25 de Junio de 1911), Tomo I* (Madrid: Eduardo Arias, 1911), pp. 51–88.

704 'Cuarto Congreso de la Unión Internacional para la investigación solar', *Boletín de la Sociedad Astronómica de Barcelona*, 1 (1910–1912), 33–34.

705 Cirera, 'Discurso relativo al establecimiento de la nueva sección de astronomía y física del globo', in *Asociación Española para el Progreso de las Ciencias, Primer Congreso (Zaragoza, 22 al 29 de Octubre de 1908), Tomo I* (Madrid: Eduardo Arias, 1908), pp. 61–66.

1912.[706] These eclipses are included in Table 11-1, which lists the solar eclipses occurring in the years after that of 1905.[707]

Table 11-1: Total solar eclipses from 1907 to 1914

Date	**Location**	**Duration (approximate)**
14 January 1907	Ural Mountains, Central Asia, China	2 minutes
3 January 1908	Pacific Ocean, ending in Panama	4 minutes
23 December 1908	South Atlantic	Annular—short total phase
17 June 1909	Greenland, Arctic regions, northern Siberia	1 minute
9 May 1910	Antarctic Ocean, Tasmania	4 minutes
28 April 1911	Australasia, Pacific Ocean	5 minutes
17 April 1912	Spain	Annular—short total phase
10 October 1912	Venezuela, Brazil, South Atlantic	1 minute
21 August 1914	Greenland, Norway, Sweden, Russia, Persia	2 minutes

11.2. The eclipse of 1912

As Table 11-1 shows, the eclipse of 17 April 1912 was the first solar eclipse observable from Europe after the 1905 eclipse. The eclipse was observed under mostly excellent conditions, and many papers on it were published. As Lànderer explained, it was an exceptional opportunity to measure the uncertain value of the apparent semi-diameter of the Moon.[708] There were discrepancies in calculated results based on observations made at totality during the late nineteenth and early twentieth centuries because different values were used for the Moon's semi-diameter.[709] At both the

[706] Cirera, 'Los eclipses de 8 de mayo de 1910 y 28 de abril de 1911', in *Asociación Española para el Progreso de las Ciencias (Valencia), Tomo III* (Madrid: Eduardo Arias, 1911), pp. 97–100.

[707] Shackleton, 'The coming total eclipse', *Knowledge & Scientific News*, 2 (1905), 45–47; 'El eclipse total de sol del próximo agosto', *El Mundo Científico*, 7 (1905), 289–291.

[708] J. J. Lànderer, 'L'Éclipse annulaire et totale de soleil des 16–17 avril 1912', *Bulletin de la Société Astronomique de France*, 25 (1911), 248–251.

[709] See, for instance: J. Y. Buchanan, 'Letters to the Editor: Eclipse predictions', *Nature*, 72 (1905), 603; and Downing, 'Letters to the Editor: Eclipse predictions', *Nature*, 72 (1905), 629.

1900 and 1905 eclipses the calculated duration of totality was slightly larger than the observed value,[710] but no definitive solution was provided. In 1902 Charles Thomas Whitmell (1849–1919) wrote that:

> There seems to be but little doubt that we need to use *three* different values for the Moon's semidiameter: the largest value for ordinary telescopic purposes, the intermediate for the occultations of stars, and the smallest for solar eclipses.[711]

The problem of the Moon's semi-diameter had not been solved by 1912. For example, some years before his death in 1906, Antonio Tarazona, former astronomer at the Observatory of Madrid, made the first calculations of the circumstances of the eclipse of 1912, using constants published in the *Connaissance des Temps* for the year 1897, including a lunar semi-diameter of 15′ 32″.83.[712] After the eclipse of 1905, Lànderer decided to employ a new value for the Moon's semi-diameter of 15′ 31″.62 to calculate the contact times and the duration of the eclipse for several locations.[713] Because of Lànderer's reputation his new value was seriously considered by several astronomers and institutions, particularly in France.[714] Furthermore, after examining the results of the observations of the 1912 eclipse made from several sites in Spain, Lànderer concluded that 15′ 31″.62 was extremely close to the exact value.[715]

---

[710] See, for example: Whitmell, C.T., 'The Moon's Semidiameters', *Journal of the British Astronomical Association*, 13 (1902-1903), 166-167; and Whitmell, 'Contact times of the eclipse on 30 August', *Journal of the British Astronomical Association*, 17 (1905–1906), 220–226.

[711] Whitmell, 'The duration of totality at Navalmoral', *Monthly Notices of the Royal Astronomical Society*, 62 (1902), 375–378.

[712] Cirera, 'Éclipse totale et annulaire de soleil du 17 avril 1912', *Astronomische Nachrichten*, 162 (1903), 101–102; and T. Moreux, 'Comment verrons-nous l'éclipse du 17 avril 1912?', *Revue générale des sciences pures et appliqués*, 23 (1912), 170–171.

[713] See for example: J. J. Lànderer, 'L'Éclipse annulaire et totale de soleil des 16–17 avril 1912', *Bulletin de la Société Astronomique de France*, 25 (1911), 248–251.

[714] See for instance: T. Moreux, 'Comment verrons-nous l'éclipse du 17 avril 1912?', *Revue générale des sciences pures et appliqués*, 23 (1912).

[715] Lànderer, 'Sur l'éclipse de Soleil du 17 avril 1912' [présentée par Bigourdan], *Comptes Rendus*, 154 (1912), 1279–1281; Lànderer, 'Observations de M ...', *Bureau des Longitudes. Observations de l'Éclipse de Soleil du 17 Avril 1912*, pp. 54–55.

Comas i Solà was one of the astronomers whose observations helped Lànderer to solve the problem of the Moon's semi-diameter. Comas i Solà had gone to Barco de Valdeorras (Orense), representing the Fabra Observatory on an expedition funded by the Reial Acadèmia de Ciències i Arts de Barcelona to observe the eclipse. He made visual observations of the corona, and used a cinematograph with two 60° flint prisms to study the solar spectrum before and during totality. Weather conditions were favourable and successful results were obtained. For example, the flash spectrum was filmed for about 20 seconds, giving about a hundred separate images. This technique, also employed by French astronomers in 1912,[716] allowed the exact moments when the different phenomena characteristic of total solar eclipses took place. Comas i Solà later published a paper in which one could follow the exact evolution of the solar spectrum from 16 seconds before totality to 10 seconds after.[717]

It is worth briefly explaining what the main Spanish astronomical institution, the Observatory of Madrid, did for the eclipse of 1912. A report issued on 24 February 1912 explained that it was not certain whether the eclipse would be total or annular, and discussed how favourable it would be for the study of the chromosphere.[718] The report argued that, although there was no need to send a large expedition or to buy new instruments to study the eclipse, Spain had to be represented by an official expedition, even if it was only for ten days. It was decided to send an official expedition to Cacabelos del Bierzo (León). The budget for this enterprise was set at 12,300 pesetas, including the astronomers' travel expenses. The expedition's members included the astronomers Francisco Cos Mermeria and Miguel Aguilar Cuadrado and the assistants José Tinoco y Acero (1882–1953) and Pedro Carrasco Garrorena (1883–1966). Captain Nicolás Prat Delcourt was also invited to join the party. The programme included the observation of contact times and Baily's beads, photography of the different phases of the eclipse, the study of the spectrum during totality, and the observation of shadow bands and

[716] 'Report of the Council to the Ninety-third Annual General Meeting of the Society. Notes on some points connected with the progress of astronomy during the past year', *Monthly Notices of the Royal Astronomical Society*, 73 (1913), 266–311, here p. 276.

[717] José Comas i Solà, 'Sonnenfinsternis 1912 April 17. Mitteilung von Herrn *José Comas Solá*, Direktor des Observatoriums Fabra in Barcelona', *Astronomische Nachrichten*, 192 (1912), 205–207.

[718] OAM, 'Expediente del eclipse total anular de Sol del 17 de abril de 1912', Folder 15.

temperature changes.[719] A similar programme of observations was also carried out from Madrid, but bad weather conditions there prevented successful observations of the partial eclipse. The flash spectrum was observed in Madrid, and some photographs were secured with a Zeiss camera and a spectroheliograph.

But the most interesting part of this story is how the research was pursued in Cacabelos and how the results were presented by the Observatory of Madrid. This time, the observations during totality were apparently carried out in a very precise and complete way.[720] A good example of this new approach is the publication of the spectrographic research carried out by Pedro Carrasco, who mapped the solar spectrum in detail and registered, measured, and identified more than 1900 lines, and augmented this descriptive work with an analysis of the distribution and main characteristics of several spectral lines.[721] On the basis of his analysis, Carrasco suggested that the chemical elements represented in the spectrum were not distributed homogeneously in the solar atmosphere but in several layers of different thicknesses and densities. Whether or not there was any truth in this suggestion, it shows that Carrasco was adopting the practice of the most distinguished foreign astronomers in explaining the results of observations in terms of an appropriate theoretical model of the structure of the solar atmosphere.[722] In fact, López Arroyo has suggested that from 1915 Carrasco developed a theoretical model of the spectrum of the corona based on the work of Seth Barnes Nicholson

---

[719] *Ibid.*

[720] See for instance: F. Iñiguez, 'Observation de l'éclipse de Soleil du 17 avril, faite à l'Observatoire de Madrid' [présentée par Bigourdan], *Comptes Rendus*, 154 (1912), 1066; and Iñiguez, 'Observatoire de Madrid', *Bureau des Longitudes. Observations de l'Éclipse de Soleil du 17 Avril 1912*, pp. 51–52.

[721] P. Carrasco, Observaciones espectrográficas del eclipse del eclipse de sol del 17 de abril de 1912 [*Revista de la Sociedad Española de Física y Química*] (Madrid: Imprenta de E. Arias, 1915).

[722] See, for example: Charles E. St John, 'Radial motion in sun-spots. The distribution of the elements in the solar atmosphere', *Contributions from the Mount Wilson Solar Observatory*, 74 (1913), 157–207; St John, 'On the distribution of the elements in the solar atmosphere as given by the flash spectra', *Astrophysical Journal*, 40 (1914), 356–371; and George E. Hale and Ferdinand Ellerman, 'The minute structure of the solar atmosphere', *Proceedings of the National Academy of Sciences of the United States of America*, 2 (1916), 102–108.

(1891–1963) and the new ideas of quantum physics to explain and identify coronal lines and to justify his empirical formulae.[723]

11.3. The eclipse of 1914

As in 1912, the total solar eclipse of 21 August 1914 was an opportunity to solve two important problems related to positional astronomy: the mathematical explanation of the time difference between the predicted and observed times of contacts (to be solved by deriving an equation which would satisfactorily explain the Moon's movement) and the value of the lunar semi-diameter to be used in these calculations (which was thought to be between 15′ 32″.83 and 15′ 31″.53).[724] In an effort to solve these problems, Lànderer observed the 1914 eclipse from Tortosa. He studied the differences between predicted contact times and those he measured with a Mailhat telescope of 88 mm aperture.[725] Astronomers from the Astronomical Observatory of the University of Valencia also observed the eclipse of 1914: Ignacio Tarazona and Vicent Martí measured by projection the contact times of the partial eclipse, finding like Lànderer that totality was shorter than predicted and that contacts happened before expected.[726] But because the solar eclipse was not total from Spain, these observations were apparently relegated to a secondary level of importance.

But from all Spanish observations, the more interesting ones were probably those from the official expedition sent by the Observatory of Madrid to the path of totality. Indeed, in 1914 the Observatory of Madrid sent an official astrophysical eclipse expedition abroad for the first time, to Theodosia (Russia), which included the astronomer Victoriano Fernández Ascarza, assistants Pedro Carrasco and José Tinoco, and the engineer Víctor Gosalvez, as well as Lady Ascarza and Lady Pizana. The budget of

---

723 López Arroyo, *El Real Observatorio Astronómico de Madrid (1785–1975)* (Madrid, 2004).

724 J. J. Lànderer, 'El eclipse total de sol del 21 de agosto de 1914', *Ibérica*, 1 (1914), 299–301.

725 See: Lànderer, 'Sur l'éclipse totale de Soleil des 20 et 21 août 1914' [présenteé par Bigourdan], *Comptes Rendus*, 159 (1914), 468–469. This text is exactly the same as: Lànderer, 'Observations de … à Tortosa (Espagne), *Bulletin de la Société Astronomique de France*, 28 (1914), 444.

726 I. Tarazona and V. Martí, 'Observation de l'éclipse de Soleil du 20–21 août 1914, faite à l'Observatoire astronomique de l'Université de Valence (Espagne)' [présentée par Bigourdan], *Comptes Rendus*, 159 (1914), 468–469.

7000 pesetas, including personal, material, and transport expenses, was subsidised by the Junta para Ampliación de Estudios, the institution founded in 1907 to promote Spanish scientific activity.[727] The main objective of this expedition was to obtain good pictures of the chromospheric spectra.[728] The official expedition left Marseille by boat on 23 July 1914 but, as Ascarza wrote,

> La déclaration de guerre européenne nous surprit à Constantinople, et, après avoir vaincu toutes sortes de difficultés et de périls, nous parvînmes à Théodosia le 17 août.[729]

The main instruments carried by this expedition were two spectrographs and a Grubb coelostat with two plain mirrors of 200 mm diameter. One of the spectrographs was employed to record the flash spectrum, while with the other the spectrum of the solar corona was studied.[730] But the most striking result was that obtained by Pedro Carrasco, who was in charge of a spectrograph formed by a Grubb coelostat, a Rowland plane grating of 15,000 lines to an inch (590 per mm), and a photographic objective of 80 mm aperture and 1.2 m focal length. Carrasco secured twenty-two photographs in two series, around the second and third contacts. He obtained remarkable pictures of the chromosphere, the flash spectrum, and the corona, but the most interesting was a 10-second exposure taken 13 seconds after the second contact which showed a new coronal line in the red region. The computed wavelength of the line, measured with reference to Hα, D3, and Hβ, was 6378.87 ± 0.04 Å.[731] This result agreed with the later observations by the Observatory of

---

[727] See: 'El próximo eclipse de Sol', *El Imparcial*, 12 July 1914 [Año 48, nº 17023], 1; and 'El próximo eclipse de Sol', *El Liberal*, 18 July 1914 [Año 36, nº 12569], 2.

[728] OAM, 'Eclipse de Sol de 21 de Agosto de 1914', Folder 15. The outbreak of war made the trip to Russia a very dangerous one. In fact, the British expedition to Kiev had to be abandoned and the German astronomers who went to Crimea were either ordered to leave the country or detained as prisoners of war. See: A. F., 'The Total Solar Eclipse, 1914 August 20–21', *Monthly Notices of the Royal Astronomical Society*, 75 (1915), 315–317.

[729] Victoriano F. Ascarza, 'L'Éclipse totale de Soleil du 21 août 1914, observée par la Comisión espagnole à Théodosia (Crimée)', *Comptes Rendus*, 159 (1914), 738–740.

[730] *Ibid.*

[731] Carrasco, 'Observations physiques faites à Théodosia, pendant l'éclipse totale de Soleil du 21 août 1914' [présentée par Bigourdan], *Comptes Rendus*, 159 (1914), 740–741; and Carrasco, 'Sonnenfinsternis vom 21. August 1914.

Meudon's expedition to Strömsund (Sweden), which gave a wavelength of 6374.5 ± 0.2 Å.[732] The detection of this previously unrecorded red line was one of the most striking features of eclipse results from all over the world. The significant difference between the wavelength values obtained by the two European expeditions is less important than the actual detection of the line, as the final values were obtained indirectly from the direct measurement of three reference lines.[733] This was a new addition to the spectrum of the corona and seemed connected "with the state of solar activity at the time of eclipses".[734]

Four groups of observers met in Hernösand (Sweden) to observe the eclipse of 1914. Together with Moll and Van der Bilt from the University of Utrecht, and the private astronomer Nordenmark from Stockholm, the astronomers Cortie, Whitelow, Gibbs, and O'Connor visited Sweden to make spectroscopic observations and photograph the eclipse with a 6 m focal-length camera. The German Wulff and the Spanish Rodés, both Jesuits, went also to Hernösand with an electrometer, a potassium cell, and a triple Zeiss lens for the general and ultraviolet photographical study of the corona. All these observers experienced problems with setting up aerials for communication by wireless telegraphy, because of the war, but their observations were successful.[735]

11.4. Spanish newspapers and the eclipses of 1912 and 1914

Spanish newspapers published several articles in 1912 and 1914 dealing with the total eclipses of those years, but neither the Spanish

---

Observación del espectro coronal a Teodosia', *Astronomische Nachrichten*, 200 (1914–1915), 147–148.

[732] See: H. Deslandres, 'Observation de l'éclipse totale du 21 août 1914 par la Mission de l'Observatoire de Meudon', *Comptes Rendus*, 159 (1914), 669–670; and J. Bosler and H. G. Block, 'Observations de l'éclipse de Soleil du 21 août 1914, faites à Strömsund (Suède), par la Mission de l'Observatoire de Meudon', *Comptes Rendus*, 159 (1914), 766–769.

[733] 'A. F.', 'The total solar eclipse, 1914 August 20–21', *Monthly Notices of the Royal Astronomical Society*, 75 (1915), 315–317.

[734] M. Selga, 'A new line in the spectrum of the solar corona', *Publications of the Astronomical Society of the Pacific*, 27 (1915), 93–94.

[735] Observers from Sweden pointed out that duration of the eclipse was less than had been predicted by Lànderer. See: 'Eclipse del 21 de agosto en Hernösand (Suecia)', *Ibérica*, 2 (1914), 168.

periodical press nor the general public were as enthusiastic about these events as at the time of the two previous eclipses whose path of totality crossed Spain. Obviously, the fact that the path of totality of the eclipse of 1912 again passed through Spain raised more attention than the eclipse of 1914. Several articles about the eclipse appeared in Spanish newspapers during April 1912 explaining the importance, characteristics, and expected evolution of the solar event in the country, as well as some of the research programmes that had been developed for it.[736] In some cases, periodicals summarised and reported the information on the eclipse published by official institutions such as the Observatory of Madrid.[737]

Many foreign astronomers went to Portugal rather than Spain to observe the eclipse of 1912. Consequently, Spanish newspapers mainly focused their reports on Spanish and other foreign expeditions both to Portugal and Spain, such as the French party from the University of Paris, led by Fred Vlès and Jacques Carvallo and located in Cacabelos.[738] The main objective of this expedition was to obtain a cinematographic film of all the phases of the eclipse—for scientific purposes, although the possibility of showing the film to the general public in cinemas was also considered. Spanish newspapers reported that this was the first ever attempt to film a complete eclipse.[739] The French expedition employed a Dalmeyer lens of 700 mm diameter and a camera which could register about thirty frames per second, far better than the apparatus employed by Comas i Solà in 1905, which allowed him to take only five exposures per second.[740] This work by foreign expeditions was complemented by the

---

[736] See, for example: 'El próximo eclipse de Sol', *El Liberal*, 2 April 1912 [Año 34, nº 11838], 2; Gabriel Galán, 'El eclipse de Sol del 17 de abril de 1912', *El Liberal*, 7 April 1912 [Año 34, nº 11842], 3; 'El eclipse del día 17', *El Imparcial*, 8 April 1912 [Año 46, nº 16202], 5; Galán, 'El eclipse de Sol del 17 de Abril. El parcial y el total', *El Liberal*, 15 April 1912 [Año 34, nº 11850], 1; Vicente Vera, 'El eclipse de mañana …', *El Imparcial*, 16 April 1912 [Año 46, nº 16210], 3; and 'Marcha del eclipse', *El Imparcial*, 16 April 1912 [Año 46, nº 16210], 3.

[737] 'Datos del Observatorio de Madrid', *El Liberal*, 17 April 1912 [Año 34, nº 11852], 1.

[738] Mario Roso de Luna, 'El eclipse de Sol del 17 de Abril. "*El Liberal*" en el Bierzo', *El Liberal*, 15 April 1912 [Año 34, nº 11850], 1; and 'Preparativos para la observación', *El Imparcial*, 16 April 1912 [Año 46, nº 16210], 3.

[739] See also: 'Películas del eclipse', *El Imparcial*, 16 April 1912 [Año 46, nº 16210], 3.

[740] Roso de Luna, 'El eclipse de hoy', *El Liberal*, 17 April 1912 [Año, 34, nº 11852], 1. Present-day movies have twenty-four or twenty-five frames per second.

research programmes pursued by Spanish institutions such as the Observatory of Madrid and the Observatory of San Fernando, and by observations by amateur astronomers and investigations by individuals such as Mario Roso de Luna, whose study of shadow bands from Cacabelos was funded by the Junta para Ampliación de Estudios.[741]

Tensions in the Spanish astronomical community, centred in particular on the figure of the Director of the Observatory of Madrid, also appeared after the eclipse of 1912. An anonymous reader of *El Liberal* complained about the mistakes made by the Observatory of Madrid in calculating the path of totality.[742] A couple of days later, Francisco Iñiguez published a note to justify some of the questioned decisions made by the Observatory and to elaborate the calculation.[743] His explanation was later supported by Mario Roso de Luna, who pointed out that in calculating the path of totality several astronomical observatories had used a value for the lunar semi-diameter that was too large. He argued that the calculated path published by the Observatory of Madrid was "exact but not accurate" and that this issue should not be used to attack the prestige of the main Spanish astronomical institution.[744] However, Iñiguez thought this defence from Roso de Luna was poor, and published a note criticising it and explaining how all corrections had been taken into account in making the appropriate calculation.[745] Roso de Luna replied immediately. He accused the Director of the Observatory of Madrid of being out of touch with modern astronomy and incapable of improving the services provided by the main

---

Silent movies had sixteen images per second, enough to create the impression of continuous movement in a film.

[741] On the reports of the research pursued by Spanish individuals and institutions, see not only the previous footnotes, but also: Vera, 'El eclipse de hoy. Observaciones curiosas al alcance del público', *El Imparcial*, 17 April 1912 [Año 46, nº 16211], 1; 'Una información curiosa', *El Imparcial*, 17 April 1912 [Año 46, nº 16211], 1; Roso de Luna, 'El eclipse de ayer. "*El Liberal*" en el Bierzo', *El Liberal*, 18 April 1912 [Año 34, nº 11853], 1; and Vera 'El eclipse de Sol. Observaciones científicas', *El Imparcial*, 18 April 1912 [Año 46, nº 16212], 1–2.

[742] Although the original letter was published in the newspaper *La Época*, it was reproduced in: 'Después del eclipse', *El Liberal*, 19 April 1912 [Año 34, nº 11854], 2.

[743] Francisco Iñiguez, 'El eclipse y el Observatorio de Madrid', *El Liberal*, 20 April 1912 [Año 34, nº 11855], 3.

[744] Roso de Luna, 'El proceso del eclipse', *El Liberal*, 22 April 1912 [Año 34, nº 11857], 1.

[745] Iñiguez, 'El proceso del eclipse', *El Imparcial*, 23 April 1912 [Año 46, nº 16217], 2.

Spanish astronomical institution. He also reminded readers of some of the problems that had occurred between Iñiguez and Ventosa (including those relating to the eclipse of 1900), and revealed some of the tensions amongst the Spanish astronomical community that have been considered here in previous chapters.[746]

As in 1900 and 1905, the total solar eclipse of 1912 proved to be a source of income for Spanish writers and traders. For example, in the newspaper *El Liberal* there appeared several advertisements for a "scientific novel" entitled *Eclipse de Sol*, written by "Rigel", which could be obtained at any newsstand for 30 centimes.[747] Enterprising "opticians" also made good money, even though sales of products were less than 15 per cent of the levels at the previous eclipses, and the products were four times cheaper than in 1900 and 1905.[748] This was a consequence of the low public interest in the eclipse of 1912, due mainly to the event taking place on the morning of a working day, and to the uncertainty about whether it was going to be total or annular. Nonetheless, some members of the public congregated in several cities and towns to observe the eclipse. In Madrid, for example, small groups assembled at La Moncloa, Paseo de Rosales, El Retiro, Puerta del Sol, Puerta de Atocha, Cuatro Caminos, San Gil. But the number of people was not at all comparable to that in previous years. Not even in the districts where the working classes resided—where, as some newspapers reported, people generally took any opportunity to express their rejoicing—did the solar eclipse attract much attention. In these districts (such as Armería, Lavapiés, Embajadores, and Rondas) people only went to their doors and onto their balconies to observe it. Several groups, mainly of women, gathered in Madrid around Ronda de Valencia, Ronda de Toledo, and Ronda de Atocha. The response was slightly better in cities closer to the path of totality, such as Gijón, where

---

[746] Roso de Luna, 'El Salmo del Eclipse', *El Liberal*, 24 April 1912 [Año 34, nº 11859], 2.

[747] See, for instance: 'Eclipse de Sol', *El Liberal*, 15 April 1912 [Año 34, nº 11850], 4; and 'Eclipse de Sol', *El Liberal*, 17 April 1912 [Año 34, nº 11852], 4.

[748] Some of the items on sale, such as simulated opera-glasses, were being sold for 20 or 25 centimes early in the morning of the eclipse. By the time of the eclipse itself the price had been reduced to 5 centimes. Further details are found in: 'El eclipse de ayer. En Madrid', *El Liberal*, 18 April 1912 [Año 34, nº 11853], 1–2; and 'El eclipse de hoy. La observación popular', *El Imparcial*, 18 April 1912 [Año 46, nº 16212], 2.

thousands of citizens observed the phenomena.[749] Despite the lack of interest from the people of Spain, some journalists ventured that the day on which Spain would join the group of developed countries that sent expeditions abroad to study scientific phenomena such as total solar eclipses was not too far away.[750] Indeed, as we have already seen, two years later Spain sent an official expedition to observe the total solar eclipse of 1914 from Russia.

The eclipse of 1914 attracted even less attention amongst journalists and the general public, as it was partial as seen from Spain. In fact, very few articles dealing with the eclipse were published in local newspapers, but all agreed in pointing out how important it was for a country that for the first time had sent an official expedition abroad to study a total solar eclipse to join the group of nations concerned about the development of scientific research.[751] But in general, there were almost no descriptions of the eclipse phenomena in Spanish newspapers.[752] All of them agreed on how uninterested the general public was[753] and how disappointing the event was for the "instant" astronomers who congregated next to the Observatory of Madrid and on the city's racecourse.[754] Some concerns about whether the Spanish astronomers sent to Russia would arrive in time and about the success of their observations were expressed in Spanish newspapers.[755]

---

[749] Further details of the popular observation of the eclipse of 1912, in cities such as Cádiz, Huesca, Oviedo, Valladolid, and Zaragoza, are found in: 'El eclipse de ayer. En Madrid', *El Liberal*, 18 April 1912 [Año 34, nº 11853], 1–2; and 'El eclipse de hoy. La observación popular', *El Imparcial*, 18 April 1912 [Año 46, nº 16212], 2.

[750] León Ortiz, 'Eclipses que se alejan', *El Imparcial*, 18 April 1912 [Año 34, nº 11853], 2.

[751] See: 'El próximo eclipse de Sol', *El Imparcial*, 12 July 1914 [Año 48, nº 17023], 1; and 'El próximo eclipse de Sol', *El Liberal*, 18 July 1914 [Año 36, nº 12569], 2.

[752] A brief note appeared on the day of the eclipse: 'El eclipse de Sol', *El Imparcial*, 21 August 1914 [Año 48, nº 17063], 2.

753. See for example: 'El eclipse de ayer', *El Imparcial*, 22 August 1914 [Año 48, nº 17064], 3.

[754] 'El eclipse de Sol', *El Liberal*, 22 August 1914 [Año 36, nº 12064], 3.

[755] 'El eclipse del dia 21', *El Liberal*, 24 August 1914 [Año 36, nº 12606], 3; and 'El eclipse de Sol', *El Imparcial*, 24 August 1914 [Año 48, nº 17066], 3.

11.5. Epilogue

After 1914, no Spanish official expeditions were sent to study any total solar eclipses until the eclipse of 25 February 1952. The Spanish Jesuits, through their network, made just one foray when the Spanish Josep Ubach was sent to Buenos Aires to observe the eclipse of 3 December 1918 as an official delegate from the Ebro Observatory.[756] However, it seems that the lack of official Spanish astronomical expeditions in this period was primarily due to economic problems rather than being related to not having in the country scholars with enough experience in the field or able to lead outstanding astronomical research programmes. As the Spanish Jesuit astronomer Luis Rodés wrote:

> We have in Spain a clear and pure sky as in few other nations; there is no lack of men of science able to compete with the best foreigners, as regards to personal qualities; once there is an abundance of funds to allow Spaniards to improve the observational equipment and to acquire those modern instruments for which the United States stands out, Spanish Astronomy will enter a new period of glory and our observatories will be the equals of the more prestigious abroad.[757]

The early years of the twentieth century saw the revitalisation of science in Spain and the engagement of Spanish governments with the promotion of science, with increasing amounts of money being invested in the development of science. Specific institutions were created to coordinate this development, such as the Junta para Ampliación de Estudios (1907), and the foundation of new laboratories such as the Laboratorio de Histología Normal y Patológica, the Laboratorio de Serología y Bactereología, the Laboratorio de Fisiología General, the Laboratorio de Investigaciones Físicas, the Laboratorio y Seminario Matemático, and the Laboratorio de Automática. But after 1914 no funds were made available for eclipse expeditions abroad, even when the astronomers of the Observatory of Madrid asked the Junta para Ampliación de Estudios for some financial assistance to observe the total eclipse of 3 February 1916 from Venezuela or Colombia, and the total

---

[756] 'América: Observaciones del eclipse anular del 3 de diciembre de 1918 en Buenos Aires', *Ibérica*, 12 (1919), 53.

[757] Luis Rodés, 'XXI Congreso Astronómico Norteamericano', *Ibérica*, 8 (1917), 270–271.

eclipse of 20 September 1922 from Australia. Both requests were denied because of the high expenses that such expeditions entailed.[758]

In order to understand why funding was refused, it is important to take into account firstly the stage that astronomy and astrophysics had reached internationally. The invention of the spectroheliograph independently by Hale and Deslandres in 1891, which enabled solar prominences and spectral lines to be photographed outside a total solar eclipse, opened a new era. Until then, the investigation of stellar structures had been restricted to the study of the physics and chemistry of the Sun, which had been possible only during total solar eclipses. The development of new astrophysical techniques not only for the study of the Sun, but also of the other stars, meant that solar eclipses gradually lost their importance. After the eclipse of 1905, the links between the organisation of solar eclipse expeditions and the development of science in a country vanished progressively. This suggests that, during the 1910s, the science achievable on eclipse expeditions moved away from the cutting edge of research. However, these enterprises were still fashionable and relevant from the scientific point of view. Good examples are the provisional arrangements made by the Joint Permanent Eclipse Committee of the Royal Society and the Royal Astronomical Society, together with the Royal Observatory and the Solar Physics Observatory, to observe the total solar eclipse of 21 August 1914 from near Kiev (Russia), Minsk (Russia), and Feodosia (Russia),[759] and Mitchell's description in 1918 of the important things to be learned from an eclipse "in order to warrant such an expenditure of energy and money".[760]

World War I also influenced strongly the organisation of eclipse expeditions. Eclipse expeditions in 1916 and 1918 were hit by wartime economic constraints. As Perrine wrote from the National Argentinean Observatory in Córdoba (Argentina) regarding the eclipse of 1916:

> Owing to the general financial depression resulting from the great European war, it was not possible to send so well equipped an expedition

[758] López Arroyo, *El Real Observatorio Astronómico de Madrid (1785–1975)* (Madrid, 2004).
[759] 'The total eclipse of the Sun, 1914 August 21', *The Observatory*, 37 (1914), 102–103.
[760] He did so by explaining in detail the scientific programme of the United States Naval Observatory carried out at Baker for the eclipse of that year. See: S. A. Mitchell, 'The total solar eclipse of 1918', *Natural History*, 19 (1919), 244–271.

> as those which went to Brazil in 1912 and to Russia in 1914. This was unfortunate, particularly in the matter of the large-scale photographs of the corona ... It was impossible to obtain some kinds of plates and films which in normal times are always to be had. This was a serious handicap chiefly to the spectroscopic and photometric work. No fresh orthochromatic or panchromatic plates could be found.[761]

European astronomers and astronomical institutions were experiencing many economic problems by 1918. And there had also not been time to organise any expeditions. Indeed, almost all the reports of eclipse expeditions from that year are American—partly because the shadow path crossed the United States of America.[762]

---

[761] C. D. Perrine, 'The total solar eclipse of February 3, 1916', *Publications of the Astronomical Society of the Pacific*, 28 (1916), 247–252.

[762] See: 'The total eclipse of the Sun, June 8, 1918', *Popular Astronomy*, 25 (1917), 412; 'The total solar eclipse of 1918 June 18', *The Observatory*, 41 (1918), 403–408; E. E. Barnard, 'The prominences of the total solar eclipse of 1918, June 8', *Popular Astronomy*, 26 (1918), 684; William Wallace Campbell, 'Bulletin Number 317: The total solar eclipse of June 8, 1918', *Lick Observatory Bulletin*, 10 (1918), 1–3; W. W. Campbell, 'The total solar eclipse of June 8, 1918', *Publications of the Astronomical Society of the Pacific*, 28 (1916), 86; Edwin B. Frost, 'The total solar eclipse of 1918, June 8: A reconnaissance', *Popular Astronomy*, 26 (1918), 103; E. B. Frost, 'Total solar eclipse, June 8, 1918', *Popular Astronomy*, 26 (1918), 458; Frances Ruth Hole, 'Duration times for the total solar eclipse of June 8, 1918', *Popular Astronomy*, 26 (1918), 306; H. M. Jeffers and C. S. Mundt, 'Note concerning the total solar eclipse of June 8, 1918', *Publications of the Astronomical Society of the Pacific*, 30 (1918), 171; S. A. Mitchell and H. R. Russell, 'The total solar eclipse of 1918: Painting the solar corona', *Publications of the Leander McCormick Observatory*, 2 (1921), 165–197; D. W. Morehouse, 'Observations of the total solar eclipse of June 8, 1918, at Matheson, Colorado', *Popular Astronomy*, 26 (1918), 551; R. M. Motherwell, 'The total solar eclipse, June 8, 1918', *Journal of the Royal Astronomical Society of Canada*, 12 (1918), 160; Edison Pettit, 'Circumstances of the total solar eclipse of June 8, 1918', *Popular Astronomy*, 25 (1917), 27; Pettit, 'The total solar eclipse of June 8, 1918', *Popular Astronomy*, 25 (1917), 15; William F. Rigge, 'Total eclipse of the Sun of June 8, 1918, as visible in the United States', *Popular Astronomy*, 25 (1917), 664; Rigge, 'Total eclipse of the Sun of June 8, 1918, as visible in the United States', *Popular Astronomy*, 25 (1917), 573; Morgan Sanders, 'Report of observations of total eclipse of Sun, June 8, 1918, made at Limon, Colorado', *Popular Astronomy*, 26 (1918), 681; V. M. Slipher, 'The spectrum of the corona as observed by the expedition from Lowell Observatory at the total eclipse of June 8, 1918', *Astrophysical Journal*, 55 (1922), 73; Florence Dell Stouder, 'Data for the total solar eclipse of June 8, 1918', *Popular Astronomy*, 25 (1917), 581; David Todd, 'Weather prospects for the total eclipse of 1918, June 8',

Investing in the organisation of total solar eclipses during these years became of secondary importance for countries with restricted budgets for the development of science but which aimed, like Spain, to join the elite of modern nations. Eclipse expeditions became less attractive, both scientifically and politically. In Spain, other fields of astronomical and astrophysical research, such as the application of new techniques to the regular study of different stellar structures, seemed to demand strong investments in order to reduce the gap with the most scientifically advanced countries.

Attitudes towards astronomical research had changed, growing closer to those in other European countries. The case of the Observatory of Madrid, with its strong interest in stellar astrophysics (developed in other countries during the 1870s), systematic spectroscopic observations of prominences outside eclipses from 1906, and systematic observation of the spectra of bright stars from 1909, is a good example of the increasing importance attached to astrophysical investigations during these years amongst the Spanish astronomical community. Spanish institutions supported the acquisition of new instruments, such as the spectroheliograph requested by the Observatory of Madrid in 1905 and finally obtained in 1910, and the creation of the Observatory's astrophysical laboratory, built in 1923.[763]

During the first quarter of the twentieth century the Spanish astronomical community developed several research programmes which required less economic investment. For example, the Observatory of Madrid promoted from 1909 spectroscopic studies of bright stars, novae, Cepheid variables, and eclipsing binaries, as well as photometric studies of the light variations of these and other bodies.[764] Other Spanish

---

Popular Astronomy, 26 (1918), 9; S. D. Townley, 'The total solar eclipse of June 8, 1918', *Publications of the Astronomical Society of the Pacific*, 30 (1918), 301; H. C. Wilson, 'Total solar eclipse of June 8, 1918', *Popular Astronomy*, 26 (1918), 285; Wilson, 'Total solar eclipse of June 8, 1918', *Popular Astronomy*, 26 (1918), 397; and Wilson, 'The total solar Eclipse of June 8, 1918', *Popular Astronomy*, 26 (1918), 447.

[763] López Arroyo, *El Real Observatorio Astronómico de Madrid (1785–1975)* (Madrid, 2004).

[764] *Ibid.* A good example of the success of these programmes is a report, published in 1913, in which there is explained how 'Prof. Iniguez's valuable series of spectra of Nova Geminorum were placed at the disposal of the Director [of the Solar Physics Observatory of South Kensington]' and were reduced and discussed. See:

astronomers, such as Comas i Solà, moved into observational astronomy and devoted their time to the study of comets and asteroids.[765] There was now, obviously, a wider range of possible astronomical and astrophysical research, and Spanish astronomers could pursue investigations which required less money but could still yield important results.

In fact, it was not until the Assembly of the International Union of Astronomy in August 1948 at Zurich that Spain rejoined the Commission that coordinated the organisation of research on total solar eclipses, with which it had not been involved since 1914—even when Spanish Jesuits kept their links with the international community. A proposal presented to the Commission by the Director of the Observatory of San Fernando was supported by the whole Spanish astronomical community, and individually by the directors of the Observatory of Madrid and the Ebro Observatory, as well as the Instituto Geográfico Central and the Professor of Astronomy of the Central University.[766] After the convulsive period of the Spanish Civil War (1936–39) and the years of repression which followed it, Spanish scholars could now return to the pathways for scientific and astronomical research that had been opened up during the first decades of the century.

'Observatories. Solar Physics Observatory, South Kensington', *The Observatory*, 36 (1913), 376.

765 The evolution of the astronomical research pursued by Comas i Solà can be found in: Josep Maria Oliver, 'Comas i l'observació astronòmica', in A. Roca Rosell (coord.), *Josep Comas i Solà* (Barcelona, 2004), pp. 43–79.

766 J. Oriol Cardús, 'En la Guinea Española para la observación del eclipse de sol', *Razón y Fe*, 147 (1953), 38–55.

CONCLUSION

Astronomy in Spain during the first half of the nineteenth century was very poorly developed, partly because of political instabilities. Nonetheless, as in many other countries, the mid-nineteenth century saw the revitalisation of astronomical activities in Spain with the re-foundation of the Observatory of Madrid. The Observatory developed a programme of meridian observations and provided several public services. Despite the political changes in the country, the 1860s, 1870s, and 1880s were decades in which the Observatory of Madrid began to be involved in some astrophysical tasks. However, the activity of the main Spanish astronomical institution progressively declined over the following years. This decline was part of the general backdrop against which the Spanish scientific activity of the time has to be understood.

Such was the state of the country's science that faced the group of Spanish scholars born in the middle decades of the nineteenth century. Most of them were self-taught and gradually applied to their observations the new astrophysical techniques developed in Europe. This new generation of Spanish astronomers even produced some original research which was recognised as such by the international scientific community, particularly in planetary, lunar, and solar observations. This new Spanish astronomical community was heterogeneous, including not only professional and amateur astronomers, but also a large number of members of the order of Jesuits, whose extensive network made possible communication and contributions to their scientific projects. The best example of this is the work pursued at the Ebro Observatory, an institution that by the early twentieth century was regarded at both national and international levels as a reference point for the Spanish scientific activity. Furthermore, scientific inquiry was gradually being removed from the arena of political ideology. New ideas were discussed and evaluated freely and openly, even when there were some underlying ideological agendas.

It was no coincidence, then, that the series of total solar eclipses of 1900, 1905, and 1912, together with the apparition of Halley's Comet in 1910, helped to regenerate interest in astronomy in Spain. This was the culmination of a process that was certainly accelerated by these events:

new scientific instruments were purchased to expand and improve the range of investigations that could be made by Spanish astronomers, new techniques were applied, and new astronomical and astrophysical institutions were founded. Of course, one has to bear in mind that the Spanish eclipse expeditions of these years were a success largely because they did not have to leave Spanish territory, and were consequently affordable. This was partly enhanced after the foundation in 1907 of the Junta para Ampliación de Estudios, an institution which aimed to promote scientific communication with foreign scholars. The support of this institution to astronomy meant the organisation of the first Spanish astrophysical eclipse expedition abroad in 1914.

Moreover, these total solar eclipses had the effect of consolidating the astronomical research community in Spain. And even more importantly, these events raised a huge expectation amongst Spaniards and led to a higher status for astronomy and astrophysics in the country. Indeed, the Spanish press heavily promoted science on the occasion of these eclipses, impressing on the population the importance of scientific knowledge. Many Spaniards showed an interest in astronomy and some of them even observed the eclipses as amateur astronomers, providing crucial popular support for the development of the discipline.

The visits of foreign astronomers and the Spanish efforts to compete with foreign researches increased scientific activity in Spain. The quality of Spanish astronomy improved rapidly after 1900, mainly as a result of the great effort made to study the solar eclipses visible from Spain during these years. Spanish astronomers used this series of eclipses not only to claim more official support, but also to promote their own activities amongst the international scientific community. Because local contacts were very important in the organisation of eclipse expeditions, Spanish scholars built strong links with foreign astronomers. These contacts proved very useful for the incorporation of Spanish astronomers into the international scientific community and for the development of astronomy and astrophysics in the country.

Eclipse expeditions were scientific enterprises with an opportunity to innovate and apply the latest technological advances to the study of astronomical phenomena, but they were also social and cultural expressions. They attracted tourism and business, and also provide us now with some clues about the changing role of women in society, the division of labour, and the concepts of national prestige and obligation in the late nineteenth and the early twentieth centuries. Official eclipse expeditions were a manifestation of a country's scientific maturity, and the eclipse of 1900 saw a re-emergence of astronomy in Spain. The conditions in which Spanish astronomers pursued their research improved, and although there were still the problems of lack of experience and outstanding equipment, after the eclipse of 1900 things began to change. There were new instruments, new projects, and more importantly, new attitudes to the development of astronomy in Spain.

The response to these eclipses in Spain provided a sounding-board for *regenerationist* calls for the involvement of the Government in the promotion of science. With the Government moved by the motivation of national pride and obligation, the eclipse of 1905 saw not only more complete and better organised Spanish expeditions, but also the success of a process of institutionalisation of astronomy which provided the basis for the future development of this discipline. By 1905, most of the astronomical instruments employed by Spanish astronomers and the research programmes they pursued were not far behind those in other countries. Furthermore, the eclipse of 1905 boosted the development of other disciplines in Spain such as meteorology and aeronautics.

In summary, astronomy and astrophysics in late nineteenth- and early twentieth-century Spain were developed through the building up, from the mid-nineteenth century onwards, of appropriate networks which integrated Spanish scholars into the international community and, as a consequence, boosted the development of astronomy in Spain. The prestige of the country and its representatives grew rapidly at the international astrophysical conferences of the first decade of the twentieth century, as shown by the proposals to cooperate with foreign astronomers received in Spain. This influenced governmental policies which directly supported scientific research, such as the foundation of new university observatories during the first decades of the twentieth century—a first step towards modernising the universities and transforming them into centres for scientific research. The eclipses of 1900 and 1905 opened up a new period for the development of astronomy and astrophysics in Spain: the frequency with which solar eclipses visited Spain not only raised interest in astronomical and astrophysical studies, but also led to the inclusion of Spanish scholars into an international astronomical community in the process of becoming institutionalised.

Bibliography

Primary Sources

Manuscript and archival sources

Archives de l'Observatoire de Paris. Folder: Ms 1060 II-A-3: *Les activites scientifiques de l'observatoire—éclipses 1850-1908 (1858, 1860, 1864, 1867, 1868, 1875, 1882, 1905, 1908).*

Archivo histórico del observatorio astronómico de la Universidad de Valencia (*OAUV*). Folders: 11, 12, and 13.

Archives Nationales de France (*ANF*). Folders: F^{17} I [F17/01(B)]), F^{17} II [F17/02(A)]), F^{17} 17274, and F^{17} 13571.

Archivo del Observatorio Astronómico de Madrid (*OAM*). Folders: 15 and 16.

Anonymous printed primary sources

'Académie des Sciences d'Amsterdam', *Revue générale des Sciences pures et appliquées*, 17 (1906), 55-56.

'Admissions et présentations', *Bulletin de la Société Astronomique de France*, 16 (1902), 528.

'América–Observaciones del eclipse anular del 3 de diciembre de 1918 en Buenos Aires', *Ibérica*, 12 (1919), 53.

'Annuaire Astronomique pour 1892', *L'Astronomie*, 11 (1892), 2.

'Antes de la batalla', *El Liberal*, 28 May 1900, 1.

'Astronomical Publications: A New Spanish Observatory', *Journal of the British Astronomical Association*, 17 (1906-1907), 152-153.

'Astronomical Publications: Eclipse Observations, August 1905', *Journal of the British Astronomical Association*, 18 (1907-1908), 185.

'Astronomical publications: Solar eclipse', *Journal of the British Astronomical Association*, 16 (1905-1906), 74.

'Astronomical Publications: Spanish Observations of the Eclipse of May 28', *Journal of the British Astronomical Association*, 11 (1900-1901), 127.

'Astronomical Publications: The August Total Eclipse of the Sun', *Journal of the British Astronomical Association*, 15 (1904-1905), 211.

'Astronomical Publications: The Eclipse of 30 August 1905', *Journal of the British Astronomical Association*, 15 (1904-1905), 346.

'Astronomical Publications: The Total Solar Eclipse of August 30', *Journal of the British Astronomical Association*, 16 (1905-1906), 119-120.

'Astronomical Publications: The Total Solar Eclipse of May 1900', *Journal of the British Astronomical Association*, 14 (1903-1904), 140.

'Astronomical Publications: Total Eclipse of the Sun, 30th August 1905', *Journal of the British Astronomical Association*, 16 (1905-1906), 40.

Burgos. Eclipse total de sol. 30 Agosto 1905 [Leaflet from the Bodleian Library]

'Capítulo IV. Otras observaciones fuera del Observatorio', *Memorias del Observatorio del Ebro*, 1 (1906), 43-51.

'Cuarto Congreso de la Unión Internacional para la investigación solar', *Boletín de la Sociedad Astronómica de Barcelona*, 1 (1910-1912), 33-34.

'Datos del Observatorio de Madrid', *El Liberal*, 17 April 1912 [Año 34, núm. 11852], 1.

'Después del eclipse', *El Heraldo de Madrid*, 30 May 1900, [año 11, nº 3487], 1.

'Después del eclipse', *El Imparcial*, 30 May 1900, 2.

'Después del eclipse', *El Liberal*, 19 April 1912 [Año 34, núm. 11854], p. 2.

'Día de tregua', *El Liberal*, 29 May 1900, 1.

'Discovery of comet *B* 1893', *Publications of the Astronomical Society of the Pacific*, 6 (1894), 146.

'Don José Joaquín Landerer', *Anuario de la Real Academia de Ciencias Exactas, Físicas y Naturales* (1923), 206-211.

'Eclipse de Sol', *El Liberal*, 15 April 1912 [Año 34, núm. 11850], 4.

'Eclipse de Sol', *El Liberal*, 17 April 1912 [Año 34, núm. 11852], 4.

'Eclipse del 21 de agosto en Hernösand (Suecia)', *Ibérica*, 2 (1914), 168.

'Eclipse expeditions planned for August', *New York Times*, 15 January 1905, SM3.

'Eclipse total de Sol', *El Heraldo de Madrid*, 12 May 1900 [Año 11, nº 3469], 3.

'El Eclipse de 30 agosto próximo. Expediciones yanquis', *El Mundo Científico*, 7 (1905), 465-466.

'El eclipse de ayer. En Madrid', *El Liberal*, 18 April 1912 [Año 34, núm. 11853], 1-2

'El eclipse de ayer', *El Imparcial*, 22 August 1914 [Año 48, núm. 17064], 3.

'El eclipse de hoy. La observación popular', *El Imparcial*, 18 April 1912 [Año 46, núm. 16212], 2.

'El Eclipse de Sol de 30 de Agosto', *El Mundo Científico*, 7 (1905), 544-550.

'El eclipse de Sol', *El Heraldo de Madrid*, 21 May 1900 [Año 11, nº 3478], 1.

'El Eclipse de Sol', *El Heraldo de Madrid*, 28 May 1900, [año 11, nº 3485], 1.

'El eclipse de Sol', *El Imparcial*, 21 August 1914 [Año 48, núm. 17063], 2.

'El eclipse de Sol', *El Imparcial*, 22 May 1900, 1.

'El eclipse de Sol', *El Imparcial*, 23 May 1900, 1.

'El eclipse de Sol', *El Imparcial*, 24 August 1914 [Año 48, núm. 17066], 3.

'El eclipse de Sol', *El Imparcial*, 27 July 1905, 2.

'El eclipse de Sol', *El Imparcial*, 28 July 1905, 1.

'El eclipse de sol', *El Imparcial*, 6 August 1905, 2.

'El eclipse de Sol', *El Liberal*, 22 August 1914 [Año 36, núm. 12064], 3.

'[El eclipse de Sol] En Elche', *El Heraldo de Madrid*, 26 May 190 [Año 11, nº 3483], 7.

'El eclipse del día 17', *El Imparcial*, 8 April 1912 [Año 46, núm. 16202], 5.

'El eclipse del día 21', *El Liberal*, 24 August 1914 [Año 36, núm. 12606], 3.

'El eclipse en Burgos', *El Imparcial*, 27 June 1905, 2.

'El eclipse total de Sol de este mes', *El Mundo Científico*, 7 (1905), 527-534.

'El eclipse total de sol del próximo agosto', *El Mundo Científico*, 7 (1905), 289-291.

'El eclipse total', *El Imparcial*, 29 May 1900, 1-2.

'El eclipse', *El Imparcial*, 27 May 1900, 2.

'El eclipse', *El Imparcial*, 28 May 1900, 1-2.

'El Eclipse', *El Imparcial*, 29 May 1900, 1.

'El mundo al día. El próximo eclipse y las borrascas magnéticas', *El Imparcial*, 8 July 1905, 1.

'El próximo eclipse de Sol', *El Imparcial*, 12 July 1914 [Año 48, núm. 17023], 1.

'El próximo eclipse de Sol', *El Liberal*, 18 July 1914 [Año 36, núm. 12569], 2.

'El próximo eclipse de Sol', *El Liberal*, 2 April 1912 [Año 34, núm. 11838], 2.

'El próximo eclipse total de Sol de 30 de Agosto. Recuerdos interesantes', *El Mundo Científico*, 7 (1905), 517-520.
'El próximo eclipse total de Sol. Observaciones por medio de globos', *El Mundo Científico*, 7 (1905), 210-211.
'En honor de Flammarion', *El Heraldo de Madrid*, 2 June 1900, [año 11, nº 3490], 3.
'En honor de Flammarion', *El Liberal*, 26 May 1900, 2.
'En honor de Flammarion', *El Liberal*, 3 June 1900, 1.
'En vísperas del eclipse', *El Heraldo de Madrid*, 26 May 1900 [Año 11, nº 3483], 3.
'Flammarion en España', *Blanco y Negro*, 2 June 1900, 4.
'Flammarion en España', *El Imparcial*, 18 May 1900, 1.
'Flammarion en España', *El Imparcial*, 24 May 1900, 3.
'Flammarion en España', *El Imparcial*, 25 May 1900, 1.
'Flammarion en Madrid', *El Liberal*, 1 June, 1900, 2.
'Flammarion en Madrid', *El Heraldo de Madrid*, 3 June 1900, [año 11, nº 3491], 1.
'Flammarion en Madrid', *El Heraldo de Madrid*, 31 May 1900, [año 11, nº 3488], 2.
'Flammarion en Palacio', *El Heraldo de Madrid*, 2 June 1900, [año 11, nº 3490], 1.
'Flammarion', *El Heraldo de Madrid*, 2 June 1900, [año 11, nº 3490], 1.
'Flammarion', *El Heraldo de Madrid*, 26 May 1900, [año 11, nº 3483], 2.
'Flammarion', *El Imparcial*, 2 Junio 1900, 2.
'Flammarion', *El Imparcial*, 3 Junio 1900, 1-2.
'Important. [Note about the] Total Solar Eclipse of 1905, August 30', [attached to] *Journal of the British Astronomical Association*, 15 (1904-1905) [issue 5].
'International Union for Cooperation in Solar Research', *Nature*, 72 (1905), 563.
'International Unit for Co-operation in Solar Research: Meeting at Oxford, 1905 September 27-29', *The Observatory*, 28 (1905), 382-387.
'L'eclipse de Soleil du 17 Juin', *L'Astronomie*, 9 (1890), 241-257.
'L'éclipse de soleil du 28 Mai 1900: Observations de M. de la Baume Pluvinel, à Elche', *Bulletin de la Société Astronomique de France*, 14 (1900), 331-332.
'L'éclipse de soleil du 28 Mai 1900: Observations de M. Hamy, à Elche', *Bulletin de la Société Astronomique de France*, 14 (1900), 326.
'L'éclipse de soleil du 28 Mai 1900: Observations de MM. Meslin, Bourget et Lebeuf, à Elche', *Bulletin de la Société Astronomique de France*, 14 (1900), 329-331.

'L'éclipse totale de Lune du 4 Novembre dernier', *L'Astronomie*, 12 (1893), 13-18.

'La fiesta de la ciencia', *El Imparcial*, 29 May 1900, 1.

'La Tache Rouge de Jupiter', *L'Astronomie*, 10 (1891), 321-322.

'Las fiestas del eclipse', *El Imparcial*, 29 June 1905, 2.

'Les Bandes d'ombre pendant les éclipses totales', *Revue générale des Sciences pures et appliquées*, 16 (1905), 1018.

'Los héroes de la ciencia', *El Mundo Científico*, 2 (1900), 417-418.

'Marcha del eclipse', *El Imparcial*, 16 April 1912 [Año 46, núm. 16210], 3.

'Ministerio de Fomento. Real Decreto', *Gaceta de Madrid*, 8 November 1899 [Año 238, num. 312, Tomo IV], 477.

'Ministerio de Hacienda. Reales Decretos', *Gaceta de Madrid*, 14 October 1899 [Año 238; num. 287; Tomo IV], 149.

'Ministerio de Hacienda. Real Orden', *Gaceta de Madrid*, 11 May 1905 [num. 131], 554.

'Mirando al cielo. El Eclipse de Sol', *El Universo*, 29 August 1905.

'No hay billetes', *El Heraldo de Madrid*, 27 May 1900, [año 11, nº 3484], 2.

'Notes sur les planètes intra-mercurielles. Vulcain et Pluton', *Les Mondes*, 48 (1879), 420-423.

'Notes: A New Observatory', *Journal of the British Astronomical Association*, 10 (1899-1900), 79.

'Notes: The Astronomical Society of Barcelona', *Journal of the British Astronomical Association*, 22 (1911-1912), 111-112.

'Notes: The Geographical Positions of Manzanares and Navalmoral', *Journal of the British Astronomical Association*, 11 (1900-1901), 322.

'Notes', *Nature*, 72 (1905), 490-494.

'[Notes:] Reports of the Solar Eclipse', *The Observatory*, 28 (1905), 430-431.

'[Notes:] The Coming Solar Eclipse of May 28', *The Observatory*, 23 (1900), 147.

'[Notes:] The following is a summary of work done by various parties during the Solar Eclipse', *The Observatory*, 28 (1905), 360.

'[Notes:] The Total Solar Eclipse of August 30 next', *The Observatory*, 28 (1905), 70.

'[Notes:] The Total Solar Eclipse of August 30', *The Observatory*, 28 (1905), 151.

'[Notes:] The Total Solar Eclipse of August 30', *The Observatory*, 28 (1905), 260-261.

'[Notes:] The Total Solar Eclipse', *The Observatory*, 28 (1905), 328-330.

'[Notes:] We…', *The Observatory*, 23 (1900), 261-262.
'Notices of the Association: The Eclipse of 1900, May 28', *Journal of the British Astronomical Association*, 10 (1899-1900), 235.
'Notices of the Association: Total Solar Eclipse of May 28, 1900', *Journal of the British Astronomical Association*, 10 (1899-1900), 268.
'Notiz', *Astronomische Nachrichten*, 188 (1911), 15-16.
'Nouvelles de la science.- Variétés : Une épée dans la Lune', *L'Astronomie*, 9 (1890), 75.
'Nouvelles de la Science.- Variétés. Conjonction de Saturne avec l'étoile double gamma de la Vierge', *L'Astronomie*, 12 (1893), 233-235.
'Nouvelles de la Science.- Variétés. L'éclipse totale de Soleil du 16 avril', *L'Astronomie*, 12 (1893), 230-233.
'Observaciones del eclipse de Sol de 30 de Agosto', *El Mundo Científico*, 7 (1905), 559-577.
'Observation de Jupiter faites à l'Observatoire de Juvisy, à Bruxelles et en Espagne. Observations de M. José Comas, à Barcelone', *L'Astronomie*, 12 (1893), 81-90.
'Observations de M. G. A. Quignon, à Burgos', *Bulletin de la Société Astronomique de France*, 19 (1905), 498.
'Observations de M. G. Tramblay, à Alcala de Chisvert (Espagne)', *Bulletin de la Société Astronomique de France*, 19 (1905), 499-500.
'Observations de MM. Georges, Henri et André Duval à Calatayud (Espagne)', *Bulletin de la Société Astronomique de France*, 19 (1905), 498-499.
'Observations made with selenium cells during the total solar eclipse of August 30, 1905', *Astrophysical Journal*, 23 (1906), 153-161 [Translated, condensed and diagrams redrawn by William F. Rigge from Wulf, Th. and Lucas, J.D., 'Zwei Beobachtungen mittels Selenzellen bei der totalen Sonnenfinsterniss am 30 August 1905', *Physikalische Zeitschrift*, 6 (1905), 838-847].
'[Observatories:] Solar Physics Observatory, South Kensington', *The Observatory*, 36 (1913), 376.
'[Observatories:] Tortosa…', *The Observatory*, (1907), 105.
'Observatory of the Ebro. Notes of the 1905 Eclipse', *Knowledge & Scientific News*, 4 (1907), 3-5.
'Our astronomical column: Atmospheric Origin of "Shadow Bands"', *Nature*, 72 (1905), 611.
'Our Astronomical Column: Employment of Selenium Cells during Total Solar Eclipse', *Nature*, 73 (1905-1906), 617.
'Our astronomical column: French Observations of the Total Solar Eclipse', *Nature*, 72 (1905), 518.

'Our astronomical column: Further Results obtained by the French Eclipse Expeditions', *Nature*, 72 (1905), 567.

'Our astronomical column: Further Results of the French Eclipse Expeditions', *Nature*, 72 (1905), 639-640.

'Our Astronomical Column: Magnetic Observations during the Total Eclipse', *Nature*, 73 (1905-1906), 400-401.

'Our astronomical column: Solar Eclipse Problems', *Nature*, 71 (1904-1905), 329.

'Our book shelf *A la Poursuite d'une Ombre*. By Proof. Moye. Pp. 98. (Montpelier: G. Firmin, 1905.)', *Nature*, 73 (1905-1906), 606.

'Para el eclipse de mañana. Guía práctica del observador', *El Heraldo de Madrid*, 27 May 1900, [año 11, nº 3484], 1.

'Para el eclipse: Servicio de trenes', *El Heraldo de Madrid*, 26 May 1900 [año 11; nº 3483], 2.

'Películas del eclipse', *El Imparcial*, 16 April 1912 [Año 46, núm. 16210], 3.

'Perfiles del día. El eclipse en Burgos', *El Imparcial*, 30 June 1905, 2.

'Petrography in Spain', *Nature*, 21 (1879-1880), 165.

'Preparativos para la observación', *El Imparcial*, 16 April 1912 [Año 46, núm. 16210], 3.

'Real Orden Circular', *Gaceta de Madrid*, 11 August 1905 [num. 22], 548.

'Recent publications', *Montly Notices of the Royal Astronomical Society*, 22 (1861-1862), 22.

Reflexiones sobre el sistema planetario (Santa Cruz de Tenerife: Imp. Isleña… de J.C. Hernandez, 1870).

'Report of the Council to the Eighty-fifth Annual General Meeting of the Society. Notes on some Points connected with the Progress of Astronomy during the Past Year', *Monthly Notices of the Royal Astronomical Society*, 65 (1905), 382-391.

'Report of the Council to the Eighty-sixth Annual General Meeting of the Society. Notes on some Points connected with the Progress of Astronomy during the Past Year', *Monthly Notices of the Royal Astronomical Society*, 66 (1906), 216-238.

'Report of the Council to the Eighty-sixth Annual General Meeting of the Society. Proceedings of Observatories', *Monthly Notices of the Royal Astronomical Society*, 66 (1906), 182-215.

'Report of the Council to the Ninety-third Annual General Meeting of the Society. Notes on some Points connected with the Progress of Astronomy during the Past Year', *Monthly Notices of the Royal Astronomical Society*, 73 (1913), 266-311.

'Report of the meeting of the Association held on February 22, 1899', *Journal of the British Astronomical Association*, 9 (1898-1899), 193-200.

'Report of the meeting of the Association held on June 28, 1899', *Journal of the British Astronomical Association*, 9 (1898-1899), 357-365.

'Report of the meeting of the Association held on March 29, 1899', *Journal of the British Astronomical Association*, 9 (1898-1899), 241-251.

'Report of the Meeting of the Association, held on January 25, 1905, at Sion College, Victoria Embankment', *Journal of the British Astronomical Association*, 15 (1904-1905), 155-161.

'Report of the Meeting of the Association, held on June 27, 1900, at Sion College, Victoria Embankment', *Journal of the British Astronomical Association*, 10 (1899-1900), 341-351.

'Report of the Meeting of the Association, held on June, 29, 1898', *Journal of the British Astronomical Association*, 8 (1897-1898), 363-373.

'Report of the Meeting of the Association, held on May 25, 1904, at Sion College, Victoria Embankment', *Journal of the British Astronomical Association*, 14 (1903-1904), 303-315.

'Report of the Meeting of the Association, held on May 30, 1900, at Sion College, Victoria Embankment', *Journal of the British Astronomical Association*, 10 (1899-1900), 309-315.

'Société Française de Physique' *Revue générale des Sciences pures et appliquées*, 18 (1907), 336.

'Sonnenfinsternis 1905 August 30', *Astronomische Nachrichten*, 169 (1905), 263-266.

'Telescopios a perra gorda', *El Liberal*, 29 May 1900, 2.

'The coming eclipse of the sun', *The Times*, 24 May 1900.

'The Solar Eclipse', *The Times*, 26 May 1900.

'The Spanish Provisional Programme of Balloon Observations in the coming Total Eclipse', *The Aëronautical Journal*, 9 (1905), 47-49.

'The total eclipse of the Sun, 1914 August 21', *The Observatory*, 37 (1914), 102-103.

'The Total Eclipse. Accounts from all Sources', *Knowledge & Scientific News*, 2 (1905), 246-247.

'The total solar eclipse of 1905, August 30', *The Observatory*, 28 (1905), 355-356.

'The total solar eclipse of August 30, 1905', *The Observatory*, 28 (1905), 377-382.

'The Total Solar Eclipse of August 30', *The Observatory*, 28 (1905), 260-261.

'The Total Solar Eclipse of May, 1900', *Nature*, 69 (1903-1904), 160.

'Topics of the Times', *New York Times*, 28 December 1897, 6.

'Trenes baratos', *El Heraldo de Madrid*, 23 May 1900, [año 11, nº 3480], 3.

'Una información curiosa', *El Imparcial*, 17 April 1912 [Año 46, núm. 16211], 1.

'Visita de Mr. y Mme. Flammarion a "Blanco y Negro"', *Blanco y Negro*, 9 June 1900, 7-8.

Other printed primary sources

A. A., 'El eclipse de Sol de 30 de Agosto', *El Mundo Científico*, 7 (1905), 544-547.

A. F., 'The Total Solar Eclipse, 1914 August 20-21', *Monthly Notices of the Royal Astronomical Society*, 75 (1915), 315-317.

Aguilar, Antonio, 'Mesures prises en Espagne pour l'observation de la prochaine éclipse totale de soleil' [Lettre de M. Aguilar à M. Le Verrier], *Comptes Rendus*, 50 (1860), 483-484.

—. 'Informe del Director del Observatorio de Madrid al Comisario Regio…', *Anuario del Real Observatorio de Madrid. Año VIII.-1868* (Madrid: Imprenta Nacional, 1867).

Alcalde, 'Homenaje a un astrónomo español', *El Mundo Científico*, 7 (1905), 241-245.

Almanaque náutico y efemérides astronómicas para el año... calculadas de orden de S.M. para el Observatorio Real de Cádiz (Madrid: En la Imprenta Real, 1791-1959).

André, C., 'Appareil à eclipses artificielles de Soleil', *Comptes Rendus*, 141 (1905), 168-169.

—. 'Sur l'éclipse totale du Soleil du 30 août 1905 à Tortosa', *Comptes Rendus*, 141 (1905), 867-869.

—. 'Quelques remarques sur les observations des contact dans les éclipses totales de Soleil', *Comptes Rendus*, 143 (1906), 806-809.

André, C and Rayet, G., *L'astronomie practique et les observatoires en Europe et en Amérique, despuis le milieu du XVIIe siècle jusqu'à nos jours. Vol. 1-2* (Paris: Gauthier-Villars, 1874).

Arcimis, Augusto, 'Observations of the zodiacal light at Cadiz', *Monthly Notices of the Royal Astronomical Society*, 36 (1875-1876), 48.

—. 'Observations of the lunar eclipse, 1876 September 3, made at Cadiz', *Monthly Notices of the Royal Astronomical Society*, 37 (1876-1877), 12-13.

—. 'On the visibility of the unilluminated portion of the disc of Venus', *Monthly Notices of the Royal Astronomical Society*, 37 (1876-1877), 259.

—. 'Phenomena of Jupiter's satellites observed at Cadiz', *Monthly Notices of the Royal Astronomical Society*, 37 (1876-1877), 259-260.

—. 'Observations of occultations of stars by the Moon, made at Cadiz in the year 1876', *Monthly Notices of the Royal Astronomical Society*, 37 (1876-1877), 261.

—. 'Lunar eclipse of 1877, February 27, observed at Cadiz', *Monthly Notices of the Royal Astronomical Society*, 37 (1876-1877), 400.

Armelin, Gaston, 'Société Astronomique de France. Séance du 1er octobre 1890', *L'Astronomie*, 9 (1890), 424-428.

—. 'Société Astronomique de France. Séance du 3 Mai 1893', *L'Astronomie*, 12 (1893), 226-229.

—. 'Société Astronomique de France. Séance du 3 Janvier 1894', *L'Astronomie*, 13 (1894), 70-73.

Ascarza, Victoriano F., 'El análisis espectral', *El Heraldo de Madrid*, 26 May 1900 [Año 11, nº 3483], 6.

—. 'Nota sobre la longitud de onda de la raya verde (1474 K) del espectro de la corona solar', *Astronomische Nachrichten*, 155 (1901), 23-24.

—. 'L'éclipse totale de Soleil du 21 août 1914, observée par la Comisión espagnole à Théodosia (Crimée)', *Comptes Rendus*, 159 (1914), 738-740.

B. B., 'El eclipse en Burgos', *La Ilustración Española y Americana*, 81 (1905) [Año 49, número 33], 131-134.

Barrett, Storrs B. (comp.), *The Astrophysical Journal: A General Index by authors and by subjects to volumes XXVI-L (July, 1907, to December, 1919)* (Chicago: University of Chicago, 1920).

Bauer, L.A., 'Appeal for Cooperation in Magnetic and Allied Observations during the Total Solar Eclipse of August 29-30, 1905', *Nature*, 70 (1904), 577.

—. 'Proposed Magnetic and Allied Observations during the Total Solar Eclipse on August 30', *Nature*, 72 (1905), 342.

Bentabol y Ureta, Horacio, *Preparación é instrucciones para observar con aprovechamiento el eclipse total de sol de 30 de agosto de 1905 estableciendo la existencia y condiciones de la atmósfera lunar y explicando el origen de todos los fenómenos que suelen presentar los eclipses de sol* (Madrid: Est. Tip. De la viuda é hijos de M. Tello, 1905).

Biske, F., 'Are diffraction phenomena possible at solar eclipses?', *Astrophysical Journal*, 38 (1913), 192-196.

Bosler, J. and Block, H. G., 'Observations de l'éclipse de Soleil du 21 août 1914, faites à Strömsund (Suède), par la Mission de l'Observatoire de Meudon', *Comptes Rendus*, 159 (1914), 766-769.

British Astronomical Association, 'The total solar eclipse of May 28th, 1900: Proposed Expedition to Spain and Algeria', *Journal of the British Astronomical Association*, 10 (1899-1900) [Circular attached to number 4].

Brown, E., 'A Visit to the Madrid Observatory', *Journal of the British Astronomical Association*, 4 (1893-1894), 361-362.

Bryant, W.W., *A history of astronomy* (London: Methuen & Co., 1907).

Buchanan, J. Y., 'Letters to the Editor: Eclipse Predictions', *Nature*, 72 (1905), 603.

Burton-Brown, Alexander, 'Notes on Stations for viewing the Total Eclipse of the Sun. 29-30 August 1905', *Journal of the British Astronomical Association*, 15 (1904-1905), 333-334.

Callendar, H. L. and Fowler, A., 'Report of the Expedition to Castellón de la Plana, Spain', *Proceedings of the Royal Society of London. Series A, Containing Papers of a Mathematical and Physical Character*, 77 (1906), 1-28.

Campbell, W. W., 'The Wave-length of the Green Coronal Line and Other Data Resulting from an Attempt to Determine the Law of Rotation of the Solar Corona', *Astrophysical Journal*, 10 (1899), 186-192.

—. 'The Crocker Eclipse Expeditions in 1905', *Astrophysical Journal*, 20 (1904), 235-236.

Carpenter, Capt. Alfred, 'Correctness of Positions at Manzanares, Spain', *Journal of the British Astronomical Association*, 12 (1901-1902), 27-28.

Carrasco, P., 'Observations physiques faites à Théodosia, pendant l'éclipse totale de Soleil du 21 août 1914' [présentée par Bigourdan], *Comptes Rendus*, 159 (1914), 740-741.

—. 'Sonnenfinsternis vom 21. August 1914. Observación del espectro coronal a Teodosia', *Astronomische Nachrichten*, 200 (1914-1915), 147-148.

—. *Observaciones espectrográficas del eclipse del eclipse de sol del 17 de abril de 1912* [*Revista de la Sociedad Española de Física y Química*] (Madrid: Imprenta de E. Arias, 1915).

Cazador, Manuel, 'El eclipse de Sol de 30 de Agosto de 1905 en el Colegio de San Julián de Vilatorta (Vich)', *El Mundo Científico*, 7 (1905), 696-697.

Chambers, George F., *The Story of the Eclipses* (London: George Newnes, 1899).

—. 'The Total Eclipse of the Sun of 1905', *Journal of the British Astronomical Association*, 12 (1901-1902), 167-169 [Read at the Meeting of the Association held on January 29, 1902].

—. 'The Total Eclipse of the Sun of 1905', *Journal of the British Astronomical Association*, 12 (1901-1902), 280-281.

—. 'The Total Eclipse of the Sun, 30th August 1905', *Journal of the British Astronomical Association*, 15 (1904-1905), 93-98.

—. 'The Coming Total Eclipse of the Sun', *Journal of the British Astronomical Association*, 15 (1904-1905), 164-166.

—. 'Proposed Eclipse Expedition to the Mediterranean', *Journal of the British Astronomical Association*, 15 (1904-1905), 384-385.

—. 'The Total Eclipse of the Sun of April 17, 1912', *Journal of the British Astronomical Association*, 22 (1911-1912), 92-99.

Chester, Colby M., 'The Total Solar Eclipse of August 29-30, 1905', *Science*, 21 (1905), 635-637.

—. 'Preliminary Results of the United States Naval Observatory Eclipse Expedition in 1905', *Astrophysical Journal*, 23 (1906), 128-143.

Cirera, Ricardo, 'Eclipse totale et annulaire de soleil du 17 avril 1912', *Astronomische Nachrichten*, 162 (1903), 101-102.

—. 'Observations magnétiques de l'Observatoire de l'Ebre à l'occasion de l'éclipse de Soleil du 30 août 1905', *Comptes Rendus*, 141 (1905), 1270-1271 [presentée par M. Mascart].

—. *Notice sur l'observatorie et sur quelques observations de l'éclipse du 30 Août 1905* (Barcelona: Gustavo Gili, 1906) [Édition Française; Traduction par E. Merveille; *Mémoires de l'Observatoire de l'Ebre*, 1 (1906)].

—. 'Prólogo', in Merveille, E. *La Section Magnétique. Memorias del Observatorio del Ebro*, 3 (1908), 5-8.

—. 'Discurso relativo al establecimiento de la nueva sección de astronomía y física del globo', in *Asociación Española para el Progreso de las Ciencias, Primer Congreso (Zaragoza, 22 al 29 de Octubre de 1908), Tomo I* (Madrid: Eduardo Arias, 1908), pp. 61-66.

—. 'Prólogo. El Observatorio del Ebro', *Boletín del Observatorio del Ebro*, 1 (1910), 7-14.

—. 'Recientes progresos de las ciencias astronómicas en España', in *Asociación Española para el Progreso de las Ciencias, Tercer*

Congreso (Granada, del 20 al 25 de Junio de 1911), Tomo I (Madrid: Eduardo Arias, 1911), pp. 51-88.

—. 'Los eclipses de 8 de mayo de 1910 y 28 de abril de 1911', in *Asociación Española para el Progreso de las Ciencias (Valencia), Tomo III* (Madrid: Eduardo Arias, 1911), pp. 97-100.

Clerke, Agnes M., *A Popular History of Astronomy during the Nineteenth Century* (Edinburgh: Alan & Charles Black, 1885 [4th ed., London: A. & C. Black, 1902]).

Cobham, A. B., 'Astronomical Drawing', *Journal of the British Astronomical Association*, 15 (1904-1905), 102-103.

Comas i Solà, Josep, José, 'Nouvelles de la Science.- Variétés : L'étoile multiple σ d'Orion', *L'Astronomie*, 9 (1890), 154.

—. 'Nouvelles de la Science.- Variétés : Antarès', *L'Astronomie*, 9 (1890), 311-312.

—. 'Nouvelles de la Science.- Variétés : Saturne, ζ du Cancer, Antarès', *L'Astronomie*, 9 (1890), 353-354.

—. 'El eclipse de Sol del 28 mayo 1900', *Astronomische Nachrichten*, 153 (1900), 85-88.

—. 'Observaciones de Júpiter durante la Oposición de 1900', *Astronomische Nachrichten*, 153 (1900), 417-430.

—. 'Observations de MM. José Comas Sola, directeur de l'Observatoire Fabra et Salvador Raurich, à Vinaroz (Espagne)', *Bulletin de la Société Astronomique de France*, 19 (1905), 495.

—. 'Observations sur l'eclipse totale du Soleil du 30 août 1905 [présentée par M. Loewy]', *Comptes Rendus des Séances de l'Académie des Sciences de Paris*, 141 (1905), 616-618.

—. 'La corona solar del eclipse de 1905', *El Mundo Científico*, 7 (1905), 818-819.

—. 'El eclipse total de sol del 30 de agosto de 1905', *Memorias de la Real Academia de Ciencias y Artes de Barcelona*, 5 [24] (1906), 479-512.

—. 'Quelques considérations sur mes observations de Mars', *Journal of the British Astronomical Association*, 21 (1910-1911), 199-201.

—. 'Observaciones de Marte', *Astronomische Nachrichten*, 191 (1912), 5-8.

—. 'Sonnenfinsternis 1912 April 17. Mitteilung von Herrn *José Comas Solá*, Direktor des Observatoriums Fabra in Barcelona', *Astronomische Nachrichten*, 192 (1912), 205-207.

—. 'Sur l'accélération de la Tache Rouge de Jupiter pendant les conjonctions avec la tache grise tempérée australe', *Astronomische Nachrichten*, 195 (1913), 413-414.

Compañía de Jesús, *Observaciones del eclipse total de Sol del 30 de agosto de 1905, hechas por los padres de la... en el Colegio de Oña* (Oña: Imprenta privada del Colegio, 1906).

Copeland, Ralph, 'Preliminary Note on Observations of the Total Solar Eclipse of 1900 May 28, made at Santa Pola (Casa del Pleito), Spain', *Proceedings of the Royal Society of London*, 67 (1900), 385-391.

Crawford, R. T., 'Students' Opinions of the Educational Advantages of the Study of Astronomy', *Publications of the Astronomical Society of the Pacific*, 29 (1917), 174-176.

Crommelin, A. C. D., 'The Weather Prospects for the Total Solar Eclipse of May 28, 1900, in Spain, Portugal, and Algiers', *Journal of the British Astronomical Association*, 9 (1898-1899), 204-207.

—. 'The Coming Spanish Eclipse. Probable Weather Conditions' *Journal of the British Astronomical Association*, 15 (1904-1905), 138-141.

—. 'Further Note on the Eclipse of 1912', *Journal of the British Astronomical Association*, 22 (1911-1912), 226.

de Cuenca, Carlos Luis, 'Del, con, en, por, sin, sobre el eclipse', *La Ilustración Española y Americana*, 81 (1905) [Año 49, número 32], 123.

de la Baume Pluvinel, A., 'Fotografías del eclipse', *El Heraldo de Madrid*, 26 May 1900 [Año 11, nº 3483].

—. 'L'éclipse totale de Soleil du 28 Mai 1900', *Bulletin de la Société Astronomique de France*, 14 (1900), 23-33.

de la Grye, Bouquet, 'Les Progrès de l'Astronomie en 1891', *Bulletin de la Société Astronomique de France*, 6 (1892), 151-153.

de Montessus de Ballore, R. (dir.) *Index generalis: annuaire général des universités, grandes écoles, académies, archives, bibliothèques...* (Paris: Gauthier-Villars, 1920).

Deslandres, H., 'Observations of the Total Solar Eclipse of May 28, 1900, at Argamasilla, Spain', *Astrophysical Journal*, 12 (1900), 287-290.

—. 'Origen eléctrico de las protuberancias, de la cromosfera y de la corona solar', *El Heraldo de Madrid*, 26 May 1900 [Año 11, nº 3483], 5.

—. 'Spectres ultra-violets de la couche renversante pendant l'éclipse totale du 28 mai 1900', *Comptes Rendus*, 141 (1905), 409-414.

—. 'Note prélimiaire sur l'observation de l'éclipse totale du Soleil du 30 août 1905, à Burgos', *Comptes Rendus*, 141 (1905), 517-519.

—. 'Observation de l'éclipse totale du 21 août 1914 par la Mission de l'Observatoire de Meudon', *Comptes Rendus*, 159 (1914), 669-670.

Dicenta, Joaquín, 'Flammarion. Instantánea', *El Liberal*, 29 May 1900, 1.

Dirección del Instituto Geográfico y Estadístico, *Coordenadas geográficas de puntos comprendidos en la zona de la totalidad del eclipse de sol de*

30 de agosto de 1905 (Madrid: Instituto geográfico y estadístico, 1905).

—. *Eclipse total de sol de 30 de Agosto de 1905. Reseña de los trabajos efectuados para su observación* (Madrid: Imprenta de la Dirección General del Instituto Geográfico y Estadístico, 1906).

Donitch, Nicolas, 'Observations de l'éclipse totale du Soleil du 29-30 août 1905', *Bulletin de l'Académie Impériale des Sciences de St Pétersbourg*, 23 (1905), 283-288.

Downing, A. M. W., 'Total Solar Eclipse of May 28th, 1900, in central Spain', *Journal of the British Astronomical Association*, 10 (1899-1900), 263-264.

—. 'The Total Solar Eclipse of August 1905', *Journal of the British Astronomical Association*, 14 (1903-1904), 367-368.

—. 'Letters to the Editor: Eclipse Predictions', *Nature*, 72 (1905), 629.

Draper, John William, *History of the Conflict Between Religion and Science* (London: H. S. King, 1875).

—. *Historia de los conflictos entre la religión y la ciencia* (Madrid: Imprenta de Aribau y Cª, 1876).

E. H. H., 'Council Notes on some points connected with the progress of Astronomy', *Monthly notices of the Royal Astronomical Society*, 59 (1899), 288.

—. 'Council Notes on some points connected with the progress of Astronomy', *Monthly notices of the Royal Astronomical Society*, 60 (1900), 376.

—. 'Council Notes on some points connected with the progress of Astronomy', *Monthly notices of the Royal Astronomical Society*, 61 (1901), 251.

Echegaray, José, 'El Sol y el eclipse', *El Heraldo de Madrid*, 26 May 1900 [Año 11, nº 3483], 1.

Esquirol, J., 'Éclipse totale de Soleil du 30 août 1905. Protubérances solaires à deux couleurs [présentée par Deslandres]', *Comptes Rendus*, 142 (1906), 757.

Evershed, J., 'Wave-Length Determinations and General Results obtained from Detailed Examination of Spectra photographed at the Solar Eclipse of January 22, 1898', *Proceedings of the Royal Society of London*, 68 (1901), 6-9.

—. 'Report of the Expedition to Pineda de la Sierra, Spain', *Proceedings of the Royal Society of London. Series A, Containing Papers of a Mathematical and Physical Character*, 77 (1906), 42-55.

Fabry, Charles, 'Sur l'intensité lumineuse de la couronne solaire pendant l'éclipse totale du 30 août 1905 [présentée par Deslandres]', *Comptes Rendus*, 141 (1905), 870-873.

Fernández Ascarza, Victoriano, 'Eclipse total de sol de 1900.- Nota sobre la longitud de onda de la raya verde (1474 K), del espectro de la corona solar', *Revista General de Marina* (Abril 1901), 1-15.

Flammarion, Camille, 'Nouvelles observations sur la planète Mars', *L'Astronomie*, 11 (1892), 36-382.

—. 'Les Progrès de la Société Astronomique de France', *Bulletin de la Société Astronomique de France*, 6 (1892), 159-163.

—. 'El eclipse total de Sol del 28 de Mayo', *El Liberal*, 18 May 1900, 6.

—. 'L'éclipse totale de soleil', *Bulletin de la Société Astronomique de France*, 14 (1900), 289-297.

—. '[L'éclipse totale de soleil] Notes complémentaires', *Bulletin de la Société Astronomique de France*, 14 (1900), 298-307.

—. 'La Société Astronomique de France', *Bulletin de la Société Astronomique de France*, 15 (1901), 205-217.

—. 'El eclipse total de Sol del 30 de agosto', *El Imparcial*, 14 August, 1905, 1.

Fontserè, Eduard, 'El eclipse total de sol del 28 de mayo de 1900', *El Mundo Científico*, 2 (1900), 259-263.

—. 'Eclipse total de 28 de mayo de 1900', *El Mundo Científico*, 2 (1900), 403-410.

—. 'Resultado generales del eclipse de Sol de 28 de Mayo de 1900', *El Mundo Científico*, 2 (1900), 419-422.

Frost, E. B., 'Astronomical Publications: The Total Solar Eclipse of 1900, May 28', *Journal of the British Astronomical Association*, 10 (1899-1900), 224.

Galán, Gabriel, 'El eclipse de Sol del 17 de abril de 1912', *El Liberal*, 7 April 1912 [Año 34, núm. 11842], 3.

—. 'El eclipse de Sol del 17 de Abril. El parcial y el total', *El Liberal*, 15 April 1912 [Año 34, núm. 11850], 1.

Garrido, Ricardo, 'El estudio espectral de las manchas del sol' in *Asociación Española para el Progreso de las Ciencias, Tercer congreso (Granada, del 20 al 25 de junio de 1911). Tomo 1-3* (Madrid: Eduardo Arias, 1911), pp. 73-92.

—. 'Estadística solar. Enero-diciembre 1920', in *La Estación Sismológica y el Observatorio Astronómico y Meteorológico de Cartuja (Granada). Memorias y trabajos de vulgarización científica* (Granada: Imprenta Gráfica Granadina, 1921).

Gaussin, Louis, 'Lois concernant la distribution des astres du système solaire', *Comptes Rendus*, 90 (1880), 518-520.

—. 'Lois concernant la distribution des astres du système solaire', *Comptes Rendus*, 90 (1880), 593-596.

Gautier, Raoul, 'L'éclipse totale de soleil du 30 aout 1905: observations de la Mission Astronomique Suisse a Santa Ponza (Ile de Majorque)', *Archives des Sciences Physiques et Naturelles*, 20 (1905), 650-674.

Gérigny, Philippe, 'Société Astronomique de France. Séance du 8 janvier 1890', *L'Astronomie*, 9 (1890), 69-73.

—. 'Société Astronomique de France. Séance du 5 février 1890', *L'Astronomie*, 9 (1890), 103-105.

—. 'Société Astronomique de France. Séance du 5 mars 1890', *L'Astronomie*, 9 (1890), 148-150.

—. 'Société Astronomique de France. Séance du 3 décembre 1890', *L'Astronomie*, 10 (1891), 70-72.

Gila y Fidalgo, Félix, *El eclipse total de Sol del día 30 de Agosto de 1905 en la provincia de Segovia* (Segovia: Diario de Aviso, 1905).

Grau Gadea, J. M., 'Desde Elche', *El Heraldo de Madrid*, 13 May 1900 [Año 11, nº 3470], 1.

Grubb, Howard, 'The New Spectroheliogaph for the Madrid Observatory', *Report of the British Association for the Advancement of Science* (1908, Dublin), 603-604.

Hale, George E. and Perrine, C. D., 'International Co-operation in Solar Research', *Astrophysical Journal*, 20 (1904), 301-305.

Hale, George E. and Ellerman, Ferdinand, 'The Minute Structure of the Solar Atmosphere', *Proceedings of the National Academy of Sciences of the United States of America*, 2 (1916), 102-108.

Herrera y Linares, Emilio, 'Observación aerostática de las sombras volantes en el eclipse de sol del 30 de agosto de 1905', *Asociación Española para el Progreso de las Ciencias, Tercer Congreso (Granada del 20 al 25 de Junio de 1911). Tomo III* (Madrid: Eduardo Arias, 1911), pp. 65-71.

—. *Memorias*, ed. Thomas F. Glick and José María Sánchez Ron (Madrid: Universidad autónoma de Madrid, 1988).

Huggins, William, *The Royal Society* (London: Methuen & Co., 1906).

—. (ed.) *The Scientific Papers of Sir William Huggins* (London: William Wesley and Son, 1909).

Ibarra, Pere, 'Desde Elche. Multitud de sucesos y escasez de cronistas...', *Las Provincias*, 23 May 1900, in in Vicent F. Soler Selva (coord.) *L'eclipsi total de sol de 1900 al Baix Vinalopó* (Elx: Ajuntament d'Elx, 2000).

—. 'Desde Elche. La fotografía y el eclipse...', *Las Provincias*, 27 May 1900, in Vicent F. Soler Selva (coord.) *L'eclipsi total de sol de 1900 al Baix Vinalopó* (Elx: Ajuntament d'Elx, 2000).

—. 'Desde Elche', *Las Provincias*, 27 May 1900, in Vicent F. Soler Selva (coord.) *L'eclipsi total de sol de 1900 al Baix Vinalopó* (Elx: Ajuntament d'Elx, 2000).

Instituto Solar Internacional, *Eclipses de sol interior: comunicación científica* (Montevideo: Instituto Solar Internacional, 1900).

Iñiguez, Francisco, 'Observaciones del Eclipse Total de Sol del 28 de Mayo de 1900', *Journal of the British Astronomical Association*, 11 (1900-1901), 84.

—. 'Observation de l'éclipse de Soleil du 17 avril, faite à l'Observatoire de Madrid' [présentée par Bigourdan], *Comptes Rendus*, 154 (1912), 1066.

—. 'Observatoire de Madrid', *Bureau des Longitudes. Observations de l'Éclipse de Soleil du 17 Avril 1912*, 51-52.

—. 'El eclipse y el Observatorio de Madrid', *El Liberal*, 20 April 1912 [Año 34, núm. 11855], 3.

—. 'El proceso del eclipse', *El Imparcial*, 23 April 1912 [Año 46, núm. 16217], 2.

Janssen, J., 'Sur l'éclipse solaire totale du 30 août 1905', *Comptes Rendus*, 141 (1905), 233-234.

—. 'Observation de l'éclipse totale du 30 août 1905 à Alcosebre (Espagne)', *Comptes Rendus*, 141 (1905), 569-571.

—. 'Observations de M. J. Janssen, directeur de l'Observatoire de Meudon, à Alcosèbre (Espagne)', *Bulletin de la Société Astronomique de France*, 19 (1905), 491.

Jiménez Landi, Pedro, *Fotografía Celeste. Tesis de doctorado* (Madrid: Bailly-Bailliere, 1906).

Johnson, Rev. S. J., 'Remarks on the Total Eclipse of the Sun, 1900 May 28, observed at Navalmoral (Spain)', *Monthly Notices of the Royal Astronomical Society*, 60 (1900), 590-591.

Kononowitsch, A., 'Beobachtung der Sonnenfinsterniss 1900 Mai 28', 154 (1901), 15-16.

Krebs, Wilhelm, 'Fliegende Schatten bei der Sonnenfinsternis vom 30. August 1905', *Astronomische Nachrichten*, 170 (1906), 283-284.

Lànderer, Josep Joaquim., 'Rapport entre les distances mutuelles des planètes et des satellites', *Les Mondes*, 15 (1867), 61-62.

—. *Principios de Geología y Paleontología* (Barcelona: Librería Religiosa, 1878).

—. 'Essai de sélénologie', *Les Mondes*, 51 (1880), 854-856.

—. 'Histoire ancienne de la lune', *Les Mondes*, 54 (1881), 184-192.
—. 'Las revoluciones del globo lunar', *Anales de la Sociedad Española de Historia Natural*, 11 (1882), 153-191.
—. 'Sur l'angle de polarisation de lune', *Comptes Rendus*, 109 (1889), 360-362.
—. 'Sur l'angle de polarisation des roches ignées et sur les premières déductions sélénologiques qui s'y rapportent', *Comptes Rendus*, 111 (1890), 210-212.
—. 'Demonstración matemática del cristianismo', *Crónica Científica*, 14 (1891), 229-230.
—. 'La tache rouge de Jupiter', *L'Astronomie*, 11 (1892), 394-395.
—. 'Diàmeter des satellites de Jupiter', *L'Astronomie*, 12 (1893), 255-256.
—. 'Sur la théorie des Satellites de Jupiter', *L'Astronomie*, 13 (1894), 256-257.
—. 'La scintillation et les climats', *L'Astronomie*, 13 (1894), 341-343.
—. 'Sur un passage de l'ombre du quatrième satellite de Jupiter', *Comptes Rendus*, 120 (1895), 248-250.
—. 'Sur la proportion de lumière polarisée de la couronne solaire', *Comptes Rendus*, 130 (1900), 1524-1525.
—. 'El eclipse total de sol del 30 de agosto de 1905', *La Ilustración Española y Americana*, 22 (1902), 375-378.
—. 'L'eclipse totale de soleil des 29-30 aôut 1905', *Astronomische Nachrichten*, 159 (1902), 125-130.
—. 'Le problème cosmologique', *Bulletin de la Société Astronomique de France*, 16 (1902), 270-274.
—. 'L'eclipse totale de soleil des 29-30 aôut 1905 (Duexième Note)', *Astronomische Nachtrichten*, 160 (1903), 309-312.
—. 'L'Eclipse totale de soleil du 30 aout 1905', *Bulletin de la Societé Astronomique de France*, 17 (1903), 105-108.
—. 'L'éclipse totale de Soleil du 30 août 1905 au nord de l'Afrique', *Bulletin de la Société Astronomique de France*, 17 (1903), 374-376.
—. 'L'éclipse totale de Soleil du 30 août 1905', *Bulletin de la Société Astronomique de France*, 17 (1903), 405-408.
—. 'En vísperas del eclipse', *La Ilustración Española y Americana*, 81 (1905) [Año 49, número 31], 93.
—. 'Antes del eclipse', *La Ilustración Española y Americana*, 80 (1905) [Año 49, número 14], 219-222.
—. 'Sur la lumière polarisée de la couronne solaire' [presentée par Janssen], *Comptes Rends*, 141 (1905), 589-590.
—. 'Sur la lumière polarisée de la couronne solaire', *Bulletin de la Société Astronomique de France*, 19 (1905), 495-497.

—. 'Sur la polarisation de la lumière lunaire', *Comptes Rendus*, 150 (1910), 1164-1165.
—. *La evolución del globo lunar* (Barcelona: Herederos de Juan Gili, 1910).
—. 'L'éclipse annulaire et totale de soleil des 16-17 avril 1912', *Bulletin de la Société Astronomique de France*, 25 (1911), 248-251.
—. 'Sur l'éclipse de Soleil du 17 avril 1912' [présentée par Bigourdan], *Comptes Rendus*, 154 (1912), 1279-1281.
—. 'Observations de M. Landerer', *Bureau des Longitudes. Observations de l'Éclipse de Soleil du 17 Avril 1912*, 54-55.
—. 'Observations de … à Tortosa (Espagne), *Bulletin de la Société Astronomique de France*, 28 (1914), 444.
—. 'Sur l'éclipse totale de Soleil des 20 et 21 aôut 1914' [présenteé par Bigourdan], *Comptes Rendus*, 159 (1914), 468-469.
—. 'El eclipse total de sol del 21 de agosto de 1914', *Ibérica*, 1 (1914), 299-301.
Langley, S.P., 'The New Spectrum', *Philosophical Magazine*, 2 (1901), 119-130.
Le Cadet, G., 'Mesures de l'intensité du champ électrique terrestre et de l'ionisation de l'atmosphère pendant l'éclipse totale de Soleil du 30 août 1905 [présentée par Loewy]', *Comptes Rendus*, 141 (1905), 925-928.
Lebeuf, A. and Chofardet, P., 'Observation de l'éclipse totale de Soleil du 29-30 août 1905 à Cistierna, province de Léon (Espagne). Rapport des membres de la mission de l'Observatoire de Besançon', *Annales du Bureau des Longitudes*, 8 (1905), 1-47.
—. 'Résultats des observations faites, pendant l'éclipse totale de soleil du 30 août 1905, à Cistierna, Espagne', *Astronomische Nachrichten*, 176 (1907), 215-218.
Levander, F. W., *General Index to the Journal of the British Astronomical Association, volumes I.-XVIII* (Edinburgh: Neill and Co., 1909).
Llorente, Teodor, 'El eclipse en Elche', *Las Provincias*, 30 May 1900.
Lockyer, Norman, *Recent and Coming Eclipses* (London: MacMillan and Co., 1900).
—. 'Total Eclipse of the Sun, May 28, 1900. Account of the Observations Made by the Solar Physics Observatory Eclipse Expedition and the Officers and Men of H.M.S. "Theseus," at Santa Pola, Spain', *Philosophical Transactions of the Royal Society of London. Series A*, 198 (1902), 375-415.
Lockyer, William J. S., 'The approaching total solar eclipse of August 30', *Nature*, 71 (1905), 393-395.

—. 'The Forthcoming Total Solar Eclipse', *Nature*, 72 (1905), 399.

—. 'Total Solar Eclipse, August 30. The Solar Physics Observatory Expedition', *Nature*, 72 (1905), 457-458.

—. 'The Solar Physics Observatory Eclipse Expedition', *Nature*, 72 (1905), 508-510.

—. 'Thorp Gratings and Small Cameras in Eclipse Work', *Knowledge & Scientific News*, 2 (1905), 117-119.

Maunder, E. Walter, 'Eclipse Suggestions', *Journal of the British Astronomical Association*, 10 (1899-1900), 157-169.

—. 'The coming eclipse of the sun', *Knowledge*, 23 (1900), 49-51.

—. 'The total solar eclipse of May 28, 1900', *Knowledge*, 23 (1900), 145-147.

—. 'The Total Solar Eclipse of 1900, May 28 (Second Paper)', *Knowledge*, 23 (1900), 175-179.

—. 'The Total Solar Eclipse of August 30, 1905', *Journal of the British Astronomical Association*, 15 (1904-1905), 240-242.

—. 'Eclipse Suggestions', *Journal of the British Astronomical Association*, 15 (1904-1905), 317-332.

Menéndez y Pelayo, Marcelino, *Historia de los heterodoxos españoles* (México: Porrúa, 1995) [First edition: Madrid, 1882].

Merino, Miguel, 'Discurso de contestación de D. Miguel Merino' in Acisclo Fernández Vallín, *Cultura científica en España en el siglo XVI* (Sevilla: Padilla Libros, 1989) [*Discursos leidos ante la Real Academia de Ciencias Exactas, Físicas y Naturales en la recepción pública del Excmo. Sr. D. Acisclo Fernández Vallín* (Madrid: Sucesores de Rivadeneyra, 1893)].

Meslin, George, 'Sur l'eclipse du 30 août 1905 et sur la polarisation de la couronne solaire', *Comptes Rendus*, 141 (1905), 493-496.

Mier y Terán, José, *Eclipse total de sol del 30 de agosto de 1905. Observaciones hechas en Carrión de los Condes (Palencia) por la sección astronómica del Observatorio de Cartuja (Granada) dirigido por padres de la Compañía de Jesús* (Granada: L. Guevara, 1905).

Miller, John A., 'Solar Eclipse of August 30, 1905', *Astrophysical Journal*, 23 (1906), 93-95.

Millochau, G., 'Sur l'observation de l'éclipse totale du 30 août 1905 à Alcosebre (Espagne)' [presentée par J. Janssen], *Comptes Rendus*, 141 (1905), 586-588.

—. 'Sur l'observation de l'éclipse totale du 30 août 1905 à Alcosebre (Espagne) [présentée par Janssen]', *Comptes Rendus*, 141 (1905), 815-818.

Mitchell, S. A., 'Preliminary Account of Flash Spectra taken 1905 30 August', *Monthly Notices of the Royal Astronomical Society*, 66 (1906), 382-391.

—. 'Wave-lengths of the chromosphere from spectra obtained at the 1905 eclipse', *Astrophysical Journal*, 38 (1913), 407-495.

—. 'The Total Solar Eclipse of 1918', *Natural History*, 19 (1919), 244-271.

—. *Eclipses of the sun* (New York: Columbia University Press, 1923).

Moreux, T., 'Autour de l'éclipse. Notes de voyage', *Bulletin de la Société Astronomique de France*, 14 (1900), 308-324.

—. 'Comment verrons-nous l'éclipse du 17 avril 1912?', *Revue générale des sciences pures et appliqués*, 23 (1912), 170-171.

Morford, Augustin, 'The Total Eclipse of 1905. At Vinaroz, Spain', *Knowledge & Scientific News*, 2 (1905), 273.

Moye, Marcel, '[Solar Eclipse of May 28:] Naked-eye Observations', *The Observatory*, 23 (1900), 279-281

—. *A la poursuite d'une ombre: l'éclipse totale de soleil du 30 août 1905* (Montpellier : G. Firmin, Montane et Sicardi, 1905).

—. 'Observation de l'éclipse du 30 août à Alcalá de Chisbert (Espagne)', *Comptes Rendus*, 141 (1905), 458.

—. 'The Total Eclipse of 1905. At Alcala de Chisvert', *Knowledge & Scientific News*, 2 (1905), 271-272.

Navarro, Manuel S., 'Nouvelles de la science.- Variétés : Le cirque Archimède', *L'Astronomie*, 9 (1890), 74.

—. 'Un nouveau cratère dans l'arène de Plinius', *L'Astronomie*, 9 (1890), 187.

—. 'Gutenberg et Godenius', *L'Astronomie*, 9 (1890), 187-188.

Navás, Longino, 'El segundo Congreso de la Asociación Española para el Progreso de las Ciencias celebrado en Valencia los días 15-20 de Mayo de 1910', *Razón y Fe*, 27(1910).

Newall, H.F., 'Note on polarisation phenomena in the solar corona, 1905 August 30', *Monthly Notices of the Royal Astronomical Society*, 66 (1906), 475-481.

Newcomb, Simon, *et al.*, 'Plans of American Eclipse Parties', *Astrophysical Journal*, 11 (1900), 314-321.

Nielsen, C., 'The Total Eclipse of 1905. At Campo de la Isla, Burgos', *Knowledge & Scientific News*, 2 (1905), 272.

Nippoldt, 'Zum Einfluβ der totalen Sonnenfinsternis vom 30. August 1905 auf die erdmagnetischen Variationen', *Physikalische Zeitschrift*, 7 (1906) 242-248.

Nordmann, C., 'Fondation d'un Observatoire astrophysique en Espagne', *Revue generale des sciences pures et appliquees*, 15 (1904), 474-475.

—. 'Recherches sur les effets magnétiques de l'éclipse totale du 30 aout 1905', *Bulletin Astronomique*, 24 (1907), 93-106.

Observatorio Astronómico y Meteorológico de Madrid [Francisco Iñiguez], *Observaciones del eclipse total de sol del 28 de Mayo de 1900 verificadas en Plasencia por la Comisión Oficial* (Madrid: Sucesores de Rivadeneyra, 1900).

Observatorio de Física Cósmica del Ebro S. J., *Instrucciones para la observación del eclipse total de sol del 30 de agosto de 1905 (Colegio de Jesús, Tortosa)* (Tortosa: Imprenta de José L. Foguet, 1904).

Ortiz, León, 'Eclipses que se alejan', *El Imparcial*, 18 April 1912 [Año 34, núm. 11853], 2.

Pereira Cabrera, Ángel, 'Eclipse total de Sol', *El Heraldo de Madrid*, 17 May 1900 [Año 11, nº 3474], 3.

—. 'Acerca del Eclipse. En Argamansilla de Alba', *El Heraldo de Madrid*, 22 May 1900, 3.

Pérez de Guzman, Juan, 'Alfonso XIII en Londres y las relaciones políticas anglo-españolas', *La Ilustración Española y Americana*, 80 (1905) [Año 49, número 22], 355-359.

—. 'El Rey', *La Ilustración Española y Americana*, 80 (1905) [Año 49, número 24], 357-358.

Perkins, Clarence, 'The Social and Economic Problems of Modern Spain', *Political Science Quarterly*, 27 (1912), 92-108.

Perrine, C.D., 'Some Total Solar Eclipse Problems', *Astrophysical Journal*, 20 (1904), 331-337 [Read at the International Congress of Arts and Science, St. Louis, September 23, 1904].

—. 'The Total Solar Eclipse of February 3, 1916', *Publications of the Astronomical Society of the Pacific*, 28 (1916), 247-252.

Pickering, W. H., 'The Total Eclipse of 1900, May 28', *Journal of the British Astronomical Association*, 10 (1899-1900), 333.

Pritchard, Ada, *The Life and Work of Professor Pritchard* (London: Seeley and Co. Limited, 1897).

Quignon, G.A., 'The Eclipse of the Sun, April 17, 1912 (Translation of Letter to the Secretary)', *Journal of the British Astronomical Association*, 22 (1911-1912), 198.

Ramón y Cajal, Santiago, *Advice for a Young Investigator* (Cambridge, Mass.: MIT Press, 1999).

Raurich, S., 'Correspondence: The Astronomical Society of Barcelona', *Journal of the British Astronomica Association*, 22 (1912), 239-240.

Rayet, G., 'L'éclipse totale de Soleil du 30 août 1905', *Comptes Rendus*, 141 (1905), 490-491.

Real Observatorio Astronómico de Lisboa (Tapada), *O eclipse de Sol de 1900 Maio 28 em Portugal* (Lisboa: Imprensa Nacional, 1900).

Rodés, Luis, 'XXI Congreso Astronómico Norteamericano', *Ibérica*, 8 (1917), 270-271.

Romeo, Octaviano, 'Observaciones espectroscópicas verificadas durante los eclipses de sol de 28 de Mayo de 1900 y 30 de Agosto de 1905', in *Asociación Española para el Progreso de las Ciencias (Valencia). Tomo III* (Madrid: Eduardo Arias, 1911), pp. 9-18.

Roso de Luna, M., 'Eclipses totales de Sol [boceto de un aparato para investigar la posición de los elementos productores de las »shadow bands« (bandas oscilantes) en el espacio]', *Astronomische Nachrichten*, 171 (1906), 87-90.

—. 'El eclipse de Sol del 17 de Abril. "El Liberal" en el Bierzo', *El Liberal*, 15 April 1912 [Año 34, núm. 11850], 1.

—. 'El eclipse de hoy', *El Liberal*, 17 April 1912 [Año, 34, núm. 11852], 1.

—. 'El eclipse de ayer. "El Liberal" en el Bierzo', *El Liberal*, 18 April 1912 [Año 34, núm. 11853], 1.

—. 'El proceso del eclipse', *El Liberal*, 22 April 1912 [Año 34, núm. 11857], 1.

—. 'El Salmo del Eclipse', *El Liberal*, 24 April 1912 [Año 34, núm. 11859], 2.

Rotch, A. Lawrence, 'Eclipse Shadow Bands', *Nature*, 72 (1905), 307-308.

Royal Astronomical Society, *General Index to the Monthly Notices of the Royal Astronomical Society, volumes LIII to LXX, 1892-1910, together with General Index to Illustrations in the Memoirs, volumes I to LIX, and the Monthly Notices, volumes I to LXX, 1822-1910* (London: Burlington House, 1911).

Ruiz Castizo, José and Galán, Gabriel, *El eclipse de Sol de 1905* (Madrid: Adrian Romo, 1905).

Selga, M., 'The Ebro Observatory, Tortosa, Spain', *Publications of the Astronomical Society of the Pacific*, 27 (1915), 21-27.

—. 'A New Line in the Spectrum of the Solar Corona', *Publications of the Astronomical Society of the Pacific*, 27 (1915), 93-94.

Shackleton, W., 'The Coming Total Eclipse', *Knowledge & Scientific News*, 2 (1905), 45-47.

—. 'The Coming Total Eclipse', *Knowledge & Scientific News*, 2 (1905), 71-73.

Soler de Morell, J., 'El día 30 de Agosto en el Observatorio del Ebro', *Memorias del Observatorio del Ebro* (1906), 53-56.

Stefanik, Milan, 'Recherches spectroscopiques pendant l'éclipse du 30 août 1905 à Alcosebre (Espagne)' [présentée par J. Janssen], *Comptes Rendus*, 141 (1905), 585-586.

Stevens, Catharine O., 'Letters to the Editor: The Problem of "Shadow-bands"', *Nature*, 72 (1905), 631.

—. 'The Problem of "Shadow-Bands"', *Journal of the British Astronomical Association*, 16 (1905-1906), 60-62.

St. John, Charles E., 'Radial motion in sun-spots. The distribution of the elements in the solar atmosphere', *Contributions from the Mount Wilson Solar Observatory*, 74 (1913), 157-207.

—. 'On the distribution of the elements in the solar atmosphere as given by the flash spectra', *Astrophysical Journal*, 40 (1914), 356-371.

Tarazona y Blanch, Antonio, *Memoria sobre el eclipse total de sol del día 28 de Mayo de 1900* (Madrid: Observatorio de Madrid, 1899).

—. *Memoria sobre el eclipse total de sol del día 30 de agosto de 1905* (Madrid: Bailly-Bailliere e hijos, 1904).

Tarazona y Blanch, Ignacio, *La fotografía solar. Discurso leído en la Universidad Literaria de Valencia en la solemne apertura del curso 1909-1910* (Valencia: Miguel Gimeno, 1909).

—. 'Observation de l'éclipse de Soleil du 22 novembre 1919, à l'Observatoire astronomique de l'Université de Valence (Espagne)', *Comptes Rendus*, 169 (1919), 1086-1087.

—. 'Observation de l'éclipse annulaire du Soleil du 27-28 mars 1922, faite à l'Observatoire astronomique de Valence (Espagne)', *Comptes Rendus*, 174 (1922), 1004-1005.

Tarazona, Ignacio and Martí Ortells, Vicente, 'Observation de l'éclipse de Soleil du 20-21 août 1914, faite à l'Observatoire astronomique de l'Université de Valence (Espagne)' *Comptes Rendus*, 159 (1914), 468-469.

—. 'Observation de l'éclipse de Soleil du 3 février 1916, faite à Valence (Espagne)', *Comptes Rendus*, 162 (1916), 313.

Todd, D. and Baker, R. H., 'The solar eclipse of 1905 in Spain', *The Observatory*, 28 (1905), 180-181.

Touchet, E., 'Société Astronomique de France', *Bulletin de la Société Astronomique de France*, 16 (1902), 518-528.

Turner, H. H. (ed.) *Miscellaneous papers of the University Observatory Oxford. Vol. 2, 1902-1905* (Oxford: University Observatory, 1905).

—. *Miscellaneous papers of the University Observatory Oxford. Vol. 3, 1905-1909* (Oxford: University Observatory, 1910).

—. *Miscellaneous papers of the University Observatory Oxford. Vol. 4, 1910-1914* (Oxford: University Observatory, 1914).

Valderrama, Juan, 'Nouvelles de la Science.–Variétés. Conjonction d'Uranus avec α de la Balance', *L'Astronomie*, 13 (1894), 234-235.

Vela y Herranz, Antonio, 'El eclipse de sol del 30 de agosto', *El Imparcial*, 10 July 1905, 1.

—. *Estudio del teodolito y del anteojo de pasos Salmoiraghi y determinación de la latitud de la hora en las estaciones de Plasencia y Burgos* (Madrid: Imprenta de la Dirección General del Instituto Geográfico y Estadístico, 1906).

Ventosa, Vicente, 'The Dark Fringes observed during Total Solar Eclipse', *Journal of the British Astronomical Association*, 10 (1899-1900), 333.

—. *Memoria del primer astrónomo del Observatorio de Madrid al director de mismo establecimiento. 1899-1900* (Madrid, 1902).

Vera, Vicente, 'El eclipse de mañana…', *El Imparcial*, 16 April 1912 [Año 46, núm. 16210], 3.

—. 'El eclipse de hoy. Observaciones curiosas al alcance del público', *El Imparcial*, 17 April 1912 [Año 46, núm. 16211], 1.

—. 'El eclipse de Sol. Observaciones científicas', *El Imparcial*, 18 April 1912 [Año 46, núm. 16212], 1-2.

Vives y Vich, Pedro, *Observaciones del eclipse total de sol de 30 de agosto de 1905 por medios de globos* (Madrid: Imprenta del Memorial de Ingenieros, 1905).

—. *Avance de los resultados obtenidos en las observaciones del eclipse total de sol de 30 de agosto de 1905* (Madrid: Imprenta del Memorial de Ingenieros, 1906).

Whitmell, C.T., 'Correspondence: The Total Eclipse at Navalmoral', *Journal of the British Astronomical Association*, 11 (1900-1901), 351-352.

—. 'The duration of Totality at Navalmoral', *Monthly Notices of the Royal Astronomical Society*, 62 (1902), 375-378.

—. 'The Moon's Semidiameters', *Journal of the British Astronomical Association*, 13 (1902-1903), 166-167.

—. 'Contact Times of the Eclipse on 30 August', *Journal of the British Astronomical Association*, 16 (1905-1906), 220-226.

Zelt, 'Flammarion en Valencia', *El Heraldo de Madrid*, 24 May 1900, [año 11, nº 3481], 1.

—. 'Flammarion', *El Heraldo de Madrid*, 25 May 1900, [año 11, nº 3482], 2.

Secondary Works

Printed secondary works

Abbott, D. (ed.) *The Biographical Dictionary of Scientists: Astronomers* (London: Blond Educational, 1984).

Abetti, Giorgio, *Historia de la astronomía* (México: Fondo de Cultura Económica, 1992).

Alter, Peter, *The Reluctant Patron: Science and the State in Britain, 1850-1920* (Oxford, Hamburg, New York: Berg, 1987).

Álvarez Junco, José, 'La conformación de una identidad', in José María Jover Zamora (dir.) *Historia de España Menéndez Pidal. Tomo XXXVI, La época de la Restauración (1875-1902). Volumen II, Civilización y cultura* (Madrid: Espasa Calpe, 2002); pp. 5-45.

Artola, Miguel, 'La burguesía revolucionaria (1808-1874)' in Miguel Artola (dir.) *Historia de España* (Madrid: Alianza Editorial, 1990).

Balfour, Sebastian, *El fin del imperio español (1898-1923)* (Barcelona: Crítica, 1997).

—, 'War, Nationalism and the Masses in Spain, 1898-1936', in Edward Acton and Ismael Saz (eds.) *La transición a la política de masas* (Valencia: Universitat de València, 2001), pp. 75-91.

Ballarín, Pilar, 'La construcción de un modelo educativo de «utilidad doméstica»', in Georges Duby and Michelle Perrot (dirs.) *Historia de las mujeres en Occidente.* [Volume 4, Geneviève Fraisse and Michelle Perrot (dirs.) *El siglo XIX*] (Madrid: Taurus, 1993), pp. 599-611.

Baratas Díaz, Alfredo, 'La cultura científica en la Restauración', in Manuel Suárez Cortina (ed.) *La cultura española en la Restauración* (Santander: Sociedad Menéndez Pelayo, 1999), pp. 279-295.

Bennett, T., *The birth of the Museum: History, theory, politics* (London, New York: Routledge, 1995).

Bensaude-Vincent, B., 'In the Name of Science', in J. Krige and D. Preste, *Science in the Twentieth Century* (Amsterdam: Harwood Academic Publishers, 1997); pp. 319-338.

Bensaude-Vincent, B. and Rasmussen, A., *La science populaire dans la presse et l'édition: XIXe et XXe siècles* (Paris: CNRS, 1997).

Bethel, Leslie (ed.) *The Cambridge History of Latin America. Volume III, From independence to c.1870* (Cambridge: Cambridge University Press, 1985).

Brain, R., *Going to the Fair. Readings in the Culture of Nineteenth-Century Exhibitions* (Cambridge: Whipple Museum of the History of Science, 1993).

Braudel, Fernand, *Capitalism and Material Life* (London: Weidenfeld and Nicolson, 1973).

British Astronomical Association, 'The history of the British Astronomical Association. The first fifty years', *British Astronomical Association Memoirs*, 42 (1989).

Brück, M.T., *Agnes Mary Clerke and the rise of Astrophysics* (Cambridge: Cambridge University Press, 2002)

Butler, S. V. F., *Science and Technology Museums* (Leicester, London & New York: Leicester University Press, 1992).

Calvo Roy, Antonio, *Cajal: triunfar a toda costa* (Madrid: Alianza, 1999).

Carr, Raymond, *Spain 1808-1939* (Oxford: Clarendon Press, 1966).

—. *Modern Spain 1875-1980* (Oxford: Oxford University Press, 1980).

Carreras, Albert, 'An Annual Index of Spanish Industrial Output' in Sánchez Albornoz (ed.) *La modernización económica de España* (Madrid: Alianza, 1985); pp. 75-89.

Carreras, Albert and Tafunell, Xavier, *Historia económica de la España contemporánea* (Barcelona: Crítica, 2003).

Cerezo Galán, Pedro, 'El pensamiento filosófico—El 98: la generación trágica', in José María Jover Zamora (dir.) *Historia de España Menéndez Pidal. Tomo XXXIX, La edad de plata de la cultura española (1898-1936). Volumen I, Identidad, Pensamiento y Vida. Hispanidad* (Madrid: Espasa Calpe, 1993); pp. 133-152.

—. *El mal del siglo. El conflicto entre Ilustración y Romanticismo en la crisis finisecular del siglo XIX* (Madrid: Biblioteca Nueva, 2003).

Chapman, A., *The Victorian Amateur Astronomer: Independent Astronomical Research in Britain 1820-1920* (Chichester: Praxis, 1998).

Clemence, G.M. and Jenkins, Louise F., *General Index to the first fifty volumes 1849-1944* (New Haven, 1948).

Collins, Harry M., *Changing Order: Replication and Induction in Scientific Practice* (Beverly Hills and London: SAGE, 1985).

Comas Solá, José, *Astronomía* (Barcelona: Ramón Sopena, 1939).

de Benito, Enrique and Martí y Ortells, Vicente, 'Don Ignacio Tarazona y el Observatorio Astronómico de Valencia', *Anales de la Universidad de Valencia*, 34 (1924-25).

de la Cierva, Juan, *Notas de mi vida* (Madrid: Instituto Editorial Reus, 1955).

Débarbat, S., Grillot, S., and Lévy, J., *Observatoire de Paris: son histoire 1667-1963* (Paris: Observatoire de Paris, 1984)

Due Rojo, Antonio, 'El Cincuentenario del Observatorio de Cartuja', *Urania*, 38 (1953), 67-80.

Fontana, Josep, *La quiebra de la monarquía absoluta* (Barcelona: Ariel, 1971).

Fox, Robert and Weisz, George (eds.) *The organization of science and technology in France 1808-1914* (Cambridge, Paris: Cambridge University Press, Editions de la Maison des Sciences de l'Homme, 1980).

Fusi, Juan Pablo, *Un siglo de España: La cultura* (Madrid: Marcial Pons, 1999).

Fusi, Juan Pablo and Palafox, Jordi, *España: 1808-1996. El desafío de la modernidad* (Madrid: Espasa Calpe, 1997).

García Camarero, Enrique and García Camarero, Ernesto, *La polémica de la ciencia española* (Madrid: Alianza, 1970).

Garrido, Julio, 'Un precursor español de las investigaciones petrográficas de la superficie lunar: José Joaquin Landerer', *Revista de la Real Academia de Ciencias Exactas, Físicas y Naturales de Madrid,* 67 (1973), 285-287.

Gerschenkron, A., *El atraso económico en su perspectiva histórica* (Ariel, Barcelona, 1968).

Gil Cremades, Juan José, *Krausistas y liberales* (Madrid: Seminarios y Ediciones, 1975).

Gil y Zárate, Antonio, *De la instrucción pública en España* (Madrid: Imprenta del Colegio de Sordo-Mudos, 1855)

Gingerich, Owen, 'The development of astronomical theory and practice from the 17th to the 20th century', in Arthur Beer (ed.) *Vistas in Astronomy*, 20 (1976), 1-9.

Glick, Thomas F., *Darwin en España* (Barcelona: Península, 1982).

—. 'Emilio Herrera and Spanish Technology', in E. Herrera, *Flying: The Memoirs of a Spanish Aeronaut* (Albuquerque: University of New Mexico Press, 1984), pp. 173-215.

—. *Einstein y los españoles: Ciencia y sociedad en la España de entreguerras* (Madrid: Alianza, 1986).

—. *Einstein in Spain: Relativity and the Recovery of Science* (Princeton: Princeton University Press, 1988).

Goldman, Lawrence, *Dons and Workers, Oxford and Adult Education Since 1850* (Oxford: Clarendon Press, 1995).

Golinski, J., *Making natural knowledge. Constructivism and the history of science* (Cambridge: Cambridge University Press, 1998).

González González, Francisco José, *El Observatorio de San Fernando (1831-1924)* (Madrid: Ministerio de Defensa, 1992).

—. *Instrumentos científicos del Observatorio de San Fernando (siglos XVIII, XIX y XX)* (Madrid: Instituto de Historia y Cultura Naval, Ministerio de Defensa, 1995).

Gozalo Gutiérrez, R. and Navarro Brotons, V., 'Josep Joaquim Landerer i Climent (València 1841—Tortosa 1922). La recerca fora del món acadèmic: astronomia i geologia' in J.M. Camarassa and A. Roca (dirs.) *Ciència i Tècnica als Països Catalans. Una aproximació biogràfica* (Barcelona: Fundació Catalana per a la recerca, 1995), pp. 457-493.

Hentschel, Klaus, *Mapping the Spectrum: Techniques of Visual Representation in Research and Teaching* (Oxford: Oxford University Press, 2002).

Hernández Martínez, Ascensión, *Vida y obra del arquitecto Ricardo Magdalena (1849-1910)* (Zaragoza: Universidad de Zaragoza, 1999).

Hufbauer, Karl, 'Solar Physics', in J. Lankford (ed.) *History of Astronomy: An Encyclopedia* (New York, London: Garland, 1997), pp. 464-471.

Huntington, Samuel P., *El choque de las civilizaciones y la reconfiguración del orden mundial* (Barcelona: Paidós, 1997).

Hutchins, Roger, *British university observatories, c.1820-1939: ideals and resources* (D.Phil. Thesis, University of Oxford, 1999).

Jover Zamora, José María, *España en la política internacional. Siglos XVIII-XX* (Madrid: Marcial Pons, 1999).

—. 'Aspectos de la civilización española en la crisis de fin de siglo', in José María Jover Zamora (dir.) *Historia de España Menéndez Pidal. Tomo XXXVI, La época de la Restauración (1875-1902). Volumen II, Civilización y cultura* (Madrid: Espasa Calpe, 2002); pp. 747-784.

Käppeli, Anne-Marie, 'Escenarios del feminismo', in Georges Duby and Michelle Perrot (dirs.) *Historia de las mujeres en Occidente.* [Volume 4, Geneviève Fraisse and Michelle Perrot (dirs.) *El siglo XIX*] (Madrid: Taurus, 1993), pp. 497-531.

Kennedy, John Loomis, *Representations of the Spectra in Early British Astrophysics* (D.Phil. dissertation, University of Oxford, 2000).

Kennedy, Paul, *Auge y caída de las grandes potencias* (Barcelona: Plaza & Janés, 1994).

Kidwell, Peggy Aldrich, 'Women Astronomers in Britain, 1780-1930', *Isis*, 75 (1984), 534-546.

King, H.C., *Geared to the stars* (Bristol: Hilger, 1978).

Kohlstedt, Sally Gregory (ed.), *History of Women in the Sciences. Readings from* Isis (Chicago and London: The University of Chicago Press, 1999).

Kuklick, Henrika and Kohler, Robert E., 'Introduction' in H. Kuklick and R. E. Kohler (eds.) *Science in the field. Osiris*, 11 (1996), 1-14.

Lafuente, A. and Sellés, M., *El Observatorio de Cádiz (1753-1831)* (Madrid: Ministerio de Defensa, 1988).

Laín Entralgo, Pedro, 'Prólogo', in José María Jover Zamora (dir.) *Historia de España Menéndez Pidal. Tomo XXXIX, La edad de plata de la cultura española (1898-1936). Volumen I, Identidad, Pensamiento y Vida. Hispanidad* (Madrid: Espasa Calpe, 1993); pp. 11-52.

Lambert, Armando, *La Astronomía* (Madrid: Mercurio, 1927).

Lankford, John, 'Amateurs and Astrophysics: a Neglected Aspect in the Development of a Scientific Speciality', *Social Studies of Science*, 11 (1981), 275-303.

Lightman, Bernard, 'Constructing Victorian heavens: Agnes Clerke and the 'new astronomy'', in Barbara T. Gates and Ann B. Shteir (eds.) *Natural eloquence: Women reinscribe science* (Madison: Univ. of Wisconsin Press, 1997), pp. 61-75.

Linari, A.F., 'Don José Joaquín Landerer y Climent', *Ibérica*, 18 (1922), 178-180.

López Arroyo, M., 'Los dos siglos de existencia del Observatorio Astronómico de Madrid', in *Doscientos años del Observatorio Astronómico de Madrid* (Madrid: Asociación de Amigos del Observatorio Astronómico, 1992), pp. 1–26.

—. *El Real Observatorio Astronómico de Madrid (1785-1975)* (Madrid: Dirección General del Instituto Geográfico Nacional, 2004).

López Piñero, José María, *Cajal* (Madrid: Debate, 2000).

López Piñero, José María, García Ballester, Luis, and Faus Sevilla, Pilar, *Medicina y sociedad en la España del siglo XIX* (Madrid: Sociedad de Estudios y Publicaciones, 1964).

López Piñero, J.M. and Navarro Brotons, V., *Història de la Ciència al País Valencià* (Valencia: Alfons el Magnànim, 1995).

Lynch, John, *Las revoluciones hispanoamericanas 1808-1826* (Barcelona: Ariel, 1980).

Maluquer, J., *Rafael Patxot i Jubert: Mecens i científic* (Barcelona: Pòrtic, 1994).

Marcuzzo, Maria Christina, Officer, Lawrence H. and Rosselli, Annalisa (eds.), *Monetary Standards and Exchange Rates* (London: Routledge, 1997).

McNeill, William, *The Pursuit of Power: Technology, Armed Force, and Society since A.D. 1000* (Oxford: Basil Blackwell, 1983).

McNeill, J.R. and McNeill, William H., *Las redes humanas. Una historia global del mundo* (Barcelona: Crítica, 2004).

Meadows, A.J., *Early Solar Physics* (Oxford: Pergamon, 1970).

—. *Science and Controversy: A biography of Sir Norman Lockyer*, (Cambridge, Mass.: MIT Press, 1972).

—. 'The origin of astrophysics', in O. Gingerich (ed.) *Astrophysics and twentieth-century astronomy to 1950* (Cambridge: Cambridge University Press, 1984).

Menéndez Ureña, Enrique and Vázquez-Romero, José Manuel, 'El pensamiento y las ideas', in José María Jover Zamora (dir.) *Historia de España Menéndez Pidal. Tomo XXXVI, La época de la Restauración (1875-1902). Volumen II, Civilización y cultura* (Madrid: Espasa Calpe, 2002); pp. 125-176.

Molina García, Vicente J. and Oriola Lafuente, Enrique P., *Historia del eclipse total de sol del 28 de mayo de 1900 observado desde Santa Pola* (Santa Pola: Ajuntament de Santa Pola, 2000).

Montiel Llorente, 'Las consecuencias de una elección para la Filosofía de la Medicina Española: Krause frente a Schelling' in Elvira Arquiola and José Martínez Pérez (cords.) *Ciencia en Expansión: estudios sobre la difusión de las ideas científicas y médicas en España (siglos XVIII-XX)* (Madrid: Complutense, 1995).

Moreno González, 'De la física como medio a la fin. Un episodio entre la Ilustración y la crisis del 98', in José Manuel Sánchez Ron (ed.) *Ciencia y sociedad en España: de la Ilustración a la Guerra Civil* (Madrid: El Arquero/CSIC, 1988), pp. 27-70.

Nadal, Jordi, *El fracaso de la Revolución industrial en España, 1814-1913* (Barcelona: Ariel, 1975).

Nadal, Jordi and Carreras, Albert (dirs. and coords.) *Pautas regionales de la industrialización española (siglos XIX y XX)* (Barcelona: Ariel, 1990).

Nash, Mary, 'Identidad cultural de género, discurso de la domesticidad y la definición del trabajo de las mujeres en la España del siglo XIX', in Georges Duby and Michelle Perrot (dirs.) *Historia de las mujeres en Occidente.* [Volume 4, Geneviève Fraisse and Michelle Perrot (dirs.) *El siglo XIX*] (Madrid: Taurus, 1993), pp. 497-597.

Navarro Brotons, Víctor, 'El conreu de les ciències' in *Cinc Segles i un dia* (València: Universitat de València, 2000), pp. 119-126.

Navarro Brotóns, Víctor and Gozalo Gutiérrez, Rodolfo, 'Entre "amateurs" i professionals: l'obra astronòmica de Josep Joaquim Lànderer i Climent (1841-1922)' in Carles Puig-Pla, Agustí Camós, Jon Arrizabalaga and Pasqual Bernat (coords.) *Actes de les III trobades d'història de la ciència i de la tècnica als Països Catalans* (Barcelona: SCHCT, 1996), pp. 303-312.

Nieto-Galán, A. and Roca Rosell, A. (dir.), *La Reial Acadèmia de Ciències i Arts als segles XVIII i XIX. Història, ciència i societat* (Barcelona: RACAB/IEC, 2000).

North, John, *Historia Fontana de la astronomía y la cosmología* (México D.F.: Fondo de Cultura Económica, 2001).

Oliver, Josep Maria, *Historia de la astronomía amateur en España* (Madrid: Sirius, 1997).

—. 'Comas i l'observació astronòmica', in Antoni Roca Rosell (coord.) *Josep Comas i Solà, astrònom i divulgador* (Barcelona: Ajuntament de Barcelona, 2004), pp. 43-79.

Oriol Cardús, J., 'En la Guinea Española para la observación del eclipse de sol', *Razón y Fe*, 147 (1953), 38-55.

Pang, Alex Soojung-Kim 'The Social Event of the Season: Solar Eclipse Expeditions and Victorian Culture', *Isis*, 84 (1993), 252-277.

—. 'Gender, Culture, and Astrophysical Fieldwork: Elizabeth Campbell and the Lick Observatory—Crocker Eclipse Expeditions' in H. Kuklick and R.E. Kohler (eds.) *Science in the field. Osiris*, 11 (1996), 17-43.

—. *Empire and the Sun* (Stanford: Stanford University Press, 2002).

Pecker, Jean-Claude and Ringeard, Gisèle, 'Astronomy, the People and the Governments', *Vistas in Astronomy* [Edited by Arthur Beer, Pergamon Press], vol.15 (1973), 1-12.

Pelayo López, Francisco, 'Ciencia y religión en España durante el siglo XIX', *Asclepio*, 40 (1988), 187-207.

Pérez Díaz, Víctor, *Peasant ethics and religion: towards a theory of the "three cultures"* (Madrid: Instituto Juan March, 1991).

Peset, José Luis, 'Educación y ciencia en el fin del Antiguo Régimen', in José Manel Sánchez Ron (ed.) *Ciencia y sociedad en España: de la Ilustración a la Guerra Civil* (Madrid: El Arquero/CSIC, 1988), pp. 17-25.

Peset, José Luis and Hernández-Sandoica, Elena, 'La recepción de la cultura científica en la España del siglo XX: la Universidad', in Antonio Morales Moya (coord.) *Las Claves de la España del siglo XX. La cultura* (Madrid: España Nuevo Milenio, 2001), pp. 127-151.

Peset, Mariano and Peset, José Luis, 'Las universidades españolas del siglo XIX y las ciencias', in José M. López Piñero (ed.) *La ciencia en la España del siglo XIX* [Serie *Ayer*, 7] (Madrid, Marcial Pons, 1992).

Pollard, Sidney, *La conquista pacífica. La industrialización de Europa 1760-1970* (Zaragoza: Prensas Universitarias de Zaragoza, 1991).

Puig, Ignacio, *El Observatorio del Ebro* (Tortosa: Imprenta Moderna del Ebro de Algueró y Baiges, 1927).

Querol Cabot, Domingo, 'Astronomía en la Valencia de la Restauración: Ignacio Tarazona y el Observatorio Astronómico de Valencia', in G. Blanes, L. Garrigós, A. Roca, J. Assizabalaga (coords.) *Actes de les IV Trobades d'Història de la Ciència i de la Tècnica* (Alcoi, Barcelona: Societat Catalana d'Història de la Ciència i de la Tècnica, 1997), pp. 173-180.

Ríos, Sixto, 'Matemáticas y Astronomía' in José María Jover Zamora (dir.) *Historia de España Ramón Menéndez Pidal, t. XXXIX, La edad de plata de la cultura española (1898-1936). Letras. Ciencia. Arte. Sociedad y culturas* (Madrid: Espasa Calpe, 1994), pp. 475-494.

Roca Rosell, Antoni, 'Una vida consagrada a la ciència' in Antoni Roca Rosell (coord.) *Josep Comas i Solà, astrònom i divulgador* (Barcelona: Ajuntament de Barcelona, 2004), pp. 13-42.

Roca Rosell, Antoni and Navarro Brotons, Víctor, 'Reflexions sobre la ciència i la tècnica a la Restauració', in G. Blanes, L. Garrigós, A. Roca, J. Assizabalaga (coords.) Actes de les IV Trobades d'Història de la Ciència i de la Tècnica (Alcoi, Barcelona: Societat Catalana d'Història de la Ciència i de la Tècnica, 1997), pp. 181-190.

Romero, Ana, 'Dos políticas de instrumental científico: el Instituto de Material Científico y el Torres Quevedo', *Arbor*, 160 (1998), 359-386.

Rossiter, Margaret W., '"Women's Work" in Science, 1880-1910', *Isis*, 71 (1980), 258-275.

Ruiz de Azúa y Martínez de Ezquerecocha, Estíbaliz, 'El bachillerato y la Universidad', in Jover Zamora (dir.) *Historia de España Menéndez Pidal. Tomo XXXVI. Volumen II* (Madrid, 2002); pp. 337-359.

Sala Catalá, José, 'Ciencia biológica y polémica de la ciencia en la España de la Restauración', in José Manuel Sánchez Ron (ed.) *Ciencia y sociedad en España: de la Ilustración a la Guerra Civil* (Madrid: El Arquero-CSIC, 1988), 157-177.

Sánchez Ron, José Manuel (ed.) *José Echegaray* (Madrid: Biblioteca de la ciencia española, 1990).

—. *Cincel, martillo y piedra. Historia de la ciencia en España (siglos XIX y XX)* (Madrid: Taurus, 1999).

Schiebinger, Londa, 'Maria Winkelmann at the Berlin Academy. A Turning Point for Women in Science', *Isis*, 78 (1987), 174-200.

Seco Serrano, Carlos, 'Las ideologías políticas', in José María Jover Zamora (dir.) *Historia de España Menéndez Pidal. Tomo XXXIX, La edad de plata de la cultura española (1898-1936). Volumen I,*

Identidad, Pensamiento y Vida. Hispanidad (Madrid: Espasa Calpe, 1993); pp. 319-467.

—. *La España de Alfonso XIII. El Estado. La Política. Los Movimientos Sociales* (Madrid: Espasa Calpe, 2002).

Sellés, Manuel A., 'Astronomía y navegación', in Manuel Sellés, José Luis Peset, and Antonio Lafuente (comps.) *Carlos III y la ciencia de la Ilustración* (Madrid: Alianza, 1988), pp. 81-98.

—.'La academia y observatorio de la marina', in Manuel Sellés, José Luis Peset, and Antonio Lafuente (comps.) *Carlos III y la ciencia de la Ilustración* (Madrid: Alianza, 1988), pp. 173-186.

Shapin, Steven, *A Social History of Truth: Civility and Science in Seventeenth-Century England* (Chicago: University of Chicago Press, 1994).

Simón Palmer, María del Carmen, 'Mujeres rebeldes', in Georges Duby and Michelle Perrot (dirs.) *Historia de las mujeres en Occidente.* [Volume 4, Geneviève Fraisse and Michelle Perrot (dirs.) *El siglo XIX*] (Madrid: Taurus, 1993), pp. 629-641.

Teague, E.H.T., 'Carrington's method of determining sunspot positions', *Journal of the British Astronomical Association*, 106 (1996), 82-85.

Ten, Antonio E. and Moya Carcel, Teodoro, 'La formación de un astrónomo en la España del siglo XIX', in Mariano Esteban Piñeiro *et al.* (coords.) *Estudios sobre la Historia de la Ciencia y de la Técnica. Actas del IV Congreso de la Sociedad Española de Historia de las Ciencias y de las Técnicas* (Valladolid: Junta de Castilla y León, 1988), pp. 469-484.

Thackray, Arnold and Merton, Robert K., 'On Discipline Building: The Paradoxes of George Sarton', *Isis*, 63 (1972), 437-495.

Topham, J. R., 'Scientific publishing and the reading of science in nineteenth-century Britain: a historiographical survey and guide to sources', *Studies in history and philosophy of science*, 31A (2000), 559-612.

Tortella, Gabriel (dir.), *La banca española en la Restauración* (Madrid: Banco de España, 1974)

—. *The Development of Modern Spain. An Economic History of the Nineteenth and Twentieth Centuries* (Cambridge, Mass. and London: Harvard University Press, 2000)

Turner, F. M., 'Public Science in Britain, 1880-1919', *Isis*, 71 (1980), 589-608.

Udias, Agustin and Stauder, William, 'Jesuit Geophysical Observatories', *EOS*, 72 (1991), 185-189.

Udias, Agustín, *Searching the Heavens and the Earth: The History of Jesuit Observatories* (Dordrecht: Kluwer Academic Publishers, 2003).

Vaucouleurs, Gérard Henry de, *Astronomical photography: from the daguerreotype to the electron camera* (London: Faber and Faber, 1961).

Vernet, Juan, *Historia de la ciencia española* (Madrid: Instituto de España, 1975)

Viñao Frago, Antonio, *Política y educación en los orígenes de la España contemporánea* (Madrid: Siglo XXI, 1982)

De Puentes Benítez, Manuel, *Educación e ideología en la España contemporánea* (Madrid: Tecnos, 1999).

Walter, Alice N., 'English Broadsides of Solar Eclipses', *History of Science*, 37 (1999), 1-43.

Welch, Claude, 'Dispelling some myths about the split between theology and science in the nineteenth century', in W. Mark Richardson and Wesley J. Wildman (eds.) *Religion and Science: History, Method, Dialogue* (London, New York: Routledge, 1996), pp. 29-40.

Williams, Mari E. W., *The Precision Makers: A history of the instruments industry in Britain and France, 1870-1939* (London, New York: Routledge, 1994).

Other secondary sources

Espinar Moreno, Manuel, 'Fundación del Observatorio de Cartuja. Primeros años de funcionamiento (1902-1906)' in M. Espinar, J. A. Esquivel, and J. A. Peña (eds.) *Historia del observatorio de Cartuja, 1902-2002. Nuevas investigaciones* [CD-ROM] (Granada: Ayuntamiento de Granada, 2002).

—. 'Notas sobre la vida y obras del R. P. Manuel María Sánchez-Navarro Neuman, S. J. (1867-1941)' in M. Espinar, J. A. Esquivel, and J. A. Peña (eds.) *Historia del observatorio de Cartuja, 1902-2002. Nuevas investigaciones* [CD-ROM] (Granada: Ayuntamiento de Granada, 2002).

Espinar Moreno, Manuel and Morcillo Puga, Juan de Dios, 'Nacimiento y evolución del edificio del Observatorio de Cartuja desde 1902 a 2002' in M. Espinar, J. A. Esquivel, and J. A. Peña (eds.) *Historia del observatorio de Cartuja, 1902-2002. Nuevas investigaciones* [CD-ROM] (Granada: Ayuntamiento de Granada, 2002).

Espinar, Manuel *et al.* (eds.) *Historia del observatorio de Cartuja, 1902-2002. Nuevas investigaciones* [CD-ROM] (Granada: Ayuntamiento de Granada, 2002).

Ignjatovlae, S.R., 'History of the exponential form of the Titius-Bode Law' in S. Jokić, I. Milošević, A. Balaž, and Z. Nikolić (eds.), *Proceedings of the Fifth General Conference of the Balkan Physical Union. Vrnjačka Banja, Serbia and Montenegro, August 25-29, 2003.* [*Section 18: History and Philosophy of Physics / Poster Presentations*] (http://www.phy.bg.ac.yu/~bpu5/proceedings/Papers/SP18%20-%20003.pdf, 2003).

Merlo Vallejo, Manuel, 'Observatorio Universitario de Cartuja. Prefacio' in: M. Espinar, J. A. Esquivel, and J. A. Peña (eds.) *Historia del observatorio de Cartuja, 1902-2002. Nuevas investigaciones* [CD-ROM] (Granada: Ayuntamiento de Granada, 2002).

Ten Ros, Antonio E., 'Ver el cielo desde Madrid. La invención de un observatorio astronómico (1846-1860)' (http://www.uv.es/~ten/umlv/oamil.html, 2003).

INDEX